# On the Origin of Autonomy

# History, Philosophy and Theory of the Life Sciences

## Volume 5



For further volumes:
http://www.springer.com/series/8916

Bernd Rosslenbroich

# On the Origin of Autonomy

## A New Look at the Major Transitions in Evolution

Springer

Bernd Rosslenbroich
Centre for Biomedical Education and Research
Faculty of Health, School of Medicine
Institute of Evolutionary Biology and Morphology
Witten/Herdecke University
Witten, Germany

ISSN 2211-1948 ISSN 2211-1956 (electronic)
ISBN 978-3-319-35386-9 ISBN 978-3-319-04141-4 (eBook)
DOI 10.1007/978-3-319-04141-4
Springer Cham Heidelberg New York Dordrecht London



Printed on acid-free paper

Springer is part of Springer Science+Business Media (www.springer.com)

*Dedicated to Wolfgang Schad*

# Acknowledgments

I thank the various friends and colleagues with whom I have discussed the subjects presented herein over many years or who helped otherwise:

Wolfgang Schad, Volker Harlan, Susanna Kümmell, Hans Christoph Vahle, Rüdiger Sünner, Friedrich Edelhäuser, Meinhard Simon, Reinhard Wallmann, Jürgen Bereiter-Hahn, John Dupré, and Alvaro Moreno and his working group.

Stephan Stockmar and members of the Frankfurt working group "Development and Evolution", who accompanied the project over many years and contributed ideas and criticism.

Peter Heusser for his cooperative work on systems aspects and many other topics.

Walter Streffer for widening my understanding of behavioral flexibility in connection with bird songs.

Kim Sterelny and Daniel McShea for helpful comments and suggestions on earlier papers on the topic that I also included within the present work.

Kristian Köchy for a series of inspiring discussions many years ago and for insisting that I should read the writings of Paul Weiss.

Richard Everett for his long-standing support. With endless patience, he corrected the language of several drafts, always following the sense of what I was trying to say and always interested in my ideas.

My wife, Marita Rosslenbroich, for the pencil drawings within the book and for many discussions we shared on the topic over the years.

The foundations who fueled my work financially and ideally and especially Arnold Langen, who pushed in the background for many years.

Thomas Reydon and the other series editors for their interest in the manuscript and the topic.

And finally I want to thank the anonymous reviewers for their comments and critiques, which really helped me to improve the text. I am well aware that it is substantial work to write such detailed, precise, and balanced reviews.

# Contents

# Chapter 1
# What Is the Outcome of Evolution?

"When discussing organic evolution the only point of agreement seems to be: 'It happened.' Thereafter, there is little consensus." With this remark, Conway Morris (2000) begins a review, summarizing the situation in the field of evolutionary biology at the beginning of the twenty-first century and concludes from this "that our understanding of evolutionary processes and mechanisms is incomplete."

Statements such as this are now increasingly emerging in the scientific literature, after the proponents of Neo-Darwinian theory have been trying for decades to convince us that there is no need to search for other or additional factors of evolution than random mutation and selection, and that the main outcome of evolution is divergence caused by different adaptations. Most authors of this new literature do not contest natural selection as one factor of evolution. However, they contest that it "alone sets the evolutionary sails.... Many of us feel that something is missing; that selection is not enough; that the actualization of some creatures, together with the failure of others to emerge from the realm of the possible, requires something else – something internal that interacts with selection in a particular way. That is what Gould and Lewontin were saying more than twenty years ago" (Arthur 2004, pp. 10, 25).

Thus, there have been indicators during the past 15 years to the effect that the great synthesis of the mid-twentieth century is due for a major revision (Pigliucci and Müller 2010; Shapiro 2011). According to some literature, a different view of Darwinian evolution is coming forth. However, it is not clear yet in which direction this new view points. But, there are valuable pieces of theories, which have to grow together in some future.

However, this revision comes at a difficult time, as Darwinian evolution is confronted with a scientifically fruitless counterpart that is trying to bring science back to a position it held before the results of the enlightenment and the scientific revolution. Indeed, it seems to be attractive for some simple-minded contemporaries to initialize again the old debate around creationism, masqueraded as "intelligent design." At least in the public perception, there seems to be no alternative between neo-Darwinian one-sidedness on the one hand and creationism on the other. To be

B. Rosslenbroich, *On the Origin of Autonomy: A New Look at the Major Transitions in Evolution*, History, Philosophy and Theory of the Life Sciences 5,
DOI 10.1007/978-3-319-04141-4_1, © Springer International Publishing Switzerland 2014

sure, in the professional literature the views become more pluralistic, exhibiting fairly sound scientific development.

Besides the repeatedly formulated doubts that the assumed random process would be able to create order within the evolutionary process, a number of empirical findings fueled the rumblings of theory modification. One of the enigmas arose with the growing knowledge of comparative molecular biology. It became increasingly difficult to explain the immense diversity of life despite its deep and pervasively similar molecular architecture.

A crucial question is how evolutionary innovations were generated. What is the origin of new constructive principles and of new organs? What was at the beginning of the major evolutionary transitions: new structures, new genes, a new environment, a new behavior, or new ontogenetic pathways (Nitecki 1990; Thomson 1992; Wagner and Altenberg 1996; Gerhart and Kirschner 1997; Shubin and Marshall 2000; Wagner et al. 2000; Hall 2003; Arthur 2004; Kirschner and Gerhart 2005; Jablonka and Lamb 2005; Pigliucci and Müller 2010; Calcott and Sterelny 2011; Shapiro 2011)?

In recent years, increasingly tangible insights into the origin of evolutionary innovations have emerged. Although the picture is still fragmentary, these insights contain several surprises. Symbiosis, for example, delivers a new system state within a single macroevolutionary step and probably has a function in a number of transitions in addition to the generation of the eukaryotic cell (Margulis and Sagan 2002). Thus, there seem to be systemic shifts in evolution and not just gradual processes, which Gould (2002) also emphasizes from the paleontological perspective. Other examples come from cell biology, comparative genetics, and developmental biology, showing that novelties can be generated by new combinations of conserved structures and functions. The genome, at least in some parts, is obviously not so much a result of random mutations but of conservation of core functions together with new arrangements and duplications of building blocks (Carroll et al. 2005; Gerhart and Kirschner 1997; Kirschner and Gerhart 2005), and these combine with epigenetic functions (Jablonka and Lamb 2005). These results, together with the paleontological description of evolutionary patterns such as heterochrony (McKinney and McNamara 1991; McNamara 1990; Schad 1993) or convergence (Conway Morris 2003), are beginning to trigger a new stage in the evolution of evolutionary biology itself (Erwin 2000; Jablonka and Lamb 2005; West-Eberhard 2003; Pigliucci and Müller 2010).

Investigations usually are made into the origins of innovations and the mechanisms by which they were generated. However, within this discussion a central aspect continues to be neglected: Likewise, it is necessary to question the properties of these innovations and to ask what is qualitatively generated during the macroevolutionary transitions. Are evolved organisms in later time periods in some consistent way, in some aspect of their individual morphology, physiology, and behavior, different from organisms more primitive in earlier times (McShea 1998)? Or, in short, what have these changes produced? There have been some attempts to tackle these questions, but they have not as yet generated a great deal of interest within the scientific community.

In this book, I develop the proposal that a recurring central aspect of macroevolutionary innovations is an increase in individual organismal autonomy in the sense of emancipation from the environment with changes in the capacity for flexibility, self-regulation, and self-control of behavior. This concept is not new. Since the days of Darwin, it has emerged occasionally. However, comments on the principle were rare and generally cursory. Authors usually gave some few examples but did not explore the implications in any depth. A systematic inquiry has been performed only recently (Rosslenbroich 2007).

I propose that the relevance of differences in autonomy for understanding macroevolutionary innovations was underestimated in the past. In addition to the interest in environmental adaptation, the principle has been neglected. Although it is somehow "en vogue," it is not integrated within evolutionary theory.

The view presented here is neither intended to replace conventional evolutionary theories nor claimed that this is some sort of driving force. Principally, increasing autonomy is presented as a recurring pattern during macroevolutionary events. However, it is proposed that an integration of the available knowledge of dependency on and independency from environmental factors that is more complete is an important element for our further understanding of macroevolution.

Changes in autonomy are observable patterns of many major evolutionary transitions and can be described with morphological and physiological properties. The intention here is "first to define and to describe the perceivable pattern in order to help to detect and identify underlying structure and cause," as was proposed as the appropriate way to study "patterns, processes and directions in the history of life" during a Dahlem workshop (Wake 1986, p. 47). Presumably, this principle should be studied in its relation to other possible patterns in macroevolution such as complexity, size, entropy, and so on (McShea 1998; Rosslenbroich 2006).

In this context, Lewontin (2000) regards it as necessary to revise the notion of adaptation through a widened understanding of the relations between organisms and their environment. He shows that this relation is more complicated and may not be reduced to a passive principle. He states that the environment of an organism is not a given physical world outside, to which it has to fit, but that there are rather complex interactions between both sides. Thus, organisms determine which elements of the external world are relevant to them to form their environment, and they smooth out the temporally and spatially varying external conditions. Moreover, organisms actively construct a world around themselves and are in a constant process of altering this environment. Thus, he states: "The time has come when further progress in our understanding of nature requires that we reconsider the relationship between the outside and the inside, between organism and environment" (p. 47).

Also, Margulis (1990) proposes a new look at organism-environment interactions. In her search for an autopoietic concept of the organism and of the biosphere, the self-produced and -maintained boundaries play a central role that corresponds to the theory of autonomy. Turner (2007) proposes focusing more on the dynamic interaction between living organisms and their environment and the building of homeostatic units within this relationship, which he calls "Bernard machines."

Perhaps the message seems not so spectacular. The theory of increasing autonomy is a synthesis of material from several scientific disciplines, a rereading of the biological text. However, it is a rereading that opens our eyes for something that has been overseen.

It is not a theory in the sense of a specific model that produces predictions to be tested against rival models. Instead, it is a general theoretical and empirical perspective on the nature of major transitions, a framework both for conducting scientific research and for understanding the broader significance of research findings.

An understanding of what large scale evolution has generated will be an essential piece of the jigsaw puzzle, which the new evolutionary biology has to put together. How can we understand evolution if we do not even know what it produced? Of course, we know that prokaryotes and single-cell eukaryotes were the earliest organisms living in this world and that later there were bees, mammals, and birds. But, what are the general characteristics that changed? What is the qualitative difference between the nervous system of a polyp and that of an octopus, a lamprey, and a dolphin?

Diversity is not the only topic, at least during the major evolutionary transitions. Otherwise, all organisms, including ourselves, would be single cell, probably with a wide divergence of colorful and muddled variants. And then, what led to the appearance of human beings and their ability for culture and civilization, for arts and humanities? Is it just an accident of evolutionary variations, or can we find out more about this event?

In the following chapters, I first (Chap. 2) trace some of the history of the discussion on trends, directionality, and the question of progress in evolution and develop a proposal on how to deal with them in the context of modern evolutionary biology. Chapter 2 uses material published in more detail in an earlier paper (Rosslenbroich 2006).

Then, in Chap. 3, I present the work of some authors who previously mentioned the principle of autonomy. I introduce the concept of biological autonomy and its changes, give a definition, and describe its general principles as I understand them so far.

Chapters 4, 5, 6, 7, 8, 9, and 10 present biological arguments for the concept during animal evolution. Most of the evidence presented is well known, and insiders will recognize many examples from textbooks of physiology, morphology, and paleontology, complemented by some results that are more recent. The interesting point, however, is that a slight, but crucial, turn in the perspective on these facts reveals a different – and presumably more appropriate – image of the underlying process than is usually transmitted. For readers more interested in the philosophical content of the theory, this part may seem long. However, I tried to balance these chapters somewhat: From an empirical point of view, many additional arguments and facts could be presented, but I limited these descriptions and tried to make them as generally understandable as possible. The reader may judge whether I was successful in this attempt. It is possible to read these chapters selectively, although that risks missing important evidence nature itself presents.

Chapter 11 discusses features of increasing autonomy during the evolution of man. Man, however, is not the most autonomous organism on Earth. But, with his special combination of autonomous features, he has the biological prerequisites to generate a world that leads far beyond the biological realm: culture. It is proposed that the theory of autonomy can be an important component of the answer to the outstanding question of how man and his cultural capacities can be linked to the evolutionary history of life. Thus, it is suitable to build a bridge between nature and culture.

Man is neither determined by his nature nor has he dissociated himself from the biological roots. Rather, the biological underpinnings are the basis we constantly use and act on. The relative autonomy of our physical and physiological organization forms the prerequisite for all those features, which are specifically human, including certain degrees of freedom.

Chapter 12 discusses the value of the theory for understanding the major transitions in evolution. It is argued that diversity is only half of the truth. The other half is that organisms maintain and expand their capacity for self-assertion and self-regulation, culminating in high degrees of flexibility and possibilities of some organisms within the environment. At the same time, it is demonstrated that there is no such thing as a linear increase in autonomy. There is rather a bush-like course of evolution as it is described by modern evolutionary research, and there are complicated ways leading to different combinations of features of autonomy.

Chapter 12 includes a brief summary of the evolutionary theories that are presently under discussion and considers the contribution of the theory of autonomy to these new developments.

# Chapter 2
# The Problem of Macroevolutionary Trends

Our general view of the large-scale evolutionary process reveals prokaryotes as the earliest forms of life, followed by the first eukaryotic cells that formed multicellular organisms. The Cambrian "explosion" added new forms of life with hard skeletons, completely changing the fauna of the world. Within the then-existing phyla, a great variety of changes led to our present-day animals, including bees, squid, frogs, crocodiles, and horses. Every evolutionary biologist thinks that there were profound changes and innovations during this process that need to be described. Traditionally, these changes have been termed *evolutionary progress*. In recent years, this term has been criticized, and some authors claim that it has now been successfully eliminated from evolutionary biology. Nonetheless, on closer examination this seems not to be the case. There are hardly any textbooks that avoid using the terms *lower* and *higher* when referring to organisms. Furthermore, many phylogenetic reconstructions, especially at the level of phyla, include sequences that lead to "advanced" forms in the traditional sense, and in zoology or paleontology textbooks organisms are usually arranged according to this sequence. The criticism of this notion in recent decades has had the effect that scientists try to avoid using terms that refer to evolutionary progress, or they explicitly distance themselves from it, although nearly everyone still thinks of evolution in the sense of overall progression.

Ruse (1996) came to the same conclusion. His question is: Why do evolutionists continue to use such unscientific terms? However, my thesis is different from his: The term *progress* carries some historical burden, as it is problematic within the modern view of evolution; but, at its core, it expresses a central aspect of evolution that cannot be ignored if it is intended to build a fairly complete view of the evolutionary process that comes close to reality. If evolutionists cannot avoid the term, what do they see in their daily work, and how could they express their observations in a scientific manner?

The term *progress* has three predominant historical roots. One is the concept of the *scala naturae*, which until the nineteenth century was the most widely prevalent view of the general order of the world. It saw the world arranged in a linear hierarchy and was originally a static concept, but during the late eighteenth and early

B. Rosslenbroich, *On the Origin of Autonomy: A New Look at the Major Transitions in Evolution*, History, Philosophy and Theory of the Life Sciences 5,
DOI 10.1007/978-3-319-04141-4_2, © Springer International Publishing Switzerland 2014

nineteenth centuries, it was "temporalized" (Lovejoy 1982) so that its elements would appear in succession. The second root lies in the notion of *social and cultural progress*, which developed during the Enlightenment and gradually replaced the notion of the invariability of human affairs. During the late eighteenth century, this idea expressed the emerging consciousness of the capability of humankind to improve its circumstances and abilities. In these early considerations, progress included the aim of an achievable perfection, which introduced a strong teleological element. Critical reflection on the questions of change, development, and progress in human history, including its problems of linearity and teleology, took place in France and Germany during the eighteenth and early nineteenth centuries. This is the true origin of evolutionary thinking on which Darwin later could build his theory. During the nineteenth century, the general progress of society, science, technology, and industry was taken for granted, especially in England (Bowler 1983, 1989a, b).

The third root is the *theory of recapitulation*, the analogy that was drawn between embryogenesis and phylogeny. Knowledge of the embryo's development from a simple to a complex structure was intellectual help for initial ideas about the changeability of organisms (Richards 1992).

For a clear picture of the notion of progress in evolutionary biology, it is necessary to reflect on the different components and connotations that may be involved in varying combinations and derive from this historical background. At least five components must be distinguished:

1. Modifications in the living world generate increasingly higher organisms (however they are characterized).
2. These higher organisms are in certain ways better than lower ones (= "improvement").
3. This progression is essentially linear.
4. Evolution has an intrinsic force that drives this progress.
5. Progressive evolution leads eventually to some sort of perfection (end stage, culmination point, goal).

In the critical literature of recent decades, these components are often mixed together, contributing to the confusion. For example, it is often assumed that the notion of progress is always looking for related driving forces in evolution (component 4). Or, it is supposed that the view of evolution as progressive implies a goal toward which the process is moving, thus making evolution a teleological concept. This supposition would be a combination of components 4 and 5. However, these components are not necessarily involved. It is true that most biological thinking before Darwin's theory was introduced into science included these components, but afterward the picture became more varied.

How did Darwin himself deal with these components in *On the Origin of Species* (Darwin 1872)? Darwin unequivocally disapproved of any idea of an inherent force that was supposed to be driving evolution. That the process should have a goal was also incompatible with his theory and was explicitly refuted, as was the idea of linearity in evolution. He argued repeatedly throughout the book against contemporary advocates of such views, and their refutation was one of his main concerns. This was an important achievement in his time.

The problem of whether the evolutionary process might generate higher organisms is nonetheless complex in Darwin's thinking, and it hides a dichotomy. His theory is mainly an explanation of how populations adapt to their changing environments and to their biotic factors. The theory of natural selection maintains that in the struggle for existence, those individuals who best adapt to new conditions will survive and reproduce, whereas others less well adapted will become extinct. Over many generations, positive adaptive characteristics are enhanced until eventually the population becomes a new species, incapable of interbreeding with the parent form. Neither the variations in features from which an adaptive characteristic is selected nor the environmental changes include directionality. Therefore, this process can deliver only a set of meandering responses in the adaptive adjustments of organisms to local environments (Gould 2002; Bowler 1989a). Depending on the respective selection factors, this process would lead to an ever-increasing divergence of forms independently of one another and result in a network of adaptations to the respective vicissitudes of the struggle for life. According to the traditional understanding of progress, this process does not include any directionality. In addition, it does not seem to make sense to compare different "levels of organization" when the main reality is branching evolution.

Through competition between individuals and victory of some creatures over others in the struggle for limited resources, direction might be involved (Gould 2002) after all. Darwin expected an accumulation of "improvements" from the struggle, which would make organisms fitter and thus generate progress. "Now species triumph because, in some sense admittedly difficult to define, winners are 'better' than the forms they vanquish. And the more uniformitarian the larger picture – the more that macroevolutionary pattern arises as a simple summation of immediate struggles – so do we gain increasing confidence that replacement and extinction must record the differential success of globally improved species" (Gould 2002, p. 475).

In a paragraph, "On the Degree to which Organisation tends to advance," Darwin (1872, pp. 127, 228) writes:

> "Natural Selection acts exclusively by the preservation and accumulation of variations, which are beneficial under the organic and inorganic conditions to which each creature is exposed at all periods of life. The ultimate result is that each creature tends to become more and more improved in relation to its conditions. This improvement inevitably leads to the gradual advancement of the organisation of the greater number of living beings throughout the world. … Although we have no good evidence of the existence in organic beings of an innate tendency towards progressive development, yet this necessarily follows … through the continued action of natural selection. For the best definition which has ever been given of a high standard of organisation, is the degree to which the parts have been specialized or differentiated; and natural selection tends towards this end, inasmuch as the parts are thus enabled to perform their functions more efficiently."

Thus, increased specialization and differentiation of parts make their bearers superior to other ones in the struggle for life.

> If we take as the standard of high organisation, the amount of differentiation and specialization of the several organs in each being when adult (and this will include the advancement of the brain for intellectual purposes), natural selection clearly leads towards this standard: for all physiologists admit that the specialization of organs, inasmuch as in this state they perform their functions better, is an advantage to each being; and hence the accumulation of variations tending towards specialization is within the scope of natural selection. (Darwin 1872, p. 128)

Hence, by means of the selection process Darwin intends to explain not only adaptation to the immediate environment but also gradual progress.

> "Thus, from the war of nature, from famine and death, the most exalted object which we are capable of conceiving, namely, the production of the higher animals, directly follows" (Darwin 1872, p. 560).

How Darwin envisaged large-scale "improvements" of organs and higher animals corresponding with increasing fitness for survival is shown, for example, in his paragraph on eyes (Darwin 1872, p. 188).

However, there exists a gap in this extrapolation from microevolutionary adaptive processes to large-scale macroevolutionary progress: Those species, which are supposedly more advanced, do not necessarily have an enhanced capacity for survival. If continuous improvements accumulate toward progressive forms (e.g., through the generation of complex organs as Darwin expected), the bearers of these improvements must finally be the fittest organisms, which is not the case in nature. Bacteria as well as protists had enough fitness to survive for a longer time than even vertebrates. The accumulation of complex organs and functions from single cells to vertebrates (which is usually referred to as progress, also by Darwin) delivers anything but enhanced survival capacity. For this reason, there is on one hand incongruence between the microevolutionary adaptation process that leads to fitness and on the other hand what is traditionally called progress in the large-scale, macroevolutionary outcome of evolution. However, this does not question the validity of either of the two principles but only states that their relationship is not clear within the original Darwinian scenario. It also does not mean that there might not be a resolution of the incongruence, but one has not yet been definitively established as differing views prevail.

It is hard to judge how clearly Darwin saw the incongruence, but his ambivalence concerning the term *progress* may hint at his struggle with it. When Darwin's theory is read as it was presented in his "ambiguous book" (Ruse 1996, p. 172), it includes a dualistic tension between populational thinking and progressive thinking, two not necessarily irreconcilable but instead complementary perspectives of a complicated matter. In one respect, Darwin wrestled with the notions of directionality and progress, and in another respect, he wrestled with his principle of an ever-branching and diverging evolution. Darwin's theory abandons elements of linearity, an intrinsic force, and a goal of evolution, but still makes an attempt at explaining "organisational advance," as he formulated it. I maintain that Darwin saw this tension and attempted to deal with the observable differences between organisms rather than focus on a more radical and reductionistic theory, which would ignore a significant part of reality.

The view that progress might be a simple accumulation of fitness is not necessarily an element of modern Darwinian thinking, but the relationship between fitness and progress has remained unresolved since the days of Darwin (Saunders and Ho 1981; Wicken 1979; Nitecki 1988; Calcott and Sterelny 2011; McShea 1991, 1998; Jablonski 2007). Gould (1996, p. 199) points out this relationship: "I have long been entirely underwhelmed by the standard arguments for general advantages of increasing complexity in the Darwinian game – adaptive benefit of more elaborate bodily form in competition for limited resources, for example. Why should more complex conformations generally prevail? … I can envisage just as many situations where more elaborate forms might be a hindrance – more parts to fail, less flexibility because all parts must interact with precision." Remember that Darwin expected functional improvements through the building of organs that were more complex, and today complexity is often equated with progressiveness.

Darwin's ambivalence and the inconsistencies in his theory led to diverging attitudes among many scientists, thus establishing at least two fundamentally different views of evolution, which have remained relatively divergent at all times. This schism pervades all evolutionary biology with a spectrum of opinion ranging from the presupposition that evolution generated progress in some form, to a complete denial of any sort of progress in it whatsoever. The history of ideas after Darwin shows how these different perspectives and their dualistic tension have always been at work (for more details, see Rosslenbroich 2006).

## 2.1 The Epistemological Problem

During the twentieth century, the term *progress* came under pressure from two different directions: One is that it transports some historical baggage that was not compatible with modern knowledge of evolution. The other is the dominance of thinking in adaptation and population dynamics, as it was strongly favored by the synthetic theory. Scientists expected the solution of evolutionary questions exclusively from this perspective. Also, major transitions seemed to be explainable through accumulated microevolutionary events. Against this backdrop, there was no interest in general macroevolutionary questions. There were even strong attempts to discuss any general qualitative changes away or to dismiss them as epiphenomena in a world that consisted exclusively of adaptation and fitness. Good examples for this are many depictions of brain organization in vertebrates (see Chap. 10). However, this is changing dramatically today, as is discussed in Chap. 12 (McShea and Simpson 2011).

At its core, the term *progress* expresses, nonetheless, the observable qualitative differences between organisms of different evolutionary levels. However, because of the dominance of thinking in terms of adaptation, not enough thought has been put into the question of these qualitative differences. There is an aspect of organic evolution to which the term has been applied and for which it is necessary to develop an epistemologically satisfying approach. The persistence in the use of the term and of related terms proves this.

In the first place, the five components in the previous list require further explication. It should be clear that accepting the notion that large-scale trends reveal increasingly higher organisms (1) does not necessarily include an agreement with the other components. Thus, the process does not need to be linear (3). Equally, it presupposes neither some sort of inherent force (4) nor a final stage, or goal even, that supposedly drives the whole process (5). Furthermore, it does not necessarily include the idea that new forms are in some sense "improved" or "better" (2); they may just have a different lifestyle, a different adaptation strategy, or a different type of general morphological and physiological organization. This much can be definitively stated, leaving open for the moment the question of the evolutionary forces that generated such differences.

In today's use of the term *progress*, when we are simply looking into the history of organisms, reconstructing and describing the sequence of basic changes, components 2 to 5 are not necessarily taken into account. In the history of theory building, these components have been abandoned. However, component 1 is still relevant. Thus, a modern interpretation of the term *progress* would accept only that macroevolution generates forms that increasingly differ from earlier forms in such a basic way that it is necessary to provide a description and analysis of the general patterns involved. For this, the terms *higher* or *lower*, and even the term *progress* itself, have until now served as metaphors.

Another basis for my further discussion comes from a proposal by Rapp (1992) in a study of the term *progress* (including social progress), who distinguishes a *genetic* (in the philosophical sense) from a *normative* form of progress. Genetic progress is a sequence of steps in time: the succession of changes and the valueless generation of the new. In addition to this, normative progress makes value judgments in the sense that every progressive step achieves an improvement with respect to a higher goal toward which it is worth striving morally. The genetic term is the prerequisite of the normative term, but the positive value judgment of the normative term can be transferred to the genetic one, either tacitly or explicitly. However, it is not certain that genetic progress always leads to improvement. This confusion stems from the historical link between the terms when they were first used.

The historical overview shows how closely the use of the term in biology has been connected with the development of social thought and theories. The notion of normative progress helped to start thinking in terms of developments, including those in the organic world. However, the normative aspect, in the sense of moral appraisal, cannot be introduced into a scientific context. Nonetheless, even the genetic aspect may contain several different elements that at the same time are possible elements of the term *evolution*. This is summarized in Table 2.1, which is compiled from considerations by Rapp (1992), Lewontin (1968), and Simpson (1973).

*Change* is the basic feature of the evolutionary process: The current state of a system is the result of one or more changes from a former state. Pure change can be the raw material of evolution, but for most evolutionists, the term *change* does not describe evolution sufficiently, as change can also be, for example, reshuffling playing cards. Thus, a different *order* can be generated by a rearrangement of elements.

**Table 2.1** Possible elements of "progress"

| | | | |
|---|---|---|---|
| Genetic: | - Change | Progression | Progress |
| | - Order | | |
| | - Direction | | |
| | - General patterns | | |
| Normative: | | | |

Using the same basic elements, it leads to the appearance of a new state in a system that was not present in the system in its former condition. The generation of a new species from an earlier one might be the generation of a new state, a new order.

The generation of a new order and properties can shift irregularly or can change in a certain *direction*, perhaps over a long time, by one or more sequences of transformations. However, these changes need not be linear. Directionality, revealed by the fossil record, for example, is usually described as an "evolutionary trend." Thus, the evolution of early mammals from mammal-like reptiles is described in paleontology as a sequence of trends that led to mammalian characters.

These trends can be described and followed throughout evolution, but often, especially in large-scale macroevolutionary trends, more basic questions can be addressed: Did a general difference evolve? Are organisms in later periods different in some general aspect of their individual characteristics from those in earlier times (McShea 1998)? The evolution from reptiles to mammals, for example, involved the generation of a largely different physiology, allowing for a completely changed lifestyle. The two classes reveal different sets of characteristics with respect to morphology, physiology, behavior, and relation to the environment. These basic differences of systems are described here as *general patterns*, expressing integrative features of large-scale macroevolutionary trends.

Evolutionary theories can be distinguished by how many of the elements (see Table 2.1) are included (Lewontin 1968). Some evolutionists only include change and order; others add directionality. Although large-scale patterns are rarely addressed explicitly (e.g., Bonner 1988; McShea 1996, 2002; Vermeij 1999; Calcott and Sterelny 2011; Jablonski 2007), they are often embedded in general discussions and in textbooks. This clearly shows that the understanding of the term *evolution* to a large extent depends on the perspective on evolution and the paradigmatic background of the respective researcher. Much of the controversy concerning the term *progress* has its origin in these different views. In today's evolutionary biology, large-scale changes of general patterns are the heart of what is usually called progress. Because a term for this is needed, it will not be possible to eliminate it in the future, just as it has not been possible to avoid it in the past.

Simpson (1973) differentiates the term *progress*, which might include a normative undertone, from *progression*, which avoids assumptions about any kind of changes for the better (Table 2.1). I argue here that today the term is used in the sense of progression because no modern scientist would include a normative judgment.

A common criticism of the term *progress* is that it could be anthropocentric. However, a scientific description of large-scale patterns in the evolutionary process need not focus a priori on the characters of man. A large portion of the organic world went through an evolution that does not contain elements of the line toward human beings and so cannot be judged according to criteria generated from this line. Here, the perspective determines the traits observed as well as the systematic level chosen. With this prerequisite, the term *progress* is not necessarily anthropocentric. It just tries to describe large-scale patterns.

## 2.2 The Ontological Problem

The course of evolution, therefore, is not characterized by a process directed toward the generation of vertebrates and mammals. Instead, early forms of organisms were joined by forms with different general patterns. This is the case among not only vertebrates but also invertebrates and plants. What are the characteristics, then, of these lineages? What is the essential difference between a bacterium and a mammal or a squid? This is the question that remains at the center of this topic. In any case, the obstinacy, with which progress has remained, shows that a term is needed for referring to the underlying phenomena. Eliminating the term from the vocabulary of evolutionary biology is not the solution but rather a moratorium.

From the middle of the nineteenth century, there have been repeated attempts to establish standards, the first attempts stemming from Meckel (1821) and Bronn (1853, 1858). Most authors compiled lists of the patterns that should be considered valid (Rensch 1959; Remane 1967; Kämpfe 1985), but opinions diverged. Several attempts to operationalize patterns scientifically have been published, but they did not generate much interest from the scientific community (Dobzhansky et al. 1977; Kämpfe 1985; Rensch 1959; Simpson 1971, 1973; Storch and Welsch 1989; Wake 1986). On the other hand, there was always a certain general consensus regarding which organisms should be considered lower and which higher. McShea (1998) published one of the most thoughtful considerations about what might constitute "largest-scale trends."

Among the patterns mentioned most often is that of "increasing complexity," not always distinguished from "increasing differentiation." In recent decades, when the term *progress* has become the subject of criticism, the term *complexity* has often been used as a substitute. McShea (1991, 1996), however, shows that the definition of "*what everybody knows*" is unsatisfactory and predominantly based on general impressions rather than on scientific data.

Some authors just took it for granted that evolution generates complexity and saw it as a product of selective processes. Bonner (1988) and Rensch (1959), for example, argue that complexity should be favored by natural selection because organisms that are more complex are mechanically more efficient, having more parts and greater division of labor among different cell types. Others claim that relating complexity to fitness is problematic, and that it is not clear whether and how complexity contributes to fitness (Wicken 1979). Further skeptical discussions are provided by Williams (1966), Lewontin (1968), McCoy (1977), Gould (1985), and Hinegardner and Engelberg (1983). Other authors make attempts at defining complexity and making it measurable (McShea 1991, 1996; Saunders and Ho 1976, 1981; Papentin 1980; Thomas and Reif 1993; Finlay and Esteban 2009). McShea (1996) developed a conceptual basis for objective investigations and found trends of increasing complexity in some measurements but not in others. McShea and Brandon (2010) propose a concept concerning increasing complexity as a constant background condition of evolution.

Many authors see "increasing differentiation" as overlapping with complexity and use the phrase in the sense of division of labor. Formulated as "increase in the number of cell types" or "increasing specialization of cells," it may provide a measurable variable (Valentine et al. 1993; Bonner 1988). The number of cell types increased with the generation of multicellularity, but a count of cell types does not seem to be able to describe the difference between an amphibian and a mammal. Also, it is difficult to distinguish cell types using a standard for comparison. Increasing "differentiation and centralization of nervous systems" has always been a widely recognized pattern. Rensch (1959), for example, sees it as a typical characteristic of his "anagenesis" and analyzes it within mollusks, arthropods, annelids, vertebrates, and others.

Some authors observe "increasing efficiency" of tissues, organs, or the whole organism, citing examples such as performance capacities of nervous systems and sense organs, biomechanical changes in movement capacities, and enhancement of metabolic capacities. However, it may be argued that also in lower organisms tissues are structured to fulfill effectively the functions for which they have been adapted. A similar argument may be applied to an increase in the "effectiveness of adaptation," whereas efficient environmental adaptation has occurred within all different phyla and classes without providing the ability to describe their general differences.

Some authors hope that an increase in the "amount of genetic information" would be a measurable trait of progress, but correlation between the supposed level and genome size turned out to be poor. This first appeared as the "C-value paradox" and has become more complicated in recent years.

Body size, of undoubted importance for many physiological, ecological, and life-history traits, is one of the general features of organisms most often deliberated. Based on a rough scale from bacteria to whales, several authors consider that the "increase in maximal body sizes" during successive periods of life on Earth is a pattern of progress (Bonner 1988).

"Energy intensiveness" has been studied in greater detail, although not always in the context of progress (McCarthy and Enquist 2005; Milewski and Mills 2010). Vermeij (1999) proposes that interactions of individuals within the adaptation process can be expected to lead to increases in size, having a higher productivity or metabolism, engaging in a larger number of interactions, and performing more functions at a higher level. Overall, they may show a pervasive increase in energy flux.

Within these various considerations, the pattern of an increasing autonomy of organisms in the sense of an "emancipation from the environment" is often mentioned. Here, again, as with some of the other criteria, authors usually give some examples, but there is no single attempt to define the pattern more closely or to describe the phenomena systematically. More about this literature follows in Chap. 3.

## 2.3 A Biological Dilemma?

Greene (1986, 1991) states that evolutionary biologists can live neither with nor without the idea of progress, a situation he refers to as a "biological dilemma." Attempts have been made to ban the term from evolutionary biology, however, with limited success. On the other hand, there have recently been attempts to reexamine the question of a directional trend in macroevolution and its possible conditions (Calcott and Sterelny 2011; Conway Morris 2003). A recurring question is how it can be detoxified from the connotations the term *progress* carries and how the idea of progressive change can become empirically tractable. This has motivated attempts to decouple work on large-scale trends from directional and progressivist ideas of history. McShea and Simpson (2011) insist that the project to examine the sense in which living beings constitute a series from the simplest forms at one end to humans at the other is legitimate. And, books are still published with titles such as, *Life Ascending* (Lane 2009), in which some of the major innovations along the path are presented.

Sterelny (2011) essentially formulates the same point when he states that evolution is a history of extraordinarily fecund changes. He describes that many lineages have seen the evolution of complex morphological innovations in the macrobes (sensory systems, locomotion, internal structural systems for circulation and support) and metabolic innovations in the microbes (nitrogen fixation, photosynthesis). Other lineages have histories of extraordinary diversification. Perhaps most striking, there has been directionality in evolutionary history. There has been an evolutionary trend of a special kind: the evolution of new forms of organization, the eukaryotic cell, multicellularity, eusociality. These transitions expand the space of biological possibility. The upper bound on possibilities is not fixed; rather, it is moved by such key innovations.

Developing a way of describing general macroevolutionary changes within a sound epistemological framework should be valuable for understanding evolution. The study of large-scale general patterns is a necessary element of the new evolutionary biology presently under construction. The study of directional qualitative shifts in morphology, physiology, molecular biology, and other fields of interest contributes to a more complete view of evolution. Within such a program, the "biological dilemma" might rather be part of the fruitful tension as Darwin implemented it.

In such an approach, the old metaphor of "progress" may be modified or understood in a new empirical light. This is a third direction between using the term indiscriminately on one hand and on the other hand ignoring the general patterns that evolution has produced. For the rest of the book, I leave the whole topic of progress and concentrate on a contribution to the question of large-scale patterns in evolution.

# Chapter 3
# The Concept of Biological Autonomy

## 3.1 Systems Biology

The principle of autonomy as a trait of living organisms has been discussed within some approaches that can be subsumed under the field of systems biology. Although the field is heterogeneous and covers ideas that are not fully unified, it has a common agenda in the search for approaches to understand the coherence of functions within a living system. In general, it attempts to understand whole systems through an integrative view of all known regulatory and molecular processes.

Systems biology was founded in the 1930s by Paul Weiss and Ludwig von Bertalanffy and received further stimuli from cyberneticists such as Norbert Wiener and W. Ross Ashby, mathematical biophysicists Nicolas Rashevsky and Robert Rosen, systems engineer Mihajlo Mesarović, and systems theorist James Miller (O'Malley and Dupré 2005). For several decades, systems biology has been making its progress in the shadow of genetics, molecular biology, and other analytical disciplines. However, since the beginning of the 21st century, systems biology has become one of the most widely discussed fields of modern biology (Noble 2006, 2008; Soto et al. 2011; Kitano 2002a, b).

There is emerging some consensus that the analytical approaches of many fields of modern science require a move from the dissection of things to the dynamics of processes and to the question of how all these mechanisms, which are being studied in increasing detail, are integrated into a coherent whole, an organism. However, there is still need for the development of a clear account of what biological systems are and how the respective definition affects research agendas (O'Malley and Dupré 2005; Rosslenbroich 2011). The emerging consensus revolves around understanding biology as a science of systems with dynamic stability (Kather 2003). Within such an understanding, autonomy is at least implicitly present within the term *stability*. However, there are several schools of thought that can be included in the field of systems biology in a wider sense and that explicitly discuss autonomy as a fundamental characteristic of living systems in general.

B. Rosslenbroich, *On the Origin of Autonomy: A New Look at the Major Transitions in Evolution*, History, Philosophy and Theory of the Life Sciences 5,
DOI 10.1007/978-3-319-04141-4_3, 

## 3.2 Autopoietic Systems

Some of these schools are inspired by the formulations of Humberto Maturana and Francisco Varela, who introduced the term *autopoiesis* as a description of living systems (Maturana and Varela 1987; Varela 1979, 1981; Luisi 2003; Barandiaran and Ruiz-Mirazo 2008; Kauffman 2003; Margulis and Sagan 2002; Di Paolo 2004, 2005).

A living system is generally described as an autopoietic unit capable of sustaining itself because of an inner network of reactions that generate and regenerate all the system's components. All the pertinent processes needed to maintain the network within a living system have their efficient cause within the system itself. The structures, based on a flow of molecules and energy, produce the components that, in turn, continue to maintain the organized bounded structure that gives rise to these components. Self-reference and automaintenance are central notions for this approach (Luisi 2003; Roth 1981; an der Heiden et al. 1985; Ruiz-Mirazo and Moreno 2012; Kather 2003). Coherent and ordered global behavior of the system constrains or governs the behavior of the individual components so that they no longer have the same behavioral alternatives as outside the system. At the same time, the behavior of the components generates and sustains the global order (Thompson 2007). This two-sided or double determination is known as circular causality (Haken 1983).

Varela founds his considerations on the idea that a living system maintains its specific organization through the active compensation of deformations (sometimes called perturbations). Here, Varela invokes Cannon's notion of homeostasis, which he expands by making every reference to homeostasis internal to the system itself through mutual interconnections of processes and by positing this interdependence as the source of the system's identity as a unit. Thus, all homeostatic operations in organisms are efficiently caused from within the system, and it is the continued existence of the set of causally dependent processes that constitutes the continued existence of the system (Bechtel 2007).

Because autopoietic systems actively distinguish themselves from their surroundings, they are autonomous: "In fact, the notion of autopoiesis can be described as a characterization of the mechanisms which endow living systems with the property of being autonomous; autopoiesis is an explication of the autonomy of the living" (Varela 1981, p. 14). An autonomous system acquires the property of specifying its own rules of behavior (Luisi 2003). Such systems need to be seen as sources of their own activity, specifying their own domains of interaction, not as transducers or functions for converting input instructions into output products (Thompson 2007).

Thompson (2007) describes this autonomy for a single cell: The cell stands out of a molecular soup by actively creating the boundaries that set it apart from what it is not and simultaneously regulate its interactions with the environment. Metabolic processes within the cell construct these boundaries, but the metabolic processes themselves are made possible by those boundaries. In this way, the cell emerges as a figure out of a chemical background. Should this process of self-production be interrupted, the cellular components no longer form a unit, gradually diffusing back into a molecular soup.

The existence of a boundary is a central element of a living system (Luisi 2003; an der Heiden et al. 1985). Inside the boundary of a cell, many reactions and chemical transformations occur; the cellular membrane encloses a defined reaction room, thus contributing to the maintenance of the cell's identity. At the same time, the membrane establishes and regulates contact to and exchange with the environment.

Thompson (2007) qualifies the necessity of a strict physical boundary for an autonomous system. He states that a system can be autonomous without having this sort of material boundary; the members of an insect colony, for example, form an autonomous social network, but the boundary is social and territorial, not material. Autonomous systems are organizationally closed in the sense that their organization is characterized by their internal network processes, which recursively depend on each other and thus constitute the system as a unit. These processes generate a far-from-equilibrium situation as long as the system is living. Equilibrium with the processes in the environment arises when the system is dead. At the same time, living systems are materially and energetically open to their environment. They receive energy and nutrients from the environment and excrete products and waste. Luisi (2003) emphasizes that there is an interesting contradiction between biological autonomy and dependence on the external medium and that all living organisms must operate within this contradiction.

In a series of papers, Moreno and coworkers work toward an understanding of a most basic form of autonomy of living organisms (Ruiz-Mirazo and Moreno 2004, 2012; Moreno et al. 2008). They see autonomy as a fundamental characteristic of life and stress explicitly the significance of the principle for understanding the origin of early life on Earth. A motivation for their search for a basic autonomy is to provide a link between this fundamental principle of life and physics and chemistry, so that the idea of autonomy itself is naturalized and can serve as a bridge from the nonliving to the living domain. Because they are crucial for the generation of simple self-maintaining and self-constructing systems, they understand that these systems must engage in an interactive loop with their respective environment across some boundary condition (gradients, influx/outflux of different compounds, energy transduction mechanisms, etc.) to sustain the processes of generation of internal "order" in accordance with the generalized second law of thermodynamics.

Moreno et al. describe that, unlike physical or chemical dissipative structures, in which patterns of dynamic order form spontaneously but whose stability relies almost completely on externally imposed boundary conditions, autonomous systems build and actively maintain most of their own boundary conditions, making possible a robust far-from-equilibrium dynamic behavior. Thus, a central question is how a system develops the capacity to channel the flow of matter and energy through itself to achieve robust self-construction (i.e., self-construction that includes regulation loops with its immediate environment).

Thompson (2007) introduces the distinction between heteronymous and autonomous systems. Whereas heteronomy literally means other-governed, autonomy means self-governed. A heteronymous system is one whose organization is defined by input-output information flow and external mechanisms of control.

Traditional computational systems and many network views, for example, are heteronymous: They have an input layer and an output layer; the inputs are initially assigned by the observer outside the system, and output performance is evaluated in relation to an externally imposed task. An autonomous living system, however, is defined by its endogenous, self-organizing, and self-controlling dynamics and determines the domain in which it operates. It has input and output; however, these do not alone determine the system. It is the internal self-production process that controls and regulates the system's interaction with the outside environment. For Thompson, the principle of autonomy is essential for understanding principles such as intentionality and subjectivity of living entities, which in complex forms generate a continuity of life and mind. He attempts to understand the relation between these entities by his "enactive approach," focusing on the conditions of this continuity.

## 3.3 Philosophical Description of Organismic Autonomy

Fuchs (2009a) gives a description of the concept of organismic autonomy to prepare a view of the human neurophysiologic functions that is more integrative. He draws on results from ecological and philosophical biology with its main exponents J. von Uexküll (1973), Plessner (1975), and Jonas (1966) and those of system theories such as those of Bertalanffy (1973) and Maturana and Varela (1987).

Fuchs also describes living beings as complex entities or systems that maintain themselves in form and structure within time, although there is a continuing exchange of substances with the environment. This maintenance is an active self-organization as the organism subordinates the substances under its own principles and transforms and integrates them. They gain new properties, which they only have within the systemic context of the organism. Fuchs points to an example: The ferrous ion in hemoglobin behaves differently from iron in the outside world – it does not oxidize irreversibly but is able to bind oxygen reversibly, which is a crucial prerequisite for the turnover of energy in animals.

Beyond this, metabolism leads to a transformation of substances during decomposing digestion and resynthesis. The nutritional components are transformed into substances with the characteristics of the organism and integrated into its processes. By means of these dynamic processes, the living being encloses itself from the environment and gains – in different degrees – self-determination or autonomy. This means that its processes and its behavior are not primarily determined from the outside but rather depend on its internal disposition and condition. External influences predominantly are stimuli, which are answered by reactions of the whole organism, rather than causal effects as in mechanical cause-and-effect relations, as long as they are not destructive.

The basis for autonomy is the special interdependence between the whole and its parts within the organism, which include a differentiation in subsystems and organs. Although the organism consists of the sum of its macromolecules, cells, organs, and circulatory and nervous systems, it has a different relation to these component parts than a crystal to its components. The organism is itself the condition of its parts because it enables their existence. It produces and reproduces them while consisting of them. Self-maintenance is continuing self-generation. At the same time, the parts fulfill their respective functions within the organism and contribute to its overall functioning.

Of course, Fuchs also describes that the autonomy of living beings is not possible in autarky. The organism only gains its sovereignty for the price of certain requirements. The changing substances need to be available and incorporated to maintain homeostasis. Thus, organisms are always in need of factors from their environment (Jonas 1966).

According to Plessner (1975), Fuchs further describes that plants exhibit a predominantly open relation to their environment, whereas animals have a more closed form of organization. In animals, the exchange surfaces for metabolism are turned to the inside. Special internal organs and internal cavities appear, while exchange surfaces on the outside are reduced. Thus, animal life steps to a certain extent out from the direct environmental relation. The enclosure from the environment requires – on the other hand – a sensorimotor interzone, which restores the contact with the environment, however, on a new level. This condition shows separate organs for sensory and motor activity and their central nervous connections. The principle of a closed-body organization enables the independent movement of the animal.

According to Fuchs, the loss of a direct environmental relation corresponds to a gain in degrees of freedom. Whereas the mimosa reacts directly to touch, the stimulus-response relationships in animals tend to be less tightly connected. Animals tend to modulate a reaction so that the probability of a certain behavior can be modified. Signals can internally be enforced, compared to other signals, and memorized. Thus, not a rigid, but rather a flexible relation between organism and environment emerges.

## 3.4 Robustness

In recent years, a somewhat-new term developed in some areas of molecular biology. It was increasingly comprehended that many structures and functions as well as proteins and genes have certain stability in the face of environmental variations and genetic changes. Many physiological and developmental systems are resistant or "robust" to such perturbations. That is, despite these natural perturbations, the systems produce relatively invariant outputs (Masel and Siegal 2009; Masel and Trotter 2010; Stelling et al. 2004; Wagner 2012; Kitano 2004, 2007; Gerhart and Kirschner 1997; Larhlimi et al. 2011). Robustness is understood as a property that allows a system to maintain its functions against internal and external perturbations and uncertainties. It encompasses a broad range of traits, from macroscopic, visible traits to molecular traits, such as the expression level of a gene or the three-dimensional conformation of a protein.

"Biological systems maintain phenotypic stability in the face of diverse perturbations arising from environmental changes, stochastic events (or intracellular noise), and genetic variation. It has long been recognized that this robustness is an inherent property of all biological systems and is strongly favored by evolution" (Stelling et al. 2004, p. 675). Masel and Siegal (2009) see it as impossible to understand whole biological systems without understanding their robustness. Stelling et al. (2004) note that robustness encompasses a relative, not an absolute, property because no system can maintain stability for all its functions when encountering any kind of perturbation.

Robustness is concerned with maintaining the possibility of a system to function rather than maintaining an actual state of a system. Thus, Kitano (2007) differentiates it from stability and homeostasis, which predominantly describe a function that keeps a condition relatively constant. A system is robust as long as it maintains functionality, even if it transits to a new steady state or if instability actually helps the system cope with perturbations. Such transitions between states are often observed in organisms when facing stress conditions. One such condition can be extreme dehydration, to which some organisms can react with a dormant state, becoming active again on rehydration. These examples of extreme robustness under harsh stress conditions show that organisms can attain an impressive degree of robustness by switching from one steady state to another rather than trying to maintain a given state.

Wagner (2012) divides the perturbations that can affect a phenotype into two broad categories. The first consists of environmental perturbations. These include changes in an organism's exterior environment, such as changes in temperature, in available nutrients, or in the abundance of other organisms, such as potential prey. They also include changes in an organism's internal environment, such as temporal fluctuations in gene expression levels, which are caused by ubiquitous intracellular noise. The second kind of perturbations is mutations, changes in an organism's genotype. Mutations affect an organism more permanently than environmental change because the changes they cause are readily inherited from generation to generation. For this reason, Wagner states that they are especially an important object of study for students of evolution.

Because the term *autonomy* describes living systems as actively distinguishing themselves from their surroundings (see Definition 1 further in the chapter), it overlaps to some extent with the term *robustness*. However, it is not congruent with it. Robustness can be seen as a prerequisite for autonomy. Self-determination and self-maintenance need robust functions to defy perturbations from the nonbiological and biological surroundings as well as from the internal variability.

However, it is also justifiable to regard robustness as a part of autonomy itself. Robustness, also in different actual states of a system, maintains basically that the system is kept in a far-from-equilibrium state. Even dormant forms are different from their immediate surroundings in a self-organized manner, including when the metabolism is completely reduced. If the system becomes like the surroundings, this results in an equilibrium state and death.

Stelling et al. (2004) mention the important point that the primary function of a system may usually be robust to a wide range of perturbations, whereas the system can show extreme fragility toward other, even seemingly smaller, perturbations. They think that the coexistence of extremes in robustness and fragility ("robust yet fragile") perhaps constitutes the most salient feature of highly evolved complexity. Making one feature robust to a class of perturbations can make the same or other features fragile to that or other perturbations. In this sense, they expect a necessary connection between complexity and robustness.

In this discussion, several principles are seen as relevant for maintaining and establishing robustness (Stelling et al. 2004; Kitano 2004, 2007). One strategy to protect against failure of a specific component is to provide for alternative ways to carry out the function the component performs. This can be called "redundancy of components." At the genetic level, this backup strategy or "genetic buffering" (Hartman et al. 2001) might be brought about by duplicate genes with identical roles or by different genes that constitute alternative but functionally overlapping pathways. In contrast to redundant systems in engineering, however, identical genes that do not diverge in functionality or regulation would not survive in evolution. Instead, structurally different entities perform overlapping functions, which seems to be a common principle in organisms, on other levels in addition to the genetic.

A further principle discussed in this regard is that of "feedback circuits" (Stelling et al. 2004; Bechtel 2007). Control circuits play a decisive role in maintaining cellular functions in the face of internal or external uncertainties. By using the output of a function to be controlled to determine appropriate input signals, feedback enables a system to regulate the output by monitoring it. Negative feedback can reduce the difference between actual output and a given set point, thereby dampening noise and rejecting perturbations. Positive feedback can enhance sensitivity. This is primarily required for robust cellular decisions that need to be derived from noisy and graded input signals and to be maintained. Well-balanced positive and negative feedback can lead to a blend of sensitivity and stability. Another possibility for achieving higher robustness consists of combining multiple levels of regulation, for instance, controlled transcription, translation, posttranslational modification, and degradation. Often, when highly precise and reliable behavior is indispensable for overall cellular functionality, multiple intertwined feedback loops operate. The different levels of control for circadian clocks (Bechtel 2010a; Hogenesch and Herzog 2011; Mohawk et al. 2012) and developmental control circuits (Carroll 2005a, b) provide good examples of these aspects.

The principle of modularity might also contribute to the robustness of organisms. The composition of cells and of organisms from "functional units" or "modules" is under increasing discussion in the literature (Stelling et al. 2004). Modules constitute semi-independent entities that show dense internal functional connections but looser connections with their environment. Modularity, as the encapsulation of functions, can contribute to both robustness of the entire system (by confining damage to separable parts) and evolvability (by rewiring of modules or by modifications in modules that are not noticeable from the outside).

Finally, the integration of cellular functionality across hierarchies seems to be important. Stelling et al. (2004) describe that cells, which under normal operation provide a certain robustness of their behavior, can collectively reduce the impact of environmental perturbations when they are components of an organism network. Thus, the "collective of cells" inherits some of the cells' robustness, augmenting it by synergistic network-level interactions. An efficient means for coordination in such networks and in complex systems is to organize the system hierarchically, namely, to establish different layers of integration. This not only might reduce the costs of information transmission but also might further enhance robustness by different level regulations, multiplying each other.

## 3.5 Homeostasis

Homeostasis is the ability of a system to regulate its internal conditions to keep some or several functions stable. Examples are properties such as temperature or blood composition in animals.

The principle was developed by Claude Bernard and later by Walter B. Cannon. Bernard focuses on the internal organization of a living system to find causal principles that would allow a description of organisms as mechanically determined entities (Bechtel 2007). He argues that the internal parts of a living mechanism reside in an internal environment that is distinct from the external environment in which the organism as a whole dwells and that a relatively strict determinism could be found in their response to fluctuating conditions. The internal environment provides a buffer between conditions in the external environment and the reactive components of the mechanism, thus insulating component parts of the mechanism from conditions in the external environment. Bernard proposes that this buffering is achieved by individual components of the organism, each performing specific operations that serve to maintain the constancy of the internal environment. The constancy would render the organism independent from vagaries of the environment and would free the organism from environmental restrictions (Bernard 1859, 1878). Most famous is his formulation of the "milieu intérieur," a phrase he coined to refer to the extracellular fluid and its physiological capacity to ensure protective stability for the tissues and organs of multicellular organisms.

> The living body, though it has need of the surrounding environment, is nevertheless relatively independent of it. This independence which the organism has of its external environment, derives from the fact that in the living being, the tissues are in fact withdrawn from direct external influences and are protected by a veritable internal environment which is constituted, in particular, by the fluids circulating in the body. … The fixity of the milieu supposes a perfection of the organism such that the external variations are at each instant compensated for and equilibrated. … All of the vital mechanisms, however varied they may be, have always one goal, to maintain the uniformity of the conditions of life in the internal environment. … The stability of the internal environment is the condition for the free and independent life. (Bernard 1974, p. 188)

Walter Cannon (1932) introduced the term *homeostasis* for the capacity of living systems to maintain a relatively constant internal environment. He also sketched a taxonomy of strategies by which animals are capable of maintaining homeostasis (Bechtel 2007). The simplest involves storing surplus supplies in time of plenty, either by simple accumulation in selected tissues (e.g., water in muscle or skin) or by conversion to a different form (e.g., glucose into glycogen) from which reconversion in time of need is possible. A second type of homeostasis involves altering the rate of continuous processes (e.g., changing the rate of blood flow by modifying the size of capillaries to maintain uniform temperature). In this sense, it somewhat overlaps with the large field of cybernetics.

The concept of homeostasis was extremely successful in different biological and medical disciplines, such as physiology, and is elaborated and described in many details today. Homeostasis is an important means to maintain an autonomy of properties of organisms that is relatively stable against environmental fluctuations as well as internal functional variations. The organism itself sets the range at which a variable is maintained and uses several functions (e.g., positive or negative feedback) to achieve this autonomy.

Both the general formation of a difference toward factors of the environment and the buffering of fluctuations establish the autonomy of the respective function. Typically, several functions at a time can be regulated, which contributes essentially to overall autonomy of the organism and some independence from external conditions. However, organisms use this principle in different degrees and sophistication.

## 3.6 Time Autonomy

A continuous characteristic of any living entity is that it establishes its own sequences in time. Development, reproduction, metabolism, rest-activity cycles, and many other functions have their respective time order. This concerns not only the well-known circadian rhythmicity, which is an endogenous rhythm, synchronized with the daily cycle of the environment, but also all biochemical, cellular, and organic processes, with different arrangements of their duration and order of sequences (Hildebrandt 1979; Hildebrandt et al. 1987; Koukkari and Sothern 2006).

Basically, all chemical reactions need a certain time, the reaction rate. However, the cell regulates these reaction rates with the help of enzymes to integrate them into its own order of sequences. It performs extremely refined sequences by subordinating the reaction rates under its own time management. The emerging time order is typically characterized by oscillations, as chronobiology describes them. The crucial point is that the sequences in time are not adopted from the environment but directed by the rules of the organism itself. The oscillations of different frequencies are endogenous, and they are compensated for temperature. Only secondarily are they synchronized with cycles of the environment. In this sense, there is also an autonomy of time in living entities, as it is both robust and tunable (Gore and Oudenaarden 2009; Duboule 2003).

Circadian rhythmicity is an excellent example of an integrated system with interdependent functions and processes (Bechtel 2010a). After research tried for a long time to find the components of the oscillations along a linear feed-forward view, it is now becoming clear that there are multiple feedback loops between a central oscillator in the brain, several peripheral oscillators, and several sensory inputs. Thus, there are indications to the effect that it is an integrated circadian system, and that a step up to a systems level that considers interactions throughout the organism is needed to understand how circadian oscillators are entrained and influence other biological processes.

## 3.7 Organisms as Hierarchically Ordered Systems

Several of the concepts mentioned are grounded on a systems view of the organism, so it might be useful to take a closer look at the notion of a biological system. There have been several attempts to define or describe organic systems. However, in my view Paul A. Weiss, who was among the first to introduce the notion into biology, developed the most coherent and consequent definition (Rosslenbroich 2011; Drack and Wolkenhauer 2011; Drack and Apftaler 2007; Overton 1997; Köchy 1997).

Weiss (1963, 1968, 1969, 1971, 1973, 1977) sees a living system as an entity that imposes restricting (i.e., regulating) functions on its component parts so the functionality of the whole system is ensured. The system itself contains constituting properties and thus possesses information that does not stem from the parts themselves. The system must be regarded as a spatiofunctional entity that integrates the functions of its parts. It has an ontological weight of its own.

Weiss expresses this in his working definition of a system: "Pragmatically defined, a system is a rather circumscribed complex of relatively bounded phenomena, which, within those bounds, retains a relatively stationary pattern of structure in space or of sequential configuration in time in spite of a high degree of variability in the details of distribution and interrelations among its constituent units of lower order" (Weiss 1969, p. 11). Not only does the system maintain its configuration and integral operation in an essentially constant environment, but also it responds to alterations in the environment by an adaptive redirection of its componential processes in such a manner that the external changes are countered in the direction of an optimal preservation of its systemic integrity.

One such system is the cell: The cell hosts a number of components, such as organelles and molecules. However, the cellular system integrates all these components into a functional unit. It needs these components and depends heavily on them, but the cell is only able to live because of the regulation imposed on the components by the system.

"The basic characteristic of a system is its essential invariance beyond the much more variant flux and fluctuations of its elements or constituents" (Weiss 1969, p. 12). Therefore, the elementary functions of a system may be variable. This corresponds exactly with modern knowledge of the cell (Shapiro 2011): Whether and when information is transcribed from the DNA, whether certain proteins are built or which components are included in the cell membrane to keep it within an optimal stage of fluidity, constantly change according to the functional state of the cell and its environmental conditions.

> This is exactly the opposite of a machine, in which the structure of the product depends crucially on strictly predefined operations of the parts. In the system, the structure of the whole determines the operation of the parts; in the machine, the operation of the parts determines the outcome. (Weiss 1969, p. 12)

A cell has subsystems (i.e., the organelles) that perform partial processes. So, a mitochondrion can be seen as a subsystem that integrates the molecular devices for processing energy. Looking at the next-higher level beyond the cell, there is the tissue in which the cells are organized. Such a tissue is also a system in which functions of single cells are integrated and regulated. One example would be epithelium, in which a boundary is established by systemic cooperation of many cells. In this case, the system can have certain characteristics, such as a barrier, that are not characteristics of the single cells. They are a property of the association of the cells. A further possible level is constituted by the organs of an organism, such as a heart, a lung, or a liver. Finally, the organism integrates all these subsystems into a coherent whole.

Thus, the integral systems operation, whether of the body as a whole or of an organ such as the brain within it, "deals with the molecules not directly, but only through the agency of intermediate subordinate sub-systems, regarded in a hierarchical scale of orders of magnitude. … Each sub-system dominates its own subordinate smaller parts within its own orbit or domain, as it were, restraining their degrees of freedom according to its own integral portion of the overall pattern, much as its own degrees of freedom have been restrained by the pattern of activities of the higher system of which it is a part and participant" (Weiss 1969, p. 14).

Weiss describes organic systems as simultaneously relatively closed and relatively open to environmental influences. They have a certain stability and thus an organizational closure; at the same time, they are open for influences from their surroundings. For example, a cell is a well-characterized entity and can be regarded as a system. However, in a multicellular organism, it needs to be regulated, requiring it to have a certain openness to regulative influences. To guarantee this, the cells of multicellular animals have a multitude of membrane receptors that mediate signals from the surroundings. They also need to have a regulated exchange of substances with the environment to maintain their basic functions.

Coincidences of this type, with two opposing principles present simultaneously, are a typical feature of organic life and can be found in many other examples. Typically, organisms balance such contradictory demands. Organismic thinking has to take such properties into account. This is the reason why Weiss presents such a long-winded definition of a system as provided previously, using formulations such as "relatively bounded," "relatively stationary," and so on.

Now, we have the components to understand Fig. 3.1, which represents the hierarchical order of the systems of an organism. Each system has relative invariance and autonomy as well as relative openness to regulative influences from

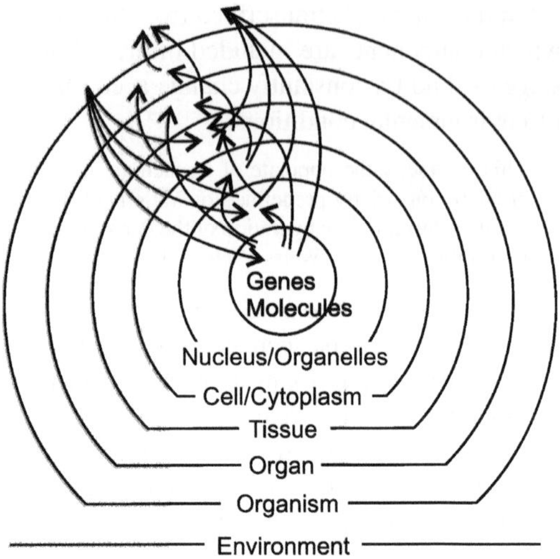


**Fig. 3.1** Schematic representation of the hierarchical concept of Paul Weiss (Redrawn from Weiss 1969 with slight changes in the levels indicated)

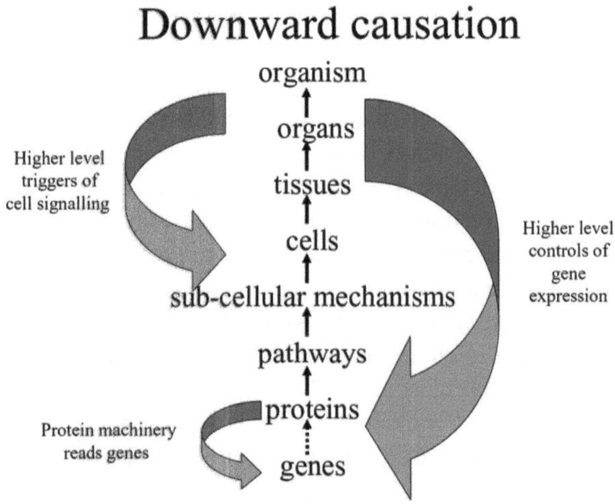


**Fig. 3.2** Schematic representation of causal relations within an organism according to Noble 2006, by permission of Oxford University Press

superimposed higher-level systems. The arrows indicate pathways of possible interactions that must be taken into account in studying such an organism. Also, the whole organism cannot be regarded as a closed system. Rather, it is integrated into its environment with many forms of exchange.

Basically, this systems view is congruent with some of the more recent views, which however show important differences among each other (O'Malley and Dupré 2005; Rosslenbroich 2011). In particular, the presently widely discussed approach of Denis Noble (Noble 2006, 2008, 2011) shows clear parallels, although it was obviously developed independently from the earlier definition of Weiss (Fig. 3.2). One could have the impression that both definitions were developed closely along the actual organic phenomena by two experienced researchers, who thus derived similar results. (Comparable approaches are to be found in Soto et al. 2011; Soto and Sonnenschein 2012; Saetzler et al. 2011; Sonnenschein and Soto 1999; Cornish-Bowden and Cardenas 2005; Cornish-Bowden 2006; Mesarovic et al. 2004; Mesarovic and Sreenath 2006; Joyner and Pedersen 2011; Bechtel 2010b; Köchy 1997.)

Even the ovum is such a system. It is not only a nutrient solution for the genome but also a real organism itself, comparable to single-cell organisms. Today, it is well known that the cytoplasm of the ovum transports many components needed for normal development. Development then takes place through continuous interactions between factors of the cytoplasm and DNA, with DNA methylation patterns introducing additional levels of information. Within these processes, the genetic information and the cytoplasmic factors are equally important. When the embryo develops into a multicellular organism, extracellular factors such as the position within the organism also become relevant. In each cell, the pertinent genetic information must be expressed at the right moment and at the appropriate place, which are dependent on a spatial order as well as a temporal order, which in turn are important in themselves and cannot simply be reduced to the genetic information.

To explain this principle, Susan Oyama developed a theory she calls the "developmental systems theory" (Oyama 2000a, b; Oyama et al. 2001; Downes 2001; Rehmann-Sutter 2002; Sterelny and Griffiths 1999). She argues that the information for the assembly of the organism can be found in neither the genome nor the environment, but it is put into effect by the process of development within the developmental system. In this context, DNA is only one of several factors for the process of development, albeit an important and necessary one. Nonetheless, sequences of DNA and any other factors cannot be privileged as bearers of ultimate causal control of the developing organism. Instead, the whole complex of factors is equally important to explain the appearance and the regularity of the steps: cellular morphology, the dynamic of biochemical processes, environmental influences, the previous history of the system, and the DNA sequences involved.

Because the embryo is "constructed" during development, Oyama calls her approach "developmental constructivism." She also expands this principle beyond the time of the development of the embryo, so that each organism can be considered as continuously "self-constructing" during its lifetime. This is a consequent systemic view applied to the ontogeny of organisms, however, basically using an organismic approach comparable to that of Paul Weiss. According to the concepts of Oyama and Weiss, it is not surprising that heredity can be found on different levels of the cell or the organism, as recent epigenetic research describes (Jablonka and Raz 2009; Jablonka and Lamb 2005).

The notion of organic systems is also applied to evolution (Riedl 2000; Wagner and Altenberg 1996; Shapiro 2011). Shapiro sees the systems view as essential for the further development of our understanding of evolution. He states that it will be possible to articulate a more interactive and information-based set of evolutionary principles without departing from the realm of established empirical observations.

I propose that the systems approach in the formulation of Paul Weiss is the most consistent one for the understanding of living entities (Rosslenbroich 2011). According to Weiss, insofar as a system can be seen as an entity that maintains its configuration within the environment and responds to alterations of the environment by an adaptive redirection of its componential processes to counter external changes, the system can be seen as the medium of the autonomy of the organism.

As a compilation of the concepts discussed so far, I propose Definition 1 for general autonomy:

Living systems are autonomous in the sense that they maintain themselves in form and function within time and achieve a self-determined flexibility.

These living systems

I. Generate, maintain, and regulate an inner network of interdependent, energy-consuming processes, which in turn generate and maintain the system;
II. Establish a boundary and actively regulate their interaction and exchange with the environment;
III. Specify their own rules of behavior and react to external stimuli in a self-determined way, according to their internal disposition and condition;
IV. Establish an interdependence between the system and its parts within the organism, which includes a differentiation in subsystems;
V. Establish a time autonomy; and
VI. Maintain a phenotypic stability (robustness) in the face of diverse perturbations arising from environmental changes, internal variability, and genetic variations.

## 3.8 Autonomy and Evolution

To this point, I have focused on a definition of autonomy as a general trait of living organisms. In the next step, I include the evolutionary view and examine changes of autonomous capacities of organisms. I try to answer the question of Ruiz-Mirazo and Moreno (2012): Is the idea of autonomy in any sense also helpful for understanding evolutionary transitions, that is, the appearance of new, more complex forms of biological organization in time?

Bechtel (2007) indicates that Moreno's notion of basic autonomy suggests additional levels of autonomy. Moreno describes that some of these may involve internal functions that enhance the system's ability to maintain itself. Others may involve ways of interacting with the environment. Basic autonomous systems, Bechtel describes further, remain highly dependent on the moment-to-moment conditions of their environment as they must continually extract energy and raw materials from it and excrete waste into it. If energy and material resources are not provided in high-enough concentration so that the osmotic or pumping functions in the membrane are able to bring them into the system or if waste accumulates, the viability of the system is undermined. By developing additional functions to ensure the needed conditions, the system can increase its ability to maintain itself.

**Table 3.1** Authors who mentioned increasing autonomy of organisms during evolution

| | |
|---|---|
| Herbert Spencer 1864 | Wolfgang Schad 1977, 1992 |
| Hermann Jordan 1908, 1913 | Verne Grant 1985 |
| Heinrich Quiring 1931 | Ludwig Kämpfe 1985 |
| Karl Beurlen 1937, 1949 | David B. Wake 1986 |
| Ivan Schmalhausen 1949 | Jeffrey S. Wicken 1987 |
| Julian Huxley 1953, 1974 | Hubert Hendrichs 1988 |
| Friedrich Kipp 1948, 1949 | Wolfgang H. Arnold 1989 |
| Ludwig v. Bertalanffy 1949 | Josef Reichholf 1992a, b |
| Klaus Günther 1950 | Jürgen Bereiter-Hahn 1996 |
| Homer Smith 1953 | Kristian Köchy 1997 |
| Maria-Josef Heuts 1953 | John Gerhart and Marc Kirschner 1997 |
| Emil Kuhn-Schnyder 1954, 1967 | Yoav Yigael 2000 |
| Edwin Hennig 1955 | Andreas Suchantke 2002 |
| Paul Overhage 1957, 1963 | Walter Streffer 2003, 2009 |
| Bernhard Rensch 1959 | Bernd Rosslenbroich 2007, 2009 |
| Conrad H. Waddington 1961 | William Bechtel 2007 |
| Erich Lange 1976 | Gerhard Neuweiler 2008 |
| | Kepa Ruiz-Mirazo and Alvaro Moreno 2012 |

Bechtel states that it makes sense to construe these additional functions as enhancing the system's autonomy.

Bechtel maintains that evolution is a process that, over time, can develop systems with greater autonomy. Although not denying the traditional accounts of evolution, he holds that the focus on autonomous systems provides a different perspective. First, it places the organism in the central role and emphasizes that an organism needs to be able to maintain itself as an autonomous system; otherwise, there is nothing to evolve. This does not mean that individual organisms must be totally self-sufficient. Organisms can evolve to rely on features of the environment that are regularly present in relation to them. However, they need to create and maintain all the mechanisms on which they rely so they can use these resources. Second, each addition to the basic system involves a cost, such that the system must generate and repair these mechanisms itself. Recognizing the organism in this sense as a subject of evolution rather than its object matches several recent approaches within the changing view of evolution (Weingarten 1993; Shapiro 2011; West-Eberhard 2003).

The idea is that during evolution the internal processes, prerequisites, and functions can change in such a way that the organisms gain increased abilities to compensate given perturbations and thus become more independent from environmental factors. Through these changes, they become more flexible and self-determined in many of their life processes, including behavior.

The principle has been noticed occasionally by scientists of relatively different provenance (Table 3.1). Deliberations on the pattern began in Darwin's time. Spencer (1864) defines life as the continuous adjustment of internal relations to external relations and formulated a "rule of increasing independence from the environment."

In the first half of the twentieth century, the pattern was occasionally included in evolutionary considerations with rather different theoretical backgrounds

(Beurlen 1937, 1949; Jordan 1908, 1913; Quiring 1931). Later, Rensch (1959) included it in a list of various factors of anagenesis, arguing that it leads to increasing plasticity of structures and functions, which allow a greater variety of reactions to the surroundings.

> "In many cases, such increased autonomy is the result of improved sensory and nervous systems. In man this autonomy finally led to control of the factors of environment. Another essential means of increasing the autonomy was the establishment of homoiothermy, by which the higher groups of vertebrates became more or less independent of the environment. ... General characters of increased autonomy, then, are a growing independence from environmental factors, and an increase of plasticity, of internal, or internally caused, physiological processes" (p. 298).

The pattern was either seen as centrally important (Bertalanffy 1949; Kipp 1948; Kuhn-Schnyder 1954, 1967; Lange 1976; Schad 1977, 1992; Schmalhausen 1949; Smith 1953) or discussed in combination with other patterns (Grant 1985; Kämpfe 1985; Overhage 1957, 1963; Waddington 1961; Köchy 1997).

The considerations of Julian Huxley (1953, 1974) are the most well known. He sees an "increased control over and independence of the environment" as a "raising of the upper level of biological efficiency" as the best definition of evolutionary progress, which was mainly achieved in the evolutionary line of the vertebrates leading to birds, mammals, and humans (Huxley 1974, p. 564). However, there are several problematic points in his discussion. One problem is the fact that relatively primitive organisms are also biologically efficient; otherwise, they would not have survived for a long time. Also, he focuses heavily on the line toward humans. I show in the forthcoming chapters that this is not necessarily the case if one assumes my definition of increasing autonomy. Beyond this, other groups with no phylogenetic relation to vertebrates – at least since the Cambrian – developed their own types of independence from the environment.

Huxley also does not define what he means by "control over the environment," especially as he does not restrict it to human beings, as Rensch does in the text cited previously. Thus, the relation of "independence" and "control" remains unclear, as McShea and Simpson (2011) indicate: They argue that it may be easy to see an exoskeleton, a shelled egg, or life cycle with a resting-cyst stage as ways to achieve some degree of independence from the external environment, but that it is difficult to see them as controls over the external environment in the same sense as a beaver or a human, building a dam, is. It seems that both criteria need their own respective definition. A somewhat clearer definition is provided by Huxley (1953), who defines biological progress as a "trend towards increased efficiency in dealing with the challenge of the environment, and an increased independence of the changes going on in it" (p. 114).

Huxley (1953, 1974) fleetingly mentions some important examples of independence from the environment. This demonstrates that he clearly saw the principle and recognized how pervasive it is. The following are some of his examples: the step to multicellularity as essential for the attainment of more-than-microscopic size and more than an elementary degree of division of labor among tissues and organs; the generation of bilateral symmetry, which allows exploration of the environment by forward movement; the capacity of higher fish to keep their internal environment chemically almost constant, while lower marine organisms have blood or body fluids identical in saline concentrations with that of the seawater in which they live, and if the composition of their fluid environment is changed, that of their blood changes correspondingly; the method of swimming in vertebrates with the aid of a tail, which gave them greater speed and power than any of their competitors and the potential to grow to a larger size; the emancipation of early land vertebrates from changes in moisture content of the air; and the ability of birds and mammals to maintain a constant temperature for their internal environment and thus be independent of a wide range of external temperature changes. These considerations belong to the most interesting ones, especially because they stand in line with the other chapters of Huxley's book with fairly pluralistic considerations about evolution in general, in fact being more pluralistic than the considerations of some other contributors to the "Evolutionary Synthesis" of Huxley's time (Witteveen 2011).

In a Dahlem workshop as presented in *Patterns and Processes in the History of Life* (Wake 1986, p. 53), "increasing autonomization" as the "degree of homeostasis or autonomous buffering of environmental variables" was included in a list of general patterns of evolution. There, it was claimed: "Across the spectrum of metazoans and metaphytes, from invertebrates through vertebrates, and algae to seed plants, autonomization and complexity obviously increase." However, the authors also stated that patterns such as this were inadequately defined and studied.

Sometimes the concept of autonomy reemerges in recent literature without sparking a broader resonance (Reichholf 1992a, b; Schad 1992, 1997; Yigael 2000; Neuweiler 2008; Bereiter-Hahn 1996). Occasionally, it appears in textbooks, especially on physiology and comparative animal morphology, again without conceptual consequences. The notion has also been discussed in the context of constructional morphology (Gutmann 1981; Weingarten 1993), and philosophical considerations of it have also been published (Jonas 1966; Spencer 1864; Steiner 1964; Fuchs 2009a).

Gerhart and Kischner (1997) argue in their inspiring book that the essential step in the transition to multicellularity of organisms was the new capacity to effectively shield themselves from the vagaries of the environment by producing their own internal conditions. "Whereas single-celled eukaryotes had little control over their environment and evolved mostly in response to it, the cells of multicellular eukaryotes could largely produce their own intercellular conditions and respond to these, as they could to the external environment" (p. 238). They call the capacity of the cell to create its own conditions "conditionality" and discuss the prerequisites for this. They regret that this has not been discussed by theoreticians, although in their view it has considerable bearing on the ability to evolve and seems to be a major evolutionary innovation.

Regularly, this idea of conditionality emerges in formulations that describe the gist of this principle without seeing the necessity for conceptualizing it further. To present just one example, the following is a passage from Vermeij (1987, p. 421):

> "It is possible, however, that species have improved in their capacity to survive in the physical environment. Many of the characteristics associated with competitive and defensive superiority – large body size, high body temperature, parental care of the young, and a tightly sealing exoskeleton, for example – also buffer individuals against short-term fluctuations in temperature and other physical factors. Consequently, individuals are able to carry on normal activity, or at least to survive, when physical conditions are temporarily unfavorable. Without such characteristics, individuals would be able to persist in a much smaller range of physical conditions."

Also within the topic of robustness cited are arguments for increases in robustness during evolution. Thus, Wagner (2012) points to the question of changes in robustness. He states that the robustness of macromolecules can change on evolutionary timescales. If robustness benefits both individuals and populations, then natural selection may favor robust phenotypes. If so, he concludes, the robustness of phenotypes might increase over time.

Even clearer about increasing robustness through evolution are some considerations of Kitano and Oda (2006). They argue that biological robustness fosters evolvability and that selection tends to favor individuals with robust traits; thus, evolvable robust systems progressively adapt to become more robust against the environment in which they are embedded. They suggest that over evolutionary time robustness against external perturbations is enhanced by adding diverse new functions to the input and output components of the organism.

However, it is conspicuous that these remarks rarely make reference to each other. This results in fairly different understandings of the topic. Usually, some examples are given, but there has been no attempt to date to define the pattern more precisely or to describe the respective phenomena systematically. This produces a rather strange situation: In some sense, one has heard of the idea. Occasionally, some people even take it for granted, so it seems unnecessary to elaborate on it in greater detail. In any case – whether it is completely overseen or is just taken for granted – evolutionists refused to integrate the principle into evolutionary theory building.

Many details of the pattern are still unclear because of the lack of further scientific endeavor on this topic. This holds true for questions on the systematic level, at which changes in autonomy can be described and whether there are autonomy-neutral and autonomy-destructive processes and events. We also know little about the relation of autonomy to adaptation. Many of the underlying details are hidden in the physiological, morphological, and paleontological literature and need to be compiled from this source under this aspect, and other questions may need to be addressed empirically.

In many considerations of large-scale evolutionary patterns, increasing autonomy is not mentioned. So, McShea (1998) did not feel compelled to include it in the overview of possible largest-scale trends in organismal evolution that are under discussion. Even in a specific chapter on this topic in the work of Rosenberg and McShea (2008), they do not take it into consideration. However, more recently McShea and Simpson (2011) saw it as a promising conceptual work to follow these lines of considerations.

## 3.9 Definition of Increasing Autonomy

A definition of increasing autonomy is attempted here in three steps. First, I present a list of features that are able to contribute to changes in autonomy of an individual organism. Second, a formal definition is developed. This definition most likely will still be a preliminary one, which could become more precise in the future. However, it is a suitable starting point to bring the phenomenon into focus. Nonetheless, it is open to further consideration. Third, more clarity is achieved by the presentation of facts and observations in the following chapters, especially for the major transitions in evolution, to which the definition and the list of features are applied.

The hypothesis is that organisms not only show autonomy as a general trait, but also that there are *differences in the degree of autonomy* within taxa. The evolutionary process generated organisms with distinguishable degrees of autonomy. Thus, there are organisms that are more subject to the direct physical, chemical, and biological conditions of their surroundings and others that can act more on their own behalf because they are more active, flexible, and selective in their interaction with the environment. Increasing autonomy can also be summarized as opening new possibilities for the organism.

I do not attempt to describe organisms as entities, which are isolated units within their environment. The inference is rather that each organism is deeply embedded in the systems of its environment. However, this inclusion can be effected either by direct physical and chemical influences that are more direct or by processes in the organism that are more emancipated, establishing organs for interactions with the environment that are more active and selected.

The term *autonomy* cannot be taken in an absolute manner but always describes *relative* autonomy. This important aspect distinguishes the concept used here from previous ones in evolutionary biology. As Ayala (1974) correctly argues: No organism can be wholly independent from the environment. In the present definition, the emphasis is, instead, on the balance of the organism-environment relationships and their changes.

A typical example is boundaries: As described previously here, the internal compartment is established within a boundary, which the system generates as a spatial separation from the environment. In its simplest form, this is realized in a single-cell organism by means of a cell membrane. However, even the simple example of the cell membrane shows that in a biological system complete separation is never obtained. Instead, we see the double function of a boundary and an exchange with the environment through and across the boundary. Each cell membrane and each integument of an animal has to perform this double function. Organisms have to balance these two requirements, and each solution looks different.

Generally, an extrinsic relation and an intrinsic relation of autonomy can be distinguished. The extrinsic relation describes the system-environment relation. The intrinsic relation describes the self-referential, internal organization within the system (e.g., homeostatic stabilization of processes, intraorganismal signaling, connectivity within neuronal systems). This is basically identical with what has been called *interactive autonomy* (how autonomously a system behaves in interaction with its environment) and *constitutive autonomy* (within the context of the biological system itself) (Bertschinger et al. 2008; Moreno et al. 2008).

This differentiation is important if we want to look for changes in autonomy because both aspects can underlie variations. Thus, changes in interactive autonomy take place when, for example, boundaries such as skins and shells are elaborated or when movement devices such as legs or wings are developed. When the capacity of

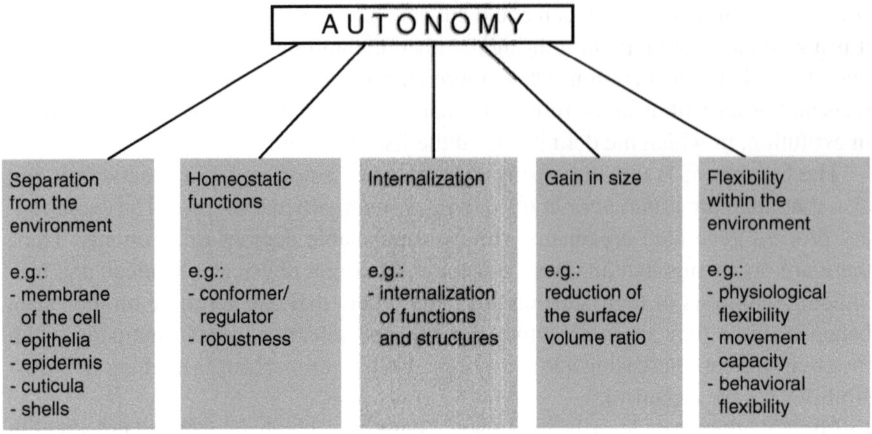


**Fig. 3.3** Set of resources to change autonomy

homeostasis in body fluids or of central nervous processing is elaborated, this is more a change in constitutive autonomy, although both also have relevance for interactive autonomy.

Several biological elements can contribute in different degrees to changes of autonomy (Fig. 3.3). They are not general rules or some sort of continuous trends. They rather function as a set of resources that can – singly or in combination with each other – increase autonomy.

These elements are probably not complete. The various relations of the somewhat-heterogeneous elements to each other will also need further examination in the future. However, they can at least be identified within the major evolutionary transitions, and changes in them can also be described. Thus, they are relevant.

One such element is *spatial separation from the environment*, such as with cell membranes, cell walls, integuments of metazoans with cuticles, shells, hairs, or feathers. To different degrees, they all serve to keep the environment outside the organism and to regulate and direct the exchange with it. Changes in their organization can contribute to an essential degree to changes in the organism-environment relation.

*Homeostatic functions* are means to establish and enhance internal functional stability. This overlaps to a large extent with changes in robustness. Another element is the displacement of morphological structures or functions from an external position into an internal position within the organism, here summarized as *internalization*. Multiple processes of internalization are involved in building up the inner anatomy of organisms, ontogenetically as well as phylogenetically. During ontogeny, gastrulation and neurulation are typical internalizations. During phylogeny, for example, the transition from prokaryotes to eukaryotes included the internalization of some organisms within others (endosymbiosis).

A *gain in size* during many transitions leads to a reduction of the surface-to-volume ratio. This means that in larger animals there is less direct contact to the immediate environment relative to the existing body mass. The smallest cells we know, bacteria, have a large surface for environmental exchange. In larger bodies, this direct exchange capacity is reduced relative to the body mass. The rates of change of state internally are much slower, giving them an "inertia" effect, which smoothes the fluctuations and gives time for regulatory functions to operate. Larger organisms may have better opportunities for storage of energy and substances, and they may have room for internal regulatory structures that are more complex. It is well known in physiology that larger animals are more likely to be regulators that stabilize their internal conditions also under fluctuating environmental circumstances (Willmer et al. 2000). Although there are no linear increases in size, evolution deals with it so that size matters and is not random.

These elements are prerequisites for establishing a certain amount of physiological flexibility within a given environment, that is, a capability of organisms to generate *flexible functional answers* to conditions and changes in their environment. Finally, this principle can be widened to include all forms of *behavioral flexibility*, emancipating organisms from mere short-term reactions to environmental factors. Together, these elements are able to generate certain degrees of physiological and behavioral freedom.

These principles can be summarized as follows in Definition 2:

Increasing autonomy is defined as an evolutionary shift in the system-environment relationship, such that the direct influences of the environment on the respective individual systems are gradually reduced (interactive autonomy) and stability and flexibility of self-referential, intrinsic functions within the systems are generated and enhanced (constitutive autonomy). This is described as relative autonomy, while, at the same time, numerous interconnections with and dependencies on the environment are retained. Thus, organisms can undergo relative emancipation from environmental fluctuations, gaining self-determination and flexibility of behavior.

A set of resources can be involved to change autonomous capacities:

I. Changes in spatial separations from the environment;
II. Changes in homeostatic capacities and robustness;
III. Internalization of structures or functions;
IV. Increase in body size; and
V. Changes in the flexibility within the environment, including behavioral flexibility.

In the following chapters, some of the major evolutionary transitions are described, and it is demonstrated that these specific elements can be identified in many of them. Thereby, their significance is outlined further. In the sense of Fuchs (2009a, p. 9), the present study is based on a combination of phenomenological thinking and approaches of organismic biology and philosophy of the living.

# Chapter 4
# The Major Transitions in Early Evolution

## 4.1 The First Cells

Traces of living organisms have been found in sediments as old as 3.5 billion years or more. The considerations of the origin of this early life are vague and hypothetical. We know virtually nothing about the first stages of life. Most of what textbooks usually try to suggest to students in their respective chapters is more wishful thinking than fact based on scientific evidence (Conway Morris 2003; Shapiro 1987). However, the essential aspect of the theory presented here is that at some time and somehow life established biochemical functions that were not identical with the processes of the surrounding inorganic environment. Just their difference from the processes around them established autonomy of an elementary manner, following their own rules by means of self-regulation and self-adjustment. They must have been able to assert a certain independence; otherwise, the components involved would have disintegrated into their surroundings. And, for this, even the first cells might have been complicated, possibly more so than sometimes is thought.

During some developments, the conditions for glycolysis, electron transport chains on membranes, protein biosynthesis, photosynthesis, replication, and transcription of DNA had to be maintained actively, consuming energy and thus defending these processes against decay. In this sense, every organic metabolism, even in its simplest form, is a form of autonomy.

Metabolism establishes dynamic disequilibrium, a dynamic stage of order within a network, that characterizes life (Margulis and Sagan 2002; Suki 2012). This is often mentioned in attempts to define life itself. A cell that drifts toward equilibrium is dying.

Autonomy is also achieved by the use of energy-rich molecules, which are assembled by metabolic processes. In the face of hydrolyzing and oxidizing influences from the environment, energy-rich bonds are maintained relatively stable. Organic molecules are always reduced compounds; thus, they are rich in energy. Energy from the environment is accumulated within these complex molecules, so an energetic gradient is maintained against the influences from the environment.

B. Rosslenbroich, *On the Origin of Autonomy: A New Look at the Major Transitions in Evolution*, History, Philosophy and Theory of the Life Sciences 5,
DOI 10.1007/978-3-319-04141-4_4, © Springer International Publishing Switzerland 2014

However, the release of the stored energy is also subject to the active regulation of the cell, as the energy in the glycolysis and the respiratory chain is transferred to adenosine triphosphate not in one, but rather in many steps. The cell prevents the energy from spontaneous sudden release and dissects this process into a succession of single steps. Thus, the cell sets free just small amounts of energy. The energetic gradient in comparison to the environment and the regulated use of energy are principles of the autonomy of the cell.

An essential part of early evolution must have been the origin of a system that carries and replicates information. Again, not much is known about the first principles of the biological storage of information. However, as soon as the principle and its transfer to descendants were established, the prerequisite was available for building an order to escape from the tendency of the environment toward entropy. Such instructions make it possible to build temporary islands of self-determined order repeatedly. These principles enable the identity of the individual as well as that of the species, that is, its characteristic identity with respect to other individuals as well as that of other species and their relative stability over many generations. This genetic self-determination is also autonomy. However, this aspect of autonomy is again relative, as changes in genetic information are necessary for evolutionary change. Relative stability must be combined with limited variability.

The essential characteristic of life is not so much the interacting matter but the information about how to use it to establish new islands of life. Information is immaterial, but it needs a material medium. Information is the source for building a higher degree of order than exists within the environment, as long as life functions are maintained.

DNA repair systems contribute to the stabilization of the genome. They are present already from the prokaryotes on and eliminate mistakes in the genetic code with high efficacy. For example, the permanently working repair systems eliminate mistakes that result just from thermic fluctuations. Thus, every day about 5,000 purine bases from the DNA of each human cell are lost as their bondings to deoxyribose are hydrolyzed (depurination). In addition, spontaneous deaminations from cytosine to uracil occur through interactions with metabolic products and environmental influences (chemicals, light). The genetic autonomy must constantly be defended against destructive influences. In particular the principle of the double helix enables repair because two copies of the information are available. One copy can be used as the backup copy to repair the other copy. Therefore, the principle of the double helix contributes to the independence and stability of the genome.

If there was a prebiotic environment with some interaction of organic molecules as forerunners of the first cells, as some hypotheses assume, at some time a closed membrane must have appeared to form the principle of the cell. It must have enclosed a compartment for the concentration of biochemical molecules and their reactions. The encapsulation of such protoplasm in membranes is an essential question for any considerations about early evolution. If metabolic networks were somehow included in some sort of capsule, there would have been problems of permeation. Capsules or vesicles would have the effect of transferring the metabolic networks without a transition from an existence within a totally open surrounding into a completely sealed situation, which would bring these networks to a standstill. The question is how the remarkable balance between a relative separation from the environment and an intensive, but regulated, exchange of substances and energy evolved. Modern cells solve this problem through a combination of the lipid bilayer with integrated membrane proteins. Not only is the transport through the membrane is regulated actively, partly through the consumption of energy, but also the properties of the lipid bilayer itself are regulated. Some authors postulate that from the beginning some sort of a protocell must have been involved.

For my purpose here, it is only necessary to look at the result, which is achieved with the generation of the first cell membrane. Seclusion from the environment was established because of the boundary, within which high concentrations of organic molecules could be accumulated so that an osmotic gradient toward the outer environment was established. A micromilieu was built and defended against environmental influences, which can be regulated and in which special conditions for the reactants of the metabolic systems are maintained. The cell membrane establishes a relatively closed compartment in which metabolism and genetic information are internalized and protected against destructive influences.

In addition, all metabolic and genetic pathways are organized within a highly structured temporal order. For example, in the same way as oxidation processes are subdivided into regulated sequences, the protein biosynthesis also only works within a controlled sequence of necessary steps. Within this sequence, the ribosomes do not catalyze the synthesis of proteins as fast as possible. They work more slowly than a protein catalyst is able to work. By this effect, fewer mistakes occur. Cell division also has a determined duration of observable single steps, each the prerequisite for the next. The cell regulates these processes not only in their appearance but also in their time sequence and thus establishes an autonomous, endogenous temporal order.

In summary, at some early point in evolution, the following characteristics were achieved:

- Metabolic networks, which are self-maintained and regulated within the inner network of functions, defended against decay and perturbations from the environment;
- The ability to use and process energy-rich molecules, thus establishing and maintaining an energetic gradient between inside and outside;
- Self-regulated processing and usage of energy;
- The processing and replication of information to relieve the organism from the tendency of the environment toward entropy (genetic and epigenetic self-regulation);
- A dynamic stage of order (situation far from equilibrium);
- The insularity vis-á-vis the environment within a membrane; and
- The endogenous time sequence of processes.

(continued)

(continued)

All these features are functions of a primal emancipation from the inorganic surroundings according to Definition 1 in Chap. 3. Thus, a first differentiation of a biological interior opposed to the outside was established. From that point, these features of a relative autonomy accompanied the history of life.

## 4.2 From Prokaryotes to Eukaryotes

In the face of some new considerations about the early evolution of the cell, the previous statement, that the principle of heredity established the identity of the individual and the succession of generations, has to be adjusted. According to sequence analyses, the genetic autonomy might have been present, but relatively weak during the early evolution of prokaryotes, as there was substantial horizontal gene transfer. This principle is known for recent bacteria. Microbes are able to absorb and discard genes as needed in response to their environment. Rather than discrete genomes, we see a continuum of genomic possibilities (Goldenfeld and Woese 2007). It is proposed that this must have been essential for the phylogenetic relations of early life (Shapiro 2010, 2011; Doolittle 1999; Martin 1999; Woese 2000; Ochman et al. 2000; Wolf et al. 2002; Koonin 2009; Boto 2010). It is possible that horizontal gene transfer was the principal driving force in early cellular evolution. Primitive cells did not carry a stable organismal genealogical trace. Thus, the evolution of primitive cells would have been basically communal, so that it was the community as a whole, the ecosystem, that evolved with a universal gene exchange pool. Instead of a tree linking life's three deepest branches (eubacteria, archaebacteria, and eukaryotes), Rivera and Lake (2004) propose a ring and Doolitle (1999) a net (Fig. 4.1).

From this pool, some cell types might have reached a level at which a more solidified, organized cellular design evolved and genomes became more stable, restricting horizontal gene transfer. This resulted in increased idiosyncrasy, which led to decreased genetic interactivity with other cell types. Woese (2002) calls this the "Darwinian Threshold," as with this stabilization a diverging phylogenetic tree came into existence. Now, vertical genetic transfer and the divergence of species became more important. Only after this threshold did real species exist; there were no species before this point. "The Darwinian Threshold truly represents the Origin of Species, in that it represents the origin of speciation as we know it" (Woese 2002, p. 8744). Genomes became more stabilized and were no longer so directly changeable through environmental demands. Eukaryotes and multicellular organisms possess higher system levels to react to environmental influences.

Thus, there was some reinforcement of genetic stability. This is consistent with the feature of autonomy of a functional stability, here as genetic robustness. The increasing autonomy within eukaryotes reduced the alienation through lateral genetic exchange. This includes the profoundly reduced mutation rates of eukaryotes compared to those of prokaryotes.

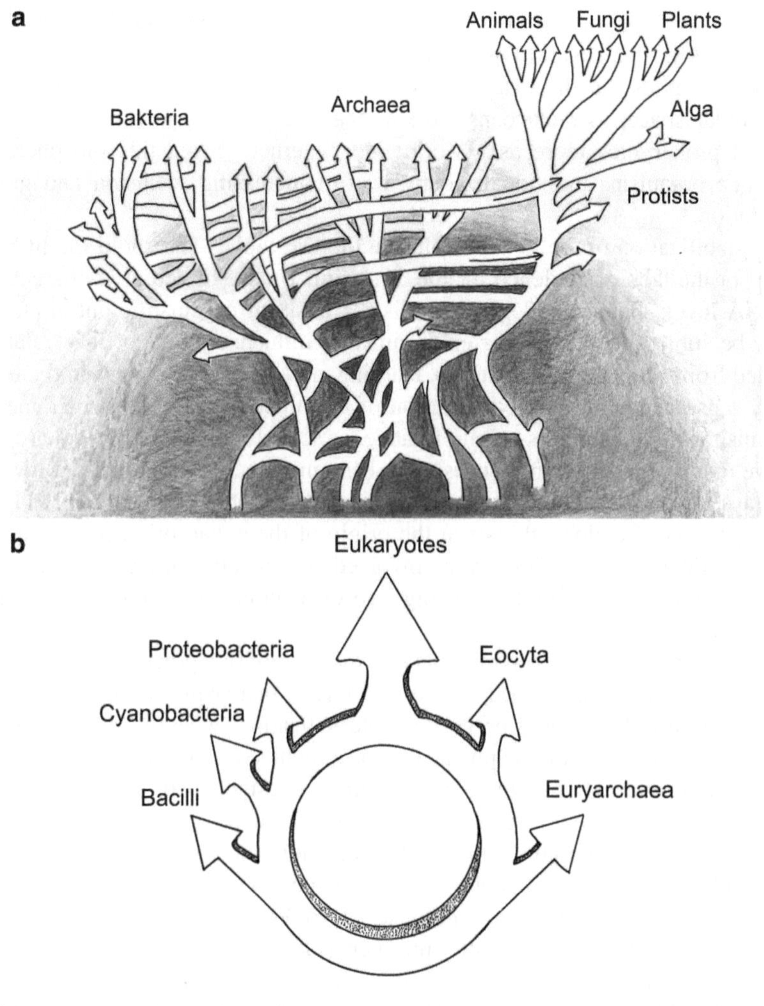


**Fig. 4.1** Modified representations of early evolution (Redrawn from Doolittle 1999, 2000; Martin and Embley 2004; Rivera and Lake 2004)

During bacterial fission, genetic changes are transmitted to all daughter cells. However, after sexual reproduction was invented in eukaryotes, especially within multicellular organisms, genetic changes are transmitted through the germ line, while somatic changes are not generally anymore heritable. This again reflects further stabilizations of the gene pool. However, the stabilization of genomes was not driven so far that horizontal gene transfer came to a standstill. It probably even played some role within further evolutionary transitions (Shapiro 2010, 2011).

The common horizontal gene transfer led to considerations that purport that there are no distinguished species within recent prokaryotes. All bacteria together might be some kind of superorganism (Sonea and Panisset 1983; Sonea 1991). Populations of bacteria in their habitat also have been described as multicellular organisms (Shapiro 1998; Shapiro and Dworkin 1997). These populations conduct all intercellular processes via the external medium rather than through an internalized extracellular space, as metazoans have it. Margulis and Sagan (2002) regard the world of prokaryotes more as a worldwide "internet." From this biosphere, the eukaryotic organisms individualized by way of increasing seclusion and gradual stabilization.

The stabilization of the genomes during the Darwinian Threshold might be the reason for the lack of evidence for the direct inheritance of acquired characters at the DNA level. Genetic independence is the basis of organismic autonomy and cannot be subject to continuous environmental influences. Rather, it is relatively secluded from short time adaptations, learning, and so on. Adaptation in the evolutionary sense needs the limited variation of the otherwise-stabilized genome as a prerequisite. But, again, this is only relative, as several epigenetic factors are more reactive to environmental influences, as is becoming increasingly clear today. This also plays a crucial role in evolution (Jablonka and Lamb 2005; Shapiro 2011).

The process of endosymbiosis at the origin of the eukaryotic cell also needed some genetic flexibility. This event involved the genetic amalgamation of two highly divergent cell lineages, causing two deep branches in the tree of life to merge outright.

This process of acquiring symbionts into the interior of a precursor cell is typical internalization, as it is described here as a feature of increasing autonomy. Margulis (1993) also describes it as a process of internalization, whereby the symbiont is removed from contact with the physical environment (Margulis 1991), and a higher complexity of internal structures and functions is established. Together with the endosymbionts, functions are internalized, which in prokaryotic precursors are associated with the external membrane. In mitochondria, it is oxidative phosphorylation, and in chloroplasts, it is photoreduction and photophosphorylation.

Typical eukaryotic cells are lengthwise 10- to 30-fold larger and 1,000–10,000 times more voluminous than a bacterium such as *Escherichia coli*. The small size of prokaryotes is responsible for many of its biological characteristics (Knoll and Hewitt 2011; Madigan et al. 2000). For example, the speed of nutrient uptake and release of waste is usually in reverse proportion to the size of the cell. The reason for this is that transportation rates to some extent are a function of the membrane surface available. Small cells have a greater surface area in relation to their volume than large cells. In a spherical body, the volume is a function of the cube of the radius ($V = 4/3\pi r^3$), whereas the surface is a function of the square of the radius ($O = 4\pi r^2$). This greatly influences the cellular metabolic rates and growth rates. However, the point to make here is that during evolution from the prokaryotic to the eukaryotic cell, tremendous reduction occurred in reference to the relative surface area and therefore of the immediate contact to the environment. The reduced area of exchange dispenses with a trait, which is often seen as an advantage of the prokaryotic cell (DeDuve 1991; Madigan et al. 2000). It was replaced or complemented by the folded system of internal membranes and the ability of endocytosis.

As prokaryotes have a large and rapid exchange with their environment in relation to their volume, they are relatively open systems. Cells that are more complex,

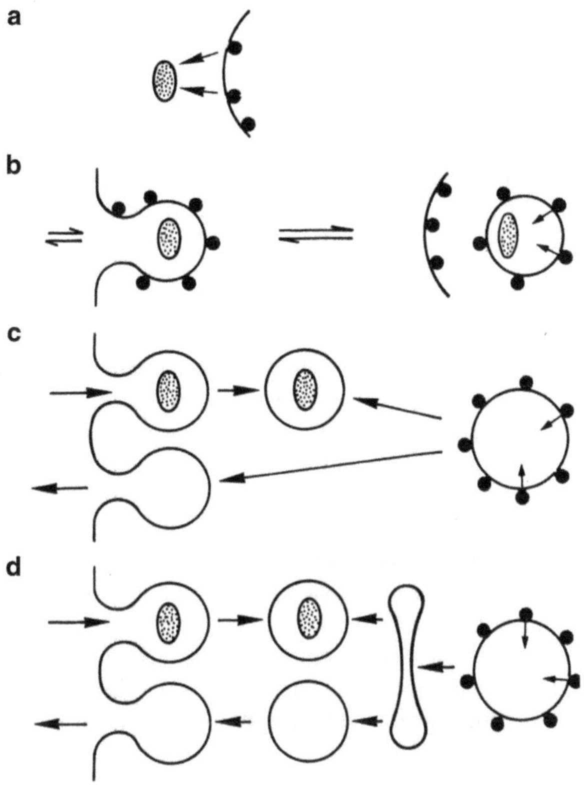


**Fig. 4.2** Hypothetical steps in the internalization of the eukaryotic system of membranes: (**a**) extracellular digestion by exoenzymes, which are secreted by membrane-bound ribosomes (prokaryotes); (**b**) temporary invaginations and generation of vesicles for a rudimentary form of internalized digestion; (**c**) membranes with ribosomes move from the surface into the cytoplasm; (**d**) some rudimentary Golgi system is inserted (From DeDuve 1991)

and later also multicellular organisms, restrict this openness gradually and thus accentuate the relative operational closure of their systems. This is accompanied by specialized areas or organs for environmental exchange, enhancing the abilities for active regulation of the transfer of substances.

Prokaryotes are dependent on the uptake of dissolved substances across their membrane. The large surface-to-volume ratio supports this (Knoll and Hewitt 2011). Therefore, many bacteria are able to prepare substances in their environment for uptake by excretion of hydrolytic enzymes into the surrounding medium. Many of them are synthesized at the cellular membrane. Thus, prokaryotes have external digestion (DeDuve 1991). Eukaryotic cells have internalized digestion, as the processes of endo- and exocytosis developed. One prerequisite for this was the generation of the elements of the cytoskeleton, which mediate the internalization of substances and the movements of the vesicles in the cell.

In a postulated precursor cell of the eukaryotes, first foldings of the membrane could have occurred. The resulting spaces would have been favored rooms for digestion of collected extracellular material by enzymes, which could have been secreted by membrane-bound systems. If such spaces would be secluded to vesicles, a primitive form of endocytosis and intracellular digestion would have developed (Fig. 4.2).

The production of hydrolytic substances, which in prokaryotes takes place at the surrounding membrane, is now shifted into the interior of the cell. This can be seen as a crucial step of emancipation from conditions of the immediate substrate, as nutrients now can be actively searched for and larger particles can be digested. External digestion was given up or reduced and integrated into the lysosomes.

However, not all eukaryotes followed this path. A spectrum of different combinations exists. Fungi, for example, generally digest molecules from the environment and still excrete their enzymes into the surroundings.

It is possible the internal membrane system of the eukaryotic cell also was generated by internalization of the prokaryotic outer membrane and was connected with the reduction of the relative surface area (Alberts et al. 2002; DeDuve 1991).

Within prokaryotic cells, the outer membrane fulfills all functions associated with membranes, such as selective transport of ions, adenosine triphosphate synthesis, secretion of proteins, and synthesis of lipids. Together with the internalization of membranes, many of their functions are also internalized.

As many characteristics of eukaryotes depend on such a membrane system, it is possible that the internalization of membranes was an early event during the transition to eukaryotes. DeDuve (1991) suspects that this stood in some relation to the loss of the cell wall. Although an enlargement of the membrane would also be possible in a cell with a wall, the changes could have required more flexibility at the surface than a rigid wall allows. As a compensation for the lack of external mechanical support, the cells could have changed to an internal support system, probably leading to a cytoskeleton. This is especially important as soon as a certain size is exceeded (Cavalier-Smith 1987). This would be an internalization of the function of mechanical support from an external wall within prokaryotes to an internal skeleton in eukaryotes, again an internalization of a function, which made the extended flexibility of the new cell type possible. We do not know much about how these transitions really occurred, but the differences that resulted from these transitions are observable in the living cells of today.

Ribosomes also might have experienced internalization together with all those processes that were internalized into intracellular vesicles and membrane systems (Fig. 4.2; DeDuve 1991; Lechner and Wieland 1989). In this way, the outer membrane was relieved of a series of functions and thus had the possibility to develop others, such as those for communication and exchange with the external world. It is equipped with a series of transport systems, pumps, and receptors, especially in multicellular animals. Generally, the eukaryotic membrane is an organelle of enormous complexity that builds external and internal boundaries on the one hand and on the other hand simultaneously mediates communication and transport to the outside. Thus, it integrates two requirements of the cell that at first seem to be incompatible. This generation of reaction rooms and compartments also extended the possibilities of a differentiation in the chronological sequence of events within the cell.

The spaces, which are enclosed by the membranous system, are in a certain way an internalized special compartment between the cytoplasm and the outside medium. Exchange processes must be controlled as well as those between the cytoplasm and the environment, although different functions are involved. This fits well with possible generation from the external membrane during early evolution (DeDuve 1991).

In line with these considerations, it is also possible that the generation of the nuclear membrane was a result of internalization of membranes (Alberts et al. 2002; DeDuve 1991). Another possibility is the generation of the nucleus by means of several endosymbiotic incidents with archaebacteria and eubacteria, possibly also together with horizontal gene transfer (Hartman 2002; Wieser 1998). However, this is hypothetical and has yet to be traced, but the definite result was that the eukaryotic cell we know today harbors DNA within a separate compartment. This is further internalized and thus has additional protection. This corresponds to another spatial separation from the environment, whereby seclusion and protection are enforced by associated histones and other proteins. Margulis (1993) assumes that these proteins might have had protective functions within extreme habitats when they evolved.

At the same time, a component more active toward the environment was generated: Prokaryotes are basically dependent on the immediate medium in which they live and from which they absorb their nutrients. On the contrary, many protozoans have a more active movement capacity within their environment (Suki 2012). Indeed, their size reduces the frictional surface and changes the relation of viscosity and inertia within the environmental fluid (Reynolds number). Through cytoskeleton and flagellum, the movement capacity is enhanced, thus also enhancing the possibilities of nutrient uptake, which can be understood as a process of emancipation. A *paramecium*, for example, swims about 10–100 times faster than bacteria with a bacterial flagellum. Margulis (1993) assumes that the enhancement of swimming rates generally might have taken place at the beginning of the transition toward eukaryotes. In this connection, she postulates that the flagellate apparatus also had an endosymbiotic origin.

In whatever manner, not just one line of development toward enhanced movement capacities exists but rather a broad spectrum of possibilities, within which active swimmers are present as well. Coming from the basic organization of the eukaryotic cell, there was a broad radiation of different adaptations within the environment.

In summary, all major innovations typically discussed in connection with the generation of the eukaryotic cell led to increases in autonomy of the individual according to Definition 2, Chap. 3:

- Relative stabilization of genomes with increasing robustness of the genetic material;
- Internalization of symbionts;
- Internalization of digestion;
- Internalization of functions into an internal membrane system and into organelles;
- Size increase with reduction of relative surface area, thus reducing the direct contact to the environment relatively;
- Internalization of mechanical support by the cytoskeleton;
- Internalization of the DNA within a separate compartment, the nucleus; and
- Enhanced movement capacities.

## 4.3 Origin of the Metazoa

Multicellularity in general appeared early and repeatedly in life's history (Rokas 2008; Butterfield 2009; Knoll and Hewitt 2011). Today, it is assumed that multicellularity in animals arose in the Precambrian about 600–1,000 million years ago. Metazoans, which may have a connection to choanoflagellates, emerged from unicellular protist ancestors in what has been called one of the "major transitions in evolution" (Holland 1998; Szathmary and Maynard Smith 1995; Maynard Smith and Szathmáry 1995; Wainright et al. 1993; DeSalle and Schierwater 2007; Calcott and Sterelny 2011). Examination of the fossil record reveals a Precambrian origin of sponge, cnidarian, and bilaterian body fossils, whereas the first fossil occurrences of the uniquely distinct bilaterian body plans of phyla such as arthropods, mollusks, echinoderms, annelids, and chordates are found in Cambrian rock strata (Valentine 2004; Erwin et al. 2011). Recently, even older traces of multicellularity have been reported, dating back as far as 2.1 billion years ago (El Albani et al. 2010; Donoghue and Antcliffe 2010).

It is interesting to look at the diverse associations of single-cell organisms in forms such as aggregates, biofilms, and many more (O'Malley and Dupré 2007; Rokas 2008). Ruiz-Mirazo and Moreno (2012) discuss this in relation to autonomy and state that since the beginning of the history of life on Earth, organisms have grouped together, constituting more or less cohesive aggregates that might increase the possibilities of the individual systems. Prokaryonts, for example, show a huge variety of this type of collective and often only temporary associations, based on processes of intercellular self-organization. These communities show features, and occasionally seem to behave, like multicellular organisms (Shapiro 1998). This would be the case, for example, of biofilms, which build a common physical border, a polymeric matrix that keeps the cells together and attached to a surface, or of myxobacteria with body-like colonies that have developed their own "life cycles."

However, as Ruiz-Mirazo and Moreno state, given the limited degree of functional differentiation within these collective associations, the cohesion and interactive capacity of the global system, as such, are rather weak. Thus, closer examination does not allow considering them as proper autonomous agents. Actually, the constitution of new composite forms of autonomy was a more difficult process than the formation of more or less cohesive colonial aggregates. This is because the creation of a full-fledged autonomous entity is not possible without stronger subordination of the constitutive elements to the new functional requirements of the emerging global autonomy. Interesting examples for such subordinations are being studied, for example, within the Volvocales (Kirk 1998, 2005; Michod 2007; Rosslenbroich 2007). In a recent study, Arnellos et al. (2013) propose a general theoretical scheme according to which a multicellular organism is an ensemble of cells that effectively regulates its own development through collective (metacellular) processes of control of cell differentiation and cell division. This theoretical result derives from a detailed study of the ontogenetic development of three multicellular systems (*Nostoc punctiforme*, *Volvox carteri*, and *Strongylocentrotus purpuratus*) and, in particular, of their corresponding cell-to-cell signaling networks. The case study supports the claim that a specific type of functional integration among the cells of a multicellular ensemble is required for it to qualify as a proper organism. They argue why a multicellular system exhibiting this type of functionally differentiated and integrated developmental organization becomes a self-determining collective entity and therefore should be considered a second-order autonomous system.

Given the limited sequence of fossils, morphology, developmental biology, and molecular and genetic aspects of the simplest extant organisms have widely been used for model building of this transition. However, understanding the origins and diversification of early animal lineages still is one of the grand challenges in evolutionary biology. As the fossil record is fragmentary and morphology leads to conflicting interpretations, there was much hope placed in the genetic sequencing of basic groups of metazoans. This contributed many new insights but the central questions remain unsolved (Ball et al. 2007; Cartwright and Collins 2007; Schierwater and DeSalle 2007; Dewel 2000).

One interesting result of recent molecular and genetic studies is the deep origin of components needed for multicellular integration. Many of them seem to have evolved before multicellularity was reached (Rokas 2008; King et al. 2003; Abedin and King 2008; Schierwater et al. 2009; Cartwright and Collins 2007). The adhesion of animal cells to their neighbors and the extracellular matrix (ECM) is a fundamental aspect of animal multicellularity. A few major classes of genes, such as the cadherins, the integrins, the selectins, and the immunoglobulin superfamily, play a key role in mediating adhesion in animal cells. Examination of the choanoflagellates suggests that the genes participating in adhesion in animals were likely well developed in the unicellular ancestor of animals and choanoflagellates. Most of the domains typically found in animals are present in choanoflagellates. However, the function of such a diverse set of adhesion molecules in a unicellular organism is not yet known.

Several genes participating in the formation of the ECM are also conserved and pre-date animal origins, including collagen, laminins, and fibronectins. Perhaps the most spectacular example of the deep, preanimal origin of these families of genes is the collagens, the most abundant protein family in mammals, homologs of which are found not only in choanoflagellates but also in the animal sister kingdom, the fungi. The same has been described for some cell-to-cell signaling pathways.

Among several possibilities for the transition from unicellular organisms to metazoans, the path involving cell divisions of unicellular ancestors staying together in a common ECM is usually favored (Rieger 1994; Rieger and Weyrer 1998; Willmer 1990; Schierwater et al. 2009). As a result, (1) several cells are embedded within an ECM, which now forms the cell's environment and thereby manifests one typical feature of a metazoan organism; (2) an outer cell layer is formed, which constitutes the boundary with the environment; (3) some cell types are internalized within the ECM and thus have no direct contact with the outside medium; (4) cells differentiate for special functions; (5) internal cavity systems for digestion and distribution are formed; and (6) the capacity to generate bigger organisms is increased. These general features of metazoan organization are now discussed in

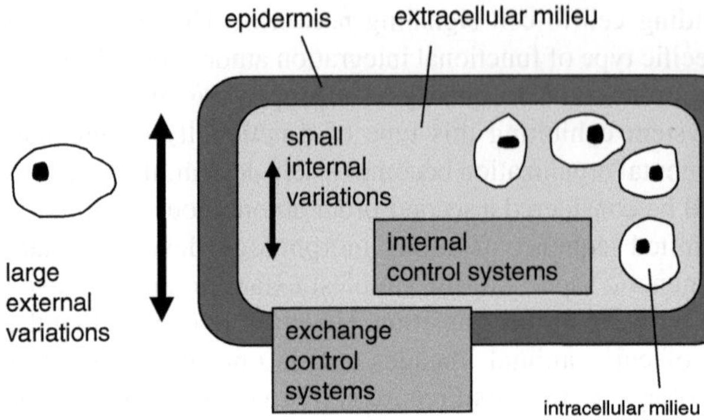


**Fig. 4.3** Schematic representation of the physiological principle of extracellular homeostasis

more detail to show that they all include traits of increasing autonomization as defined in Chap. 3.

### *4.3.1 The ECM*

Whereas the ECM was once only regarded as a substance that somehow keeps cells together, it now increasingly appears to be an essential dynamic system, by means of which cells are morphologically and functionally integrated within tissues. During development, it is an essential component of maturation processes and cell movement (Morris 1993). Rieger (1994) infers that the ECM played a central role during early radiation of the Metazoa because evolution acted on developmental programs. As an ECM is present in all phyla, including sponges, it may have originated early in the evolution of multicellularity (Gerhart and Kirschner 1997; Morris 1993; Müller 2001; Rieger and Weyrer 1998).

Özbek et al. (2010) see the ECM as a key mediator of metazoan multicellularity. They indicate it enabled the building of bridges between cells, contributed to cell spatial arrangements by binding cell-surface adhesion receptors, and supported cell survival, differentiation, and tissue organization. Thus, the ECM enabled the emergence of larger and more complex eukaryotes that could resist predation, evolve specialized tissues and higher-order biological capacities, and colonize new environments.

The ECM effectively allows cells to create their own intercellular conditions, which regulate and protect them from the external milieu, whereas the environment of single cells is itself the external medium in which the cells live (Fig. 4.3). Thus, a more or less homeostatically stabilized buffer zone is established, an extracellular space internal to the organism (Bonner 1998; Gerhart and Kirschner 1997). The exchange with the outside medium is increasingly restricted to specialized cells predominantly organized as epithelia, while other cells can be completely withdrawn from the environment and internalized into the ECM. However, there are large differences in the degrees of buffering of physical and chemical changes in this intercellular space. Some organisms have only basic regulative functions for the ECM (e.g., when its ionic composition essentially resembles that of the outside medium). Others develop organs such as nephridia, which regulate this composition.

Prerequisites for the integration of cells within the metazoa are the ECM and the relative exclusion of the outside medium. This becomes most obvious with regard to the ECM as the space through which intercellular signaling takes place. The cells react not to substances from the environment but to signaling processes within the organism, which integrate them into a system. Single-cell protist division, for example, mostly depends on factors in the outside medium, especially nutritional factors, while cells of a multicellular organism divide by way of their own signals within the system.

### 4.3.2 *The Epithelial Boundary*

A general feature of metazoa is their superficial epithelial layer, which is organized in special ways as an integument. The cells, together with occluding junctions between them, seal the internal space from the environment so that the passage of substances can be controlled. The composition of fluids in the inner spaces can be regulated, and concentration gradients compared to the environment can be created. Thus, several authors conclude that the generation of more or less tightly sealed external epithelia, enclosing inner spaces, must have been a critical step in the generation of early metazoans (Mackie 1984; Rieger 1994; Rieger and Weyrer 1998; Tyler 2003; Cereijido et al. 2004; Magie and Martindale 2008).

However, epithelia again demonstrate the double function mentioned previously: Besides their properties as boundaries, they are selectively open either via membranes of the epithelial cells or via the intercellular junctions, which are part of a dynamic and active regulation of transport and barrier functions (Madara 1998; Mitic and Anderson 1998). In addition, the epidermis has to admit information about the outside world via an integrated nervous system. Willmer (1990) called these double functions the "skin paradox." However, it reflects the inward-outward balance, which the epidermis has to fulfill, and rather than being a paradox, demonstrates well the principle of relative seclusion of the environment.

Rieger and Weyrer (1998) developed a possible scenario of successive stages in the evolution of early metazoan colonies (Fig. 4.4). In this figure, (A) demonstrates the separation of somatic cells and germ cells. Some cells would be in contact with the outside medium via cilia; others are internalized into an ECM. A portion of a multicellular organism with an incomplete epithelioid layer at the boundary of the cell colony is seen in (B). Cells in this layer adjoin, but apical junctional complexes and basal matrices are more or less absent. Epithelioid layers with such intermediate characteristics are found in extant Porifera and Placozoa. The general eumetazoan organization with an outer epithelial layer as an epidermis and an inner monociliated

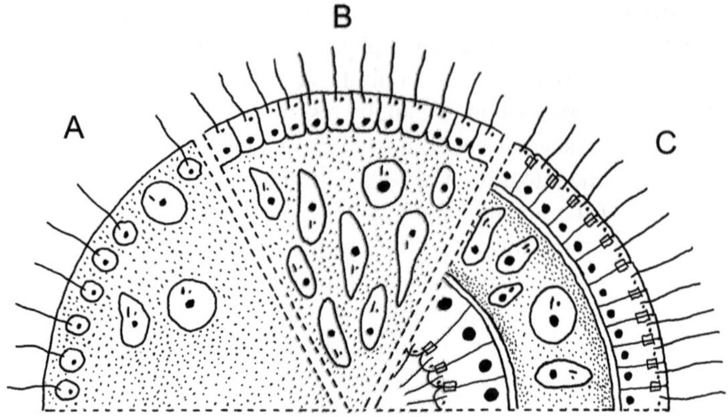


**Fig. 4.4** A hypothesis for possible stages in the evolution of early metazoan cell colonies. For explanations, see text (Redrawn from Rieger and Weyrer 1998)

gastrodermis surrounding the digestive cavity is demonstrated in (C). Both epithelial layers enclose a space in between, thus forming an internalized compartment. Provided the scenario is correct, it exhibits different stages of closure and emancipation of tissues from the environment.

The closure of the intercellular space by occluding junctions might represent a crucial point in the generation of true epithelium (Guillot and Lecuit 2013). Junctions are known from all metazoan phyla (Nielsen 2001). Septate junctions occur in invertebrates and tight junctions in vertebrates, tunicates, and some arthropods. Septate junctions have the shape of a series of parallel septa bridging the intercellular space between cells. Physiological studies increasingly demonstrate that there is a partial barrier function with septate junctions (Dan-Sohkawa et al. 1995; Tepass et al. 2001; Magie and Martindale 2008).

The double function of epithelia has been elaborated tremendously throughout the further evolution of metazoans, as epithelia also form internal sheets wherever some kind of separation is needed. Epithelia separate biological compartments with different composition. From tight junctions, it is well known today that they can form more or less tight epithelia according to their respective function and regulate transepithelial transport according to physiological needs (Cereijido et al. 2004). Thus, the general principle of environmental seclusion was more elaborated and used by further developments in animal history. Manifold internalizations of these sheets were the basis of increasingly robust physiologies.

Tyler (2003) closes his comparative study of epithelial organization with the conclusion that the structure of epithelia, including the arrangement of the proteins and other molecules, which determine epithelial function, is remarkably similar in all eumetazoans and reflects a highly complex and conserved mode of cell differentiation. Comparison with cell layers of sponges suggests that many of the same determinants are present even in this noneumetazoan phylum, although these layers do not qualify by established criteria as true epithelia. Thus, the function of separation

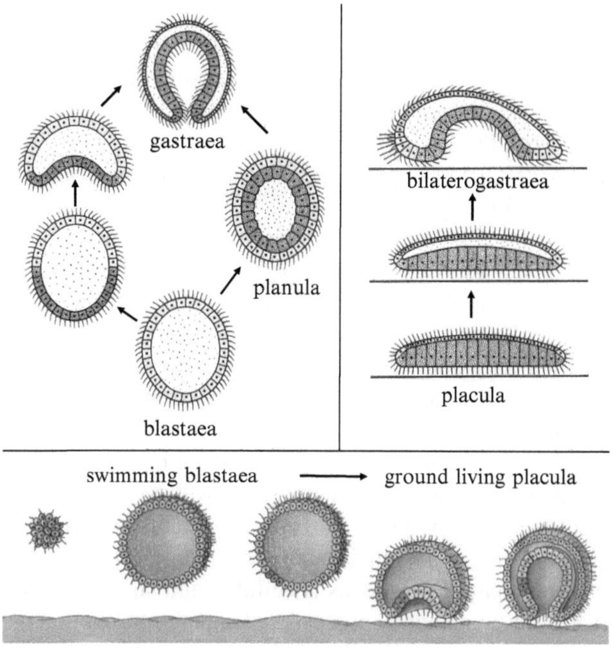


**Fig. 4.5** Recent forms of gastrulation that have been discussed as models for the origin of early metazoans (Modified from Campbell 1997; Gruner 1993)

from the environment must have been involved in the advent of early animal life and thus is a basic and general principle of metazoans.

Magie and Martindale (2008) state that the evolution of cell adhesion is intimately connected with the evolution of multicellularity. The ability for cells to adhere to one another is of importance in the development of multicellular forms, and in this sense, the epithelium can be viewed as the fundamental metazoan innovation. The early evolution of the metazoa is essentially seen as the evolution of the ability of cells to organize into epithelia, something that requires the cell-cell contacts.

## 4.3.3 *Internalized Cells*

It has been argued that in an early metazoan ancestor, all cells might have been on the surface of the organism. This would be the case in a blastaea- or bilateroblastaea-type organism as well as in a planula-type ancestor (Fig. 4.5) (Nielsen 1998, 2001; Ruppert and Barnes 1994). A widely discussed model for such an ancestor is a spherical choanoflagellate colony. It is further assumed that a differentiation occurred in feeding and nonfeeding cells as a starting point for a more complicated structure and that nonfeeding cells were displaced to the interior of the sphere. Germ cells, cells stabilizing the shape of the animal, and cells with special accumulative functions may have lost contact with the surrounding water. These internal cells may have originated from the outer layer through ingression, delamination, or invagination (Nielsen 2001).

Knoll and Hewitt (2011) express the view that the presence of exterior and interior cells is critical for the transition to multicellularity. As a consequence, cells do not all have equal access to nutrients and therefore will not accumulate biomass at a uniform rate unless a process evolves for the transfer of resources from one cell to another. Therefore, Knoll and Hewitt argue, active transfer processes evolved to free multicellular organisms with interior and exterior cells from the sharp constraints of molecular diffusion. Also, interior cells no longer receive signals directly from the environment, even though response to environmental dynamics remains decisive to growth and reproduction. Complex multicellular organisms therefore require processes by which environmental signals can be received by surficial cells and be transduced to interior cells, where genes will be up- or downregulated in response. Development in complex multicellular organisms involves regulation of genes in response to molecular signals from surrounding cells. In animals or plants, the effective environment of most cells is cellular, and signaling between adjacent cells may derive from fundamental processes of signal transduction in response to a need for life-history regulation. Knoll and Hewitt argue that the principle of development may have evolved within this context. All these features match the principle of internalization, as not only cells are internalized, but also essential functions and processes at the same time.

It is especially typical that germ cells are internalized. As this is also the case in organisms that otherwise only show few internalized cells, the removal of germ cells from direct influences of the environment may have played a crucial role in early evolution. In many cnidarians, for instance, the germ cells move into the mesogloea. Sponge oocytes lying within the mesohyl are even surrounded by a special follicular epithelium, which is built from modified archeocytes (Westheide and Rieger 2006).

### 4.3.4 Cell Differentiation

Signaling systems also are the basis for the differentiated functions of specialized cells within a multicellular organism and their integration into the whole of the system. Gerhart and Kirschner (1997) argue that one of the main innovations in the generation of a multicellular organism is its capability of controlling and regulating the functions of cells, whereas many core reactions are already present within single cells and were highly conserved during metazoan evolution. They call a cell's responsiveness to extracellular or intracellular conditions a "contingency." Most processes in metazoan cells are more contingent than those in single-cell eukaryotes. Regulation and differentiation, and thus contingency of cellular functions, are mostly introduced by imposing inhibitions on conserved reactions, only allowing them to become activated under certain conditions and thereby integrating them into the regulated networks.

As argued previously, the prerequisite for the use of signaling systems either through extracellular molecules or through neuronal pathways is the relative closure of the extracellular space. Single-cell organisms are well able to influence each other, but the permanent regulation and integration of cells within the multicellular system need the constancy of signaling pathways. The system then reacts as a whole to environmental factors (Knoll and Hewitt 2011), gaining an active component toward the environment. Gerhart and Kirschner (1997) call the capacity of the cell to create its own conditions "conditionality." An increase in cell types allows the organism to perform new functions (McCarthy and Enquist 2005), including organs for regulation and physiological robustness, which contributes to a gain in flexibility and stability of the organism toward environmental challenges.

### 4.3.5 Formation of Interior Cavities

During embryonic development, animals generally build up several body cavities. The first cavity, which is temporarily generated, is the blastocoel. As all animals, but not members of the other kingdoms of life, develop from such a hollow ball of cells, Margulis (1990) sees this as a distinguishing feature of animals. Form and size of the blastula depend on the amount of yolk in the eggs of different animals, giving different starting points for gastrulation. The blastocoel is already an interior space removed from the outside medium and enclosed by a cell layer. The process of formation of the two-layer condition is always a typical internalization because the processes involved take place in the form of either "infolding" or "inwandering" (Willmer 1990). The differences can be seen as variations on this basic theme (Wolpert 1990).

The other result of gastrulation processes is the formation of the archenteron, another internalized space, which develops into the gut. Functionally, it remains a part of the outside world, which is only gradually sealed off from the environment for digestion and distribution. In its simple organization as in the gastrocoel of cnidarians, digestive enzymes cannot be highly concentrated and predominantly contain water, identical with the surrounding medium. Thus, digestion is still partly intracellular. Also, in many turbellarians part of the digestion is intracellular; in other forms, the cavity is functionally more closed, so that extracellular digestion dominates. Generally, the early course of metazoan evolution shows a tendency to localize exchange surfaces within the protected interior of the body (Schlichter 1984). Thus, gastrulation can be seen as a basic process of the generation of internalized body cavities, which are secluded from the environment. Other internal cavities are built up in triploblastic animals as the different forms of a coelom, again as interior spaces.

The knowledge of these embryological features in extant animals led to some hypotheses about the phylogenetic origin of body cavities (Fig. 4.5) (Gruner 1993; Willmer 1990; Wolpert 1990; Dewel 2000). Haeckel's Blastaea/Gastraea theory is still discussed. Different variants of this theory exist, most of them accepting an early blastula stage and then postulating that the gut evolved either by an invagination of the outer cell layer or by other forms of inward migrations of cells. Another line of argument comes from the placula theory on the basis of the feeding behavior of *Trichoplax adhaerens*: It implies that the lower "epithelium" of a placula-type cell colony could have formed temporarily and later permanent invaginations took place for feeding (Grell 1971, 1981). Each of these theories has to consider the generation of internalized, secluded cavities and the internalization of cell material to build up the body layers. Therefore, this element of autonomization must have been a basic characteristic of animal phylogeny.

### 4.3.6 Size Increase

If it is assumed that multicellularity arose from protists, which somehow stayed together, this involves a dramatic increase in the body size of the resulting organism (Rokas 2008; Knoll and Hewitt 2011). Bonner (1998, 2003) argues that the key step in the appearance of multicellularity was the transition to larger size, which would make it possible for the organism to be isolated from the outside world. A straightforward argument for Bonner's statement is to compare the changes in surface-to-volume ratio with changing size. As noted previously, the surface area of an animal changes with the second power of its linear dimensions and the volume with the third power. Therefore, larger animals considerably reduce their area of direct contact with the environment through the surface, while the interior compartment becomes more elaborate. This is in principle included in Bonner's argument that an increase in size relates to a gain in internal complexity. Especially in larger animals, organs are necessary for functions such as osmoregulation or distribution of gases and nutrients because exchange through the surface no longer suffices. Compared with a large organism, a small organism can generally lose, for example, water faster through its large surface but can also take up water more easily (if it has the same exchange rate via the integument and the same shape). This means that increased size supports physiological stability of fluid management.

Knoll and Hewitt (2011) discuss the chicken-and-egg problem of whether the increase in size reflects or promotes the active transfer of nutrients, oxidants, and cell signals to the internalized cells. They argue that the solution might be to consider the system of size, metabolism, and differentiation as a positive-feedback loop. Under a given set of environmental conditions, the size of an emerging multicellular organism will be limited, in the first instance, by diffusion. Plasmodesmata, gap junctions, or other principles will increase the allowable thickness of tissues by facilitating the transfer of metabolites and signaling molecules between adjacent cells. The length scale on which such transfer will be effective, however, is small, and the degree to which oxygen and diffusible nutrients will be available to interior cells will remain a function of diffusion. With small sizes, direct cell-to-cell connections might be sufficient to support interior cells in three-dimensional organisms. Increasing thickness, however, will enlarge the distance between surface and interior, setting up an increasingly strong gradient of nutrients and signaling molecules. The gradient might itself promote cell differentiation, and cells that differentiate in ways that facilitate molecular transfer will potentiate further increase in size. In this way, size and cell differentiation can generate a positive-feedback loop that will eventually carry populations across a functional threshold to the size and complexity of vascular plants, for example, or of bilaterian animals.

In summary, it can be stated that the general characteristics of multicellularity, as they evolved in early metazoans, essentially enhanced the possibilities of physiological regulation of internal compartments and tissues. Elements that contribute to this achievement are as follows:

- The generation of an ECM, which allows the cells to create their own intercellular conditions that can be buffered against environmental fluctuations;
- The internalization of cells into the ECM;
- The establishment of signaling systems, internalized into the ECM;
- The generation of epithelia as effective boundaries, including enhanced possibilities for regulation of exchange processes;
- Cell differentiation that contributes to the abilities of regulation and specialized reactions to internal and external challenges;
- Formation of body cavities achieved through several processes of internalization; and
- Increase in size with reduction of surface-to-volume ratio and possibilities to support physiological stability of fluid management.

Thus, the features of increasing autonomization according to Definition 2 are matched: refinement of external boundaries, enhancement of regulation capacities, several processes of internalization, and size increase.

## 4.4 Comparison of Presumed Early Stages

So far, it has been shown that typical features of the metazoan organization exhibit elements that contribute to enhanced possibilities of self-determined and self-regulated actions within the environment. Comparing the simplest extant metazoa, it is now demonstrated that these elements are instantiated in different degrees, suggesting that metazoan autonomy evolved in successive steps.

### 4.4.1 Sponges

Sponges (Porifera) are regarded as possessing an organization that could be close to a hypothetical "urmetazoan" (Müller 2001; Brusca and Brusca 2002; Maldonado 2004; Wörheide et al. 2012). Several molecular and morphological studies have identified poriferans as the earliest diverging branch of the animal tree, a placement in agreement with observations that poriferans are the first animals to appear in the fossil record (Rokas 2008; Love et al. 2009; Srivastava et al. 2010; Dunn et al. 2008; Brain et al. 2012). This matches with their low degree of closure from the environment.

Although poriferans have evolved a remarkably complex connective tissue compartment, they do not partition and seal an internal space, and contacts between their cells are far more transitory than in other metazoans. They lack epithelia sensu stricto, which can be defined as sheets of cells forming "sealed" internal compartments within which unique internal milieus may be established (Dewel 2000).

The system of canals in adult calcareans and demosponges – with their exopinacocytes, endopinacocytes, and choanocytes – exposes large surface areas to the water flowing through them and thus to the medium that is surrounding the organism. All these cells have their distal surface in contact with the water, so that feeding and gas exchange can take place directly at the cellular level. The water is used to transport nutrients and gases to the cells and partly fulfills transport functions, which are internalized in more autonomous metazoans as circulation systems with their own body fluids. The whole sponge is pervaded by, and thus exposed to, its environment (Tyler 2003; Srivastava et al. 2010; Cereijido et al. 2004).

As pinacoderm cells and choanocytes together surround the mesohyl, which is the ECM of the sponge, the cells within the mesohyl show a basic although incomplete degree of internalization. The pinacoderm is described as a thin layer of cells that is loosely organized and "leaky" (Magie and Martindale 2008). In sponges, desmosome-like junctions have only been observed infrequently. Gap junctions and basal membranes are also lacking. Of functional importance is that no septate junctions occlude the space between the pinacocytes. Only some special intercellular connections, formed by way of thickenings of opposing cell edges and interdigitations to increase the area of contact, have been found. The physiological barrier function provided by the covering layer is relatively inefficient, and there is little homeostatic autonomy within the mesohyl. In some sponges, the pinacoderm even allows the entrance of water directly into the mesohyl (Fig. 4.6). In addition to the transfer of ingested material from choanocytes to archeocytes, the archeocytes can take up food directly from the water, which enters the mesohyl. Because of the low grade of homeostatic autonomization of the mesohyl, ionic regulation is performed by cellular contractile vacuoles. The flattened shape of the endopinacocytes is regarded as a specialization for rapid diffusion between the water and the mesohyl. Weissenfels (1983) has even demonstrated that yeast cells applied experimentally to *Ephydatia fluviatilis* enter the mesohyl and are phagocytized there. He calls this situation an "open mesenchyme."

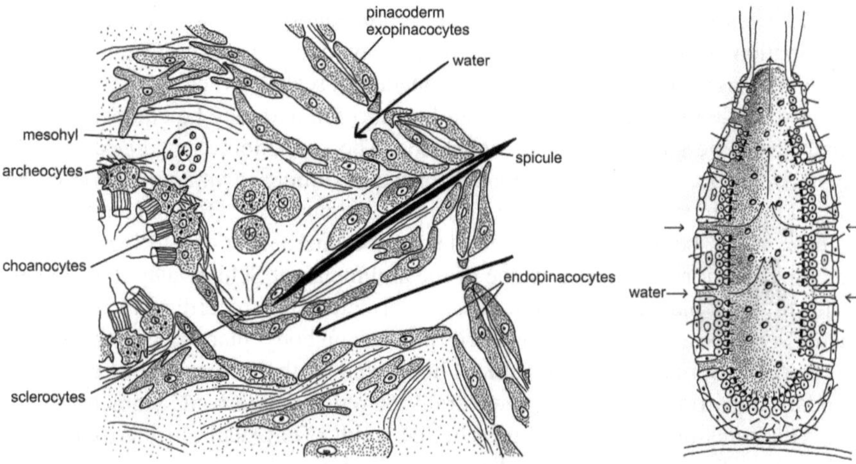


**Fig. 4.6** Sponge tissue (Redrawn from Westheide and Rieger 2006; Moore 1990)

Some observations in demosponges showed that only the areas in which the spicules are secreted seem to be more closed up because the sclerocytes are connected to each other by septate junctions (Dan-Sohkawa et al. 1995; Green and Bergquist 1979; Ledger 1975; Dewel 2000). Besides this, septate junctions in sponges have been found between choanocytes and between gemmula-building spongocytes of some demosponges. Green and Bergquist (1979) state that all cases where structures with occluding functions were observed can be associated with the need to maintain a protecting environment, as in embryonic development, or with the need to maintain a certain ionic environment, as in spicule secretion. In some species, septate junctions seem to be lacking completely.

## 4.4.2 Placozoans

*Trichoplax adhaerens*, a representative of the enigmatic phylum of placozoans, exhibits a simple body construction characterized by just four cell types, an absence of organs, and an axis of symmetry. It is based on two cell layers enclosing central cells, the fiber cells. Because of this simple multicellular organization, the placozoans are also under consideration as possible urmetazoans (Schierwater et al. 2009; Ball and Miller 2010; Schierwater and DeSalle 2007).

The two layers are more similar to the real epithelia of the Eumetazoa than the covering sheets of sponges because they possess apical junctions and septate junctions similar to those of Eumetazoa (Nielsen 2001; Rieger 1994). Ruthmann (2006) supposes that these terminal junctions provide a less-effective closure from the outside than those of eumetazoans, and that the interstitial fluid between the two layers is not too different from seawater. Both epithelia are without a basal matrix.

Thus, *Trichoplax*, a sort of "surface animal," is another organism with a low degree of emancipation from the environment. However, by exposing some features of real epithelia, it has rudimentary elements of eumetazoan sealing structures.

The ventral epithelium is responsible for nutrition and may be compared functionally with the gastrodermis of Eumetazoa. The formation of a "digestive bag" (Fig. 4.5) has been regarded as a model for the beginning of an archenteron in the sense of a "temporal gastrulation" (Grell and Ruthmann 1991; Ruthmann 2006). In this case, it can be interpreted as a model for the internalization of digestive cells and a first interior cavity. This is one of the arguments of Schierwater et al. (2009) for postulating placozoans as the possible urmetazoan. According to this "new placula hypothesis," the nonsymmetric placozoan body plan transforms into a symmetric cnidarian (or also bilaterian) body plan by the former ring of epithelial boundary transforming into the new "oral" region of the derived symmetric body plan. This transformation would then be the result of a placula lifting its feeding epithelium to form the external feeding cavity, keeping function and morphology of the epithelium unchanged. In the final stage, the oral pole would then develop specialized organs, such as a mouth and tentacles for feeding. Thus, the principle of internalization as a tool for gaining autonomy would essentially be involved here.

### 4.4.3 Cnidarians

Cnidarians are composed of two epithelia, an outer epidermis and an inner gastrodermis, the latter enclosing the gastrovascular cavity. This seemingly simple basic organization is more complicated within the different groups and morphologies. Some authors even hold that cnidarians have elements of a triploblastic organization (Boero et al. 2007; Ball et al. 2007).

However, with their morphology, cnidarians are also basically surface animals, exposing their tissue layers to a large extent to the environment, while only a few cells are enclosed between the two layers in the ECM, the mesogloea. The cells of both layers are in direct contact with the surrounding water or the fluid of the coelenteron. The direct contact with the surrounding water and the short distances for diffusion make it possible for the cells to independently fulfill their physiological functions, such as osmoregulation and gas exchange.

However, cnidarians possess fully developed intercellular junctions (Filshie and Flower 1977; Nielsen 2001; Magie and Martindale 2008) (Fig. 4.7). The apical junctional complex consists of a zonula adherens and septate junctions. Some cnidarians lack the zonula adherens. With these connecting and occluding structures, the two epithelia of cnidarians are regarded as "real epithelia," resembling those of other invertebrate groups (Rieger 1994; Magie and Martindale 2008). Proteins have been found that are typical for the basal membrane. The septate junctions seal a space between the cells and thus establish a basic physiological delimitation from the environment. The barrier function is best documented in medusae, many of which maintain a specialized internal ionic environment, which influences the

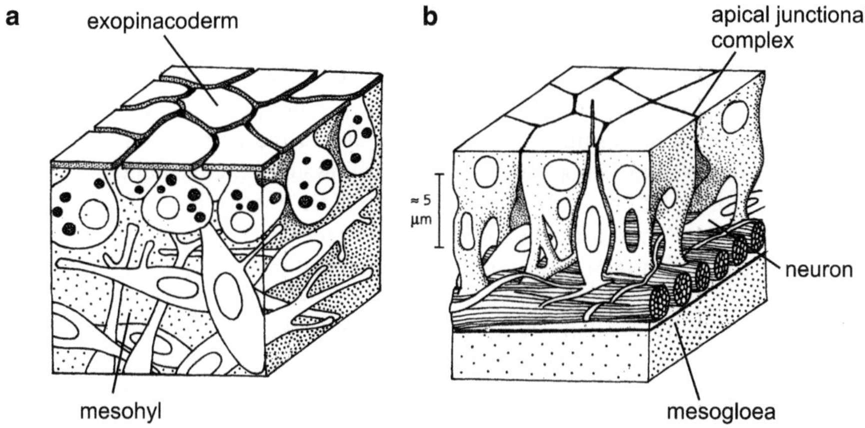


**Fig. 4.7** Comparison of surface structures in poriferans (**a**) and cnidarians (**b**). Essential is the closure between the cells in cnidarians, which is lacking in sponges (Modified from Bergquist 1978; Westheide and Rieger 2006)

buoyancy of the mesogloea. To achieve this condition, a well-sealed epithelial covering is needed (Mackie 1974, 1984).

Between the epithelial cell layers, the nerve cells are internalized as basoepithelial nerve cells or beyond the basal layer as subepithelial nerve cells (Fig. 4.7). By means of the occlusion by the apical junctional complex, these cells are not in direct contact with the water surrounding the organism but lie in a distinctive internal milieu and thus exhibit an internalized cell fraction. Rieger (1994) supposes that the ability to control the intercellular environment was the starting point for the evolution of nervous systems. Although single cells and sponges also have excitable membranes using other ions, for the first time $Na^+$ canals have been found in cnidarians (Mackie 1990).

Muscular structures are internalized in the same manner. The contractile parts of the epithelio-muscle cells lie deep under the epithelial cells in such a way that the apical part lines up with them. In some cases, the myocytes lie as fiber-muscle cells within the mesogloea. Rieger and Lombardi (1987) and Westheide and Rieger (2006) hypothesize that these fiber-type muscle cells developed from epithelio-muscle cells, which sank under the cell layer as fiber-muscle cells within the mesogloea, and that this could be a basic evolutionary process in metazoans, which might have led several times to fiber-muscle cells. This can be interpreted as another example of internalization.

In addition, the cnidarian epidermis shows other basic features of an integument. Although neither stratified nor cornified, the epithelium is frequently thick and tough, protected externally by secreted materials. These external layers range from thin mucous cuticles to the stiff chitinous perisarcs of hydroids and the calcified exoskeletons of corals, thus exhibiting diverse delimitations from the environment (Mackie 1984).

On the whole, the two layers sitting on the mesogloea surround a simple interior cavity. Digestive enzymes can be released and concentrated within the enclosed space. This permits the use of a much greater range of food sizes than is possible in protozoa and sponges (Barnes 1986). For *Hydra*, a transepithelial electrical potential between the water at the outside epithelia and in the gastric cavity has been reported as involved in osmoregulation. Water follows $Na^+$ excreted into the gastric cavity, and the solution is eliminated by periodical contractions of the animal (Schlichter 1984). Thus, the cells can eliminate water via a collective process, and the gastric cavity can be regarded as a simple internalized organ, which functions here for feeding, distribution, and osmoregulation. However, in cnidarians the space of the gastric cavity is relatively open, and in many forms, its content is not too different from the surrounding water. In certain forms, digestion has been found to be restricted to the area of contact of prey with the gastrodermis (Rieger and Weyrer 1998). In other metazoan groups, the gastric cavity is increasingly closed off from the surrounding medium to concentrate digestive enzymes and to use the whole space for digestive functions. Thus, regarding the internalized gastric cavity, cnidarians are only in part surface animals.

### *4.4.4 The Triploblastic Organization*

The origin of triploblastic organization has been discussed extensively (Boero et al. 2007; DeSalle and Schierwater 2007; Schierwater and DeSalle 2007). Some authors identify triploblastic features already within cnidarians (Boero et al. 2007). Here, only one special point is added: The third germ layer, the mesoderm, is a compartment embedded between two layers, the ectoderm and the endoderm (Fig. 4.4C). Ectoderm and endoderm form tissues establishing borders with the outside world, the epidermis and the gastric system. Especially with their organization of intercellular junctions and their basal membrane, these epithelia exhibit the full organization, which is capable of excluding the outside world or rather of regulating exchange with it. In addition, integuments build up special structures that more or less isolate the animal from its environment (Bereiter-Hahn et al. 1984; Rieger and Rieger 1976).

Mesodermal tissue with its derivatives is embedded in between the endo- and ectoderm, thus establishing the main part of a physiologically self-controlled compartment. Mesoderm may be organized as epithelia surrounding cavities, as muscles and blood vessels, and it may form compact mesenchymal tissue. Thus, the proportion of cells in an ECM without direct contact with the environment is increased. Also, other derivatives of the ectoderm (e.g., the nervous system) and of the endoderm (e.g., appendages of the gut system) are included in this internalized compartment. Characteristically, the mesoderm originates ontogenetically through internalizations by ingression or invagination as discussed previously.

The evolution of nephridial organs starts with the generation of these internalized compartments. As certain cell masses are removed from direct contact with the environment, systems for osmoregulation and excretion are needed and possible. These functions are now performed by specialized organs, which lead to phyla with extensive capabilities for homeostasis (e.g., insects and vertebrates). However, the Acoelomorpha, having no nephridia, are an exception, but platyhelminths tend to the surface principle in some parts of their organization because they are flat and thus lack the more elaborated internal differentiation of other triploblasts.

Mesoderm formation within the different phyla of metazoans is closely related to other homeostatic functions. Thus, for example, it is related to the formation of body cavities, which are involved in distributive functions, storage, and movement. Coelomic cavities also are secluded spaces without contact with the environment. Willmer (1990) sees the quantum increase in size permitted by body cavities as the most important factor for coelom formation. However, it may be argued that diploblastic animals can also grow large. Sponges 2 m in size are known. However, Willmer's argument is cogent in that the coelom allows the formation of large animals with autonomized tissues. In diploblastic animals, all functions are fulfilled by the surface principle; triploblastic animals reduce their surfaces and achieve contact with the environment via specialized structures. In this sense, larger size and the functions of the coelom deliver increased possibilities of homeostatic control and independence from the environment.

## 4.5 Closing Remarks

In summary, all these phenomena demonstrate that the transition from single cells to metazoans included different degrees of emancipation from the environment. During early evolution, cells might have been at first only weakly integrated within the ECM with a low degree of seclusion from the environment. Sponges seem to be relicts of such a state, although they already had some sealing functions without having the characteristic epithelia. Another degree of organization generated epithelia-like layers, which were able to establish basic boundaries by the closure of intercellular spaces. Pure surface organisms are conceivable, either with or without sealed epithelia and without an enclosure of a cavity. Trichoplax might be a relict of such a level. Vendozoan organisms with the enigmatic Ediacara fauna might have possessed this type of organization (Knoll and Hewitt 2011). Erwin and Tweedt (2012) assume that they may represent a variety of metazoan clades lying above sponges and below the origin of the protostomes and deuterostomes, and that their surface-to-volume ratios and morphology suggest that at least some of these organisms were osmotrophic, feeding off dissolved organic carbon in seawater.

(continued)

(continued)

A further level might have been the delimitation of a cavity as an internalized space. Cnidarians are still surface animals in their tissue organization but enclose such a cavity as the gastrocoel. The "surface stage" was overcome with the introduction of the third body layer, the mesoderm, well sealed from the environment by an ectodermal and an endodermal epithelium. The internalized tissue enabled the elaboration of an increasingly self-regulated compartment with organs. Although the evidence used is well known, the fact that the generation of multicellularity in metazoans exhibits autonomization as a central evolutionary pattern has largely been overlooked.

During the transition to multicellularity a new systemic level (in the sense of the system model mentioned in Chap. 3) emerged. The cells are integrated into an overall system that now acts and reacts as a coherent whole (Knoll and Hewitt 2011; Michod 1997; Buss 1987; Arnellos et al. 2013). Buss (1987) describes metazoans as discrete entities in which cellular differentiation routinely acts to limit the inherent potential of their constituent cells for unbounded growth. "Metazoan evolution is characterized by an increasing sophistication of cells, tissues, and organs which perform somatic duties of value to the individual as a whole, but which require the cells composing them to limit their inherent potential for proliferation. The propensity for continued self-replication has been subjugated to the interests of the whole" (p. 53). Then, Buss describes how the whole system controls this process in a direct fashion during embryonic development in metazoans and how experiments demonstrated that the organization of the unfertilized egg predetermined, to varying extents, the ontogenetic path. "The developing embryo follows a path of differentiation enforced upon it by its mother." Today, we have much more insight into many of the molecular processes involved, which generally confirms this principle.

Also, Butterfield (2009, p. 201) characterizes multicellularity "as a concept" that entails "a fundamental shift in the nature of individuality. No longer does each cell have an independent identity, but serves instead as part of some larger more inclusive individual, defined by a suite of 'emergent' characters. Such collectivization undoubtedly has its advantages, not least increased size and accompanying environmental resilience."

Contingency on the cellular level in the sense of Gerhart and Kirschner (1997) is a prerequisite for this. In terms of robustness, Stelling et al. (2004) discuss that intracellular processes provide a certain amount of stability to diverse perturbations, but that they must dispense with this robustness in certain situations within a multicellular system, as high robustness might be detrimental to the organism and lead to uncontrolled, robust growth of neoplastic cells. Thus, the regulation of the cell's robustness within the system, such as for apoptosis, is essential and becomes a new feature of the overall robustness of the multicellular system. Cancer occurs when a cell develops a new behavior that – from its own perspective – is robust (e.g., against prohibitive signals from the organism) (Sonnenschein and Soto 1999). Thus, physiological robustness shifts to the next-higher level of the hierarchical system.

Currently available molecular data support the idea that the metazoan ancestor was equipped with the molecular tools necessary for the specification of complex body plans (Cartwright and Collins 2007; Bromham 2011). This supports the origin of a complex genome pre-dating the Cambrian radiation, with the ancestral genome possessing the molecular tool kit necessary for an "explosion" of body plans and complex traits. Thus, it can be assumed that the trigger for the Cambrian radiation has to be expected on the new integrative system level (West-Eberhard 2003; Carroll 2005b; Prud'homme et al. 2007).

# Chapter 5
# The Cambrian Explosion and Thereafter

The Cambrian explosion, which occurred some 540 million years ago, marks a period in which the animal phyla we know today appear in the fossil record for the first time. However, the diversity of forms, which becomes visible at that time, makes it likely that it had a longer prehistory. Although some fossils of metazoan origin are found before the Cambrian explosion, there are no sequences of fossils that reveal the steps of morphological evolution leading to this divergence (Valentine 2002; Knoll and Carroll 1999; Butterfield 2009; Donoghue and Antcliffe 2010; Erwin et al. 2011; Morris 2006; Gaidos et al. 2007; Cartwright and Collins 2007).

To the oldest well-preserved fossils of metazoans belong findings from the Doushantou formation in China. Their age is estimated at 590–550 million years before present. The fossils are so well preserved that even the cells are partly visible. Eggs and embryos in the blastula stage also were found (Yin et al. 2004). Tissues with structures that resemble spicula may have belonged to sponges and thin tubular structures to cnidarians of that time (Xiao et al. 2000). Other fossils of a comparable age come from Newfoundland (Conway Morris 1998). However, most of the fossils, which may be small metazoans, are younger.

Other types of fossils, which possibly came from early metazoans, are traces and burrows, which have been left in the sediments for at least 570 million years (Crimes 1992; Jensen 2003; Valentine 2004; Xiao and Kaufman 2006). So far, it is not possible to identify which of the known body fossils of that time may be responsible for these traces. Nonetheless, the traces allow some estimation of the activities the animals could have displayed and of some of the structural prerequisites that must have been achieved to that point. The Precambrian traces were predominantly horizontal, and the sediments were only slightly penetrated. Thus it is assumed that the animals mainly crept over the soil or dug only superficially (Erwin and Tweedt 2012). The traces are minute, usually less than 1 mm in width. Extant animals, which leave such superficial and small traces, mostly have a mucociliary way of creeping, using their cilia to creep within a trail of mucilage over the substratum.

B. Rosslenbroich, *On the Origin of Autonomy: A New Look at the Major Transitions in Evolution*, History, Philosophy and Theory of the Life Sciences 5,
DOI 10.1007/978-3-319-04141-4_5, © Springer International Publishing Switzerland 2014

However, this type of movement does not allow more forceful digging in and penetration of the soil. Many extant marine invertebrates, which are able to dig further into the substrata, use muscles of the body wall to generate peristaltic waves or specialized organs on their front part. Because deeper traces are lacking in the Precambrian, it is assumed that hydrostatic skeletons, which are needed for forceful peristaltic movement, were lacking and that this was the case until the beginning of the Cambrian. Thus it can be presumed that the animals that left these traces must have been small. Possibly, most of the typical features of extant bilaterians were not yet present. In comparison with recent animal groups of this size and such life habits, it can be assumed that they were acoelomats or pseudocoelomats. They lacked a mineralized skeleton and might have had only a thin cuticle or none at all and only minimal organs. Such organisms only fossilize under special conditions, which might be one reason for the lack of body fossils from these animals.

The beginning Cambrian then is characterized by the huge variety of body fossils and the occurrence of larger, penetrating traces, thus marking a phase of profound innovations in the biomechanical and physiological equipment of the animals, leading to changes in biological activities (Bromham 2011; Erwin et al. 2011; Erwin and Tweedt 2012; Cartwright and Collins 2007). Looking at this fossil record, together with knowledge about the physiological and morphological characteristics of the phyla that began their evolutionary history around this time, it can be said that the resources of this set, which can increase autonomy (Definition 2), are all involved in these innovations: increasing body size; reinforcement of the environmental separation through epithelia, cuticles, and shells; extension of homeostatic possibilities; and internalization of organs and functions (e.g., respiratory organs, circulation systems, nephridial organs). The movement capacities expanded and a stronger mechanical impact on the environment became possible. The generation and extension of nervous systems enabled multicellular organisms to channel the flood of information from the environment by creating a dam between the external world and the interior milieu of the organism and to react in a modulated and more or less self-determined way.

The different body plans of the phyla are characterized by the varying combinations of this set of resources. The phyla started their way through evolution with these different prerequisites, allowing for different possibilities of independence and flexibility. In some cases, the basic plan was considerably reorganized so that special forms of flexibility arose (e.g., an octopus). Other body plans seem to have lacked these possibilities, perhaps because they were more constrained.

It is interesting that diversification during the transition to Cambrian faunae was more an explosion of body plans than of cell types and molecular processes. Most of the necessary building blocks had been invented earlier (Gerhart and Kirschner 1997). It seems that the innovations emerged more on a systemic level (according to the characterization of systems in Chap. 3) than on the level of molecular or cellular processes, which further indicates the significance of the systemic perspective for understanding such transitions. This is in accordance with recent considerations about the importance of the phenotype in general for evolutionary processes (West-Eberhard 2003).

Thus, I propose the view that the Cambrian explosion and main parts of metazoan evolution thereafter were a diversification of different means of evolution generated to gain autonomic functions on the level of body plans. In the following paragraphs, I discuss in more detail several of the features involved.

## 5.1 Body Size and Movement Capacity During the Cambrian Explosion

While Ediacara organisms maintained the surface principle as they extended their surface with minimal thickness and other faunae were smaller than 1 mm in size, the Cambrian explosion brought forth organisms that reached a new dimension of size with a closed organization and internalized tissues. Thereby, considerable reduction of the external surface contact to the environment relative to the volume took place. This made it necessary to generate increasingly specialized organs for exchange functions.

The size of the animals now reached up to 60 cm (*Anomalocaris*) or 75 cm (some trilobites), and the smaller ones also reached a new dimension, such as some mollusks of about 2–6 cm (some nautiloids) or the early chordate *Pikaia* (about 4 cm). However, these examples show that there was no linear increase in size but rather a spectrum of different sizes. According to the hypothesis that the Cambrian diversification might have been a time of experimentation of evolution with various possible body plans (Storch et al. 2001), it may also have been a time of experimentation with different body sizes. Movement capacity and body size are in close relation with each other, as the Reynolds number changes with size: Larger organisms are emancipated from the restrictions of small Reynolds numbers (Fig. 5.3).

Even more important was the generation of new instruments for movement. A primitive form of locomotion in an ancestral Precambrian metazoan may have been the propulsion by cilia as it is still used by many protists and some small metazoans today (Rieger et al. 1991). However, this functions only in extremely small organisms. If the organisms become larger, the available surface covered with cilia only increases by the square of body size, whereas the volume increases by the cube of body size. Thus the available ciliary surface must transport an increasingly larger relative body volume, setting limits to this type of locomotion.

Some new innovations seem to be necessary for larger animals. Probably one important principle was invented during the Precambrian, but it stayed small and inconspicuous. An organism with this innovation may have been something like the "roundish flatworm," postulated by Gerhart and Kirschner (1997) as the possible origin of bilateral animals. It might have had a mesoderm with some sort of a pseudocoel in it and could have had a set of *Hox* genes, which allowed for compartmentalization of its body axis. It might have been the organism that left the horizontal traces underneath the microbial biomat during the Ediacaran period and might have used a simple hydrostatic skeleton to generate the necessary forces for digging in this limited dimension.

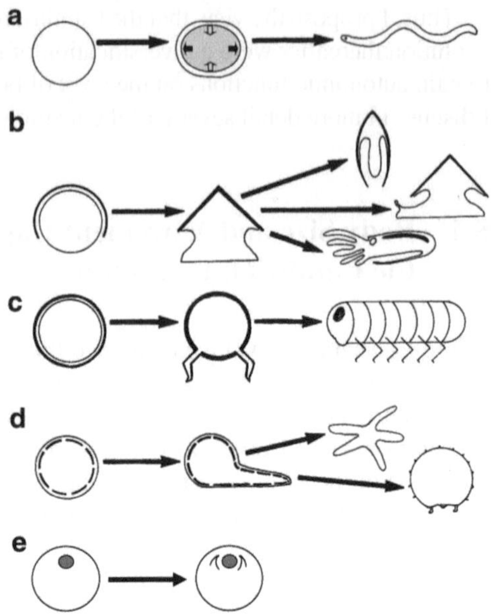


**Fig. 5.1** The five basic types of skeletons. (**a**) Hydrostatic "worms." (**b**) Incomplete exoskeletons: the "shell" of mollusks. (**c**) Complete exoskeletons of arthropods. (**d**) Endoskeletons of echinoderms. (**e**) Endoskeletons of chordates (Modified from Willmer 1990)

The hydrostatic skeleton could have been an essential prerequisite for the generation of body size in later digging animals, which would have been continued by the generation of stiff body parts as either exoskeletons or endoskeletons (Fig. 5.1). Unto these constructions muscles could attach, which made the movements more effective than just swimming with cilia. By way of size as well as by means of the connection of skeletal elements with joints and legs, stronger momentum could be generated for more directed and more varied movements.

The internal mass of muscles and thus the available power can increase together with the volume of the organism. Several bundles of muscles can perform their force relatively removed from their direct location through elongations, and they can combine the effects of multiple units. This can tremendously increase the force that can be exerted on a certain point. Thus, the transition to the propulsion by muscles allows organisms to achieve a far greater range of power that can be applied to the surroundings (Gans 1989; Scheiba 1990). This applies generally to the divergence of forms of movement, whether digging into the ground, crawling on it, or swimming in free water.

Under extant animals, the larger ciliary swimmers are also slower than comparable muscular swimmers. For example, the turbellar *Convoluta*, which is about 2 mm long, swims only 0.6 mm/s, while *Daphnia*, at the same size, swims up to 7 mm/s using muscular force and levers (McMahon and Bonner 1983).

Thus, these organisms experimented with their possibilities within the environment so that they could move more skillfully, obtain a more physical effect on the environment during digging, or exhibit a sessile lifestyle, often together with an elaboration of shells and other stiff seclusions from the outside world. Decisive for those animals that stayed mobile was the generation of a spectrum of body plans, which in different combinations enabled self-determined movement together with a certain amount of environmental seclusion.

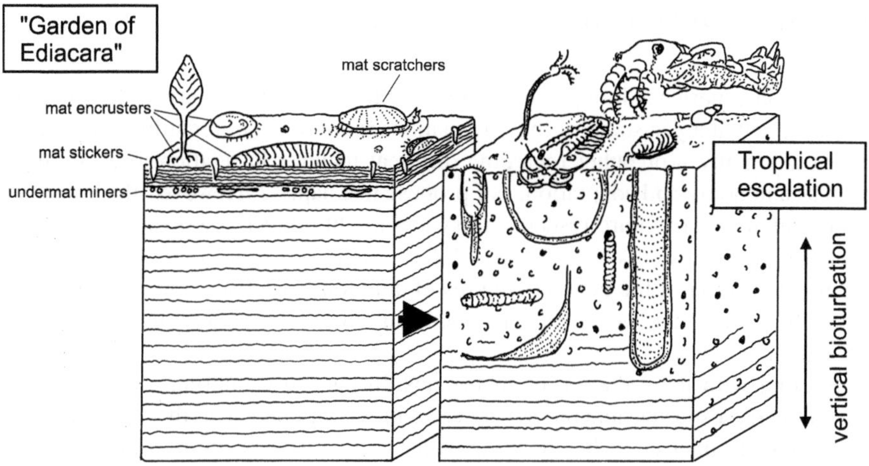


**Fig. 5.2** The Cambrian explosion as an ecological change (Redrawn from Seilacher 1999)

Many animals of the Cambrian had eyes. This is probably related to aim-directed movements (Bromham 2011). The bilateral organization of the body with a head was a starting point for a centralized nervous system and thus for the generation of widened flexibility of behavior (Conway Morris 1998).

By means of these innovations, the metazoans liberated themselves from settlement in suitable littoral seas, to which the precursor faunae seem to have been restricted. Because of their increased size and the possibility to exert more force and effects on the environment, many new biotopes became available, including the open sea and deeper sea areas.

It is assumed that the sediments of the sea during Ediacaran times were sealed by firm and erosion-resistant biomats produced by microbial life (Seilacher 1999; Seilacher et al. 2003). Most organisms living on the ground moved over the surface or were stuck on it. Only a few small burrowing animals penetrated these biomats superficially without destroying their structure. In contrast to this situation, the occurrence of metazoan life in the Cambrian must have been accompanied by a fundamental ecological change. The biomats disappeared because of intensive digging and burrowing (bioturbation) (Erwin et al. 2011; Erwin and Tweedt 2012; Cartwright and Collins 2007). Thus, trilobites ploughed through the soil of the sea and wormlike animals burrowed deep into the ground (Fig. 5.2). A fauna from which there was no sign of profound mechanical effects on the environment of mutual prey catching ("garden of Ediacara") was replaced by a fauna with intensive burrowing, digging, filtering, and prey-catching behavior. This resulted in a radical change in the complexity of behavior and in the ecological relationships. The disappearance of the biomats also led to a more intensive exchange of substances between water and the sediments, so that the feeding interrelations must have changed in a profound way. These "engineering activities" also enhanced the oxygenation of the sediments and microbial primary productivity (Erwin and Tweedt 2012).

## 5.2 Size and Movement in Extant Animals

Most phyla experienced long evolution before today's forms emerged. It would be interesting to analyze some of these transitions in more detail regarding the principle of autonomy – or rather regarding different features in relation to each other, such as autonomy and adaptation, as I propose in Chap. 12. However, this must yet be accomplished in future research work and might deliver new insights on "how evolution navigates" (Conway Morris 2003). For now, we must be content with a rough description of the general features of autonomy as observable in animals of today. In this manner, we can at least obtain an impression of how fundamental the principle really is.

Starting with the radiation of metazoans during the Cambrian, diverse trends toward increases in size and movement capacities continued up to the modern world of animals. Bonner describes how the upper limit of size in animals as well as in plants increased permanently (McMahon and Bonner 1983; Bonner 1988, 2004). However, this is only a rough generalization and may just demonstrate the size increase as it was reached from the world of single-cell organisms during Precambrian times.

Figure 5.3 depicts the relationships between body size, Reynolds number, and absolute speed. Movement capacities are generally measured in absolute numbers rather than relative to body size because the interactions between organisms and their environment as well as with other organisms are determined by the absolute, not by the relative size (Bennett 1991). Roughly larger animals move in dimensions of larger speeds than small ones (McMahon and Bonner 1983).

Looking closer, there is greater diversification within the respective dimension according to the morphological and functional prerequisites and the respective forms of adaptation. However, body size matters, and there have been systematic changes of these features during evolution, on the large scale as well as within certain groups (Roy 2008; Blanckenhorn 2000; Bonner 2004).

Many small invertebrates do not swim actively within the water but rather stay floating and drift with currents. Because of their low Reynolds numbers, they partly rely on the viscosity of the water and also move with the help of cilia (Willmer et al. 2000). In contrast to this, muscles of different forms are used in larger animals to stay actively floating or swimming. Jet propulsion, undulation of the body or parts of it, and propulsion with body appendages are the basic principles employed for swimming. Often, crawling on the ground is generated by variations of these muscular movements.

Larger animals also use the force of combined muscles to dig themselves into the ground, thus giving them the ability to generate larger forces. Some sort of anchor often penetrates into the soil first, and the rest of the body is then pulled behind. Movement frequently is a modification of the function of alternating circular and longitudinal muscle contractions, as they take place in many other forms of soft body movements. The hydrostatic skeleton enables especially effective and forceful digging in the soil, for example, in the segmented coelom of annelids.

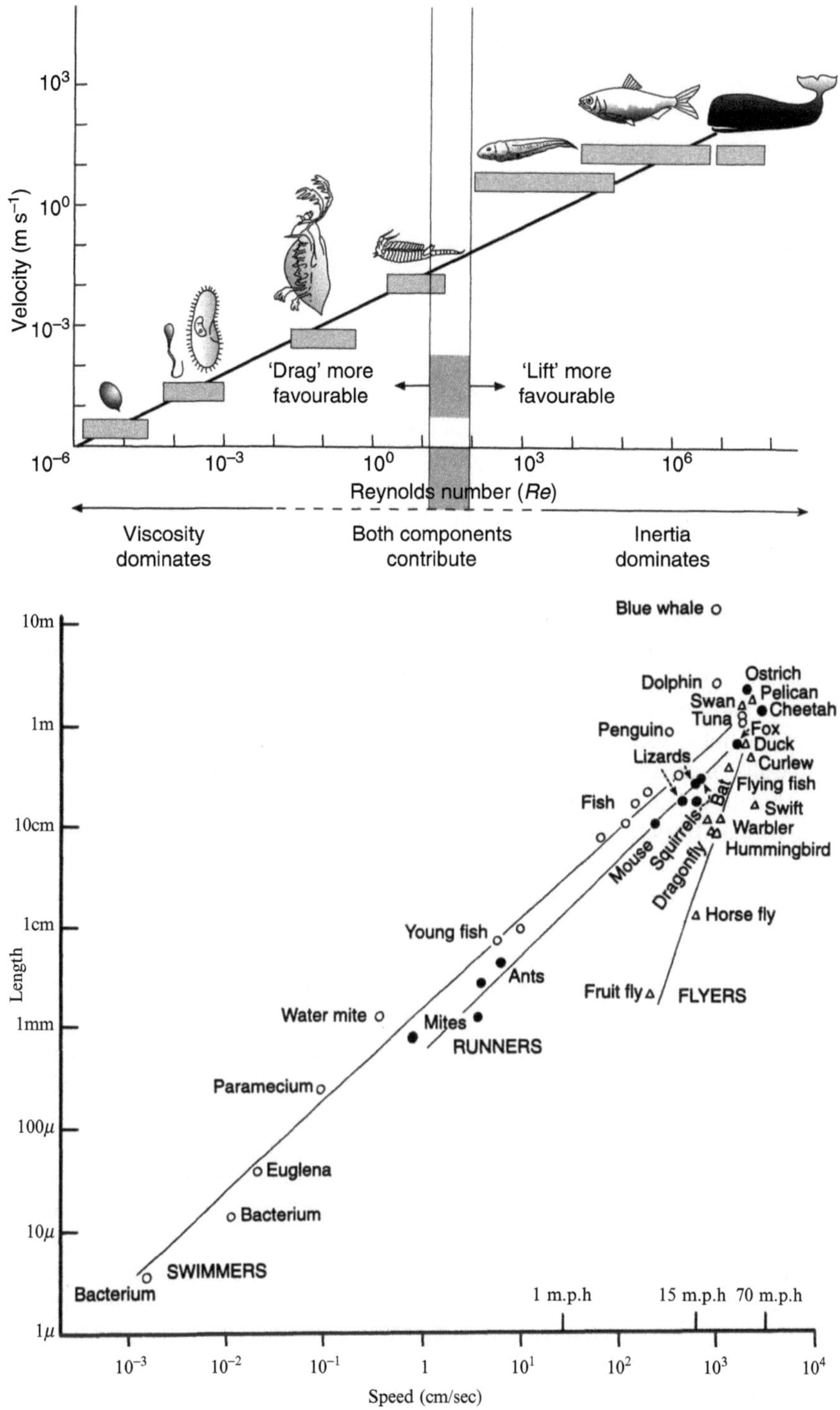


**Fig. 5.3** Relations of body size, Reynolds numbers, and speed during locomotion (*Above*: Willmer et al. 2000, with permission of Blackwell Publishing Ltd.; *below*: Bonner 1993, with permission of Princeton University Press)

Many mollusks develop a tough exoskeleton, which mainly serves as a separation from the environment. In many cases, the exoskeleton is suitable to retreat completely from environmental influences. At the same time, however, this restricts the possibilities of movement. Reductions of the shell, on the other hand, can be associated with increased movement behavior. Some groups of cephalopods, for example, reduce their shell and develop an effective mode of swimming by jet propulsion. The spectrum, which evolved in the different groups of mollusks, with different combinations of environmental seclusions on the one hand and movement capacities on the other hand, is an example of how features of autonomy are realized in a diversified manner.

In metazoans, which went on land, the morphological prerequisites for movement possibilities are even more important. Soft-bodied animals with their predominantly hydrostatic support systems now have tremendous disadvantages (Willmer et al. 2000). Without support by way of buoyancy in water, they tend to become flat under their own weight. Therefore, they need thicker muscles to maintain tonus and form. As such, they are dependent on complete hydration because dehydration restricts their activities. Also, they are often sensible to frictions on hard substrata.

In a functional sense, the real innovators within the invertebrates were the onychophorans, which might be the intermediate form between a possible annelid ancestor and euarthropods (Westheide and Rieger 2006; for a different view on theses relations, see, e.g., Bromham 2011). Many pairs of legs stick out of their body wall. The body works essentially hydrostatically with considerable changes of form, but the appendages work as real legs. Moving the legs forward and backward using own muscles results in a lever action that reduces friction with the ground.

Within the broad spectrum of forms of movement generated by metazoans, the use of the lever principle was of special significance. This is valid for movement in water, but even more so on land. Within the euarthropods, the possibilities of the cuticle are used to develop effective lever principles in legs and wings. The cuticle, which can be secreted in a fluid state and subsequently hardens to variable degrees and into any form that is needed, delivers the necessary resistance for antagonistic working muscles. The generation of legs with flexors and extensors at each joint uses the advantage of the lever action, reduced frictional contact with the substratum, and a smooth forward action for the main body mass, with only the legs themselves alternately accelerating and decelerating (Willmer et al. 2000).

Effective movement is especially reached by a reduced number of legs, which itself reduces weight. The most agile land arthropods have only five pairs (decapod crustaceans), four pairs (arachnids), or three pairs (insects). In many decapods and arachnids, one pair of legs is not even used for the fastest gaits. Muscles are moved to upper parts of the limbs so that the lower parts can be light and flexible. The advantages of legs are summarized in Table 5.1. Together, these features lead to a profound increase in movement capacity and maneuverability. The fact that legs of this sort are present in animals that live in water as well as in animals living on land makes it more likely that it is a phylogenetically generated potential rather than some sort of direct environmental adaptation.

**Table 5.1** Advantages of legs (in contrast to soft-bodied locomotion)

| |
|---|
| Lever action exists – small movement of muscle generates large movement at limb tip |
| Limited ground contact is made – reduced friction |
| Legs stop and start, body moves forward smoothly – reduced acceleration and deceleration of large masses, energy costs reduced. Aided by having main mass of leg (muscles, etc.) at the top, close to body |
| Muscles can be small instead of sheet-like, with less connective tissue strapping and increasing speed of contraction |
| No lateral sinusoidal components exist that waste energy |
| Legs largely independent of hydration state |
| Effects of muscle contraction are localized, do not affect other body wall muscles |
| Increased number of gaits and gait/speed/energy trade-offs |
| Good proprioception permitted |
| Legs can be diversified for other uses |

From Willmer et al. (2000, p. 480)

Speeds reached by land arthropods can be substantially faster than those by marine species, largely because of the reduced viscosity of the medium. Land crabs can achieve 1–2 m/s, with only 8 of the 10 legs in use and with only 3 in contact with the ground at any time. Within the chilopods, there also exist fast runners (*Scutigera* up to about 0.5 m/s), and many of them are well known for their forceful digging into the soil. Digging beetles with powerful legs and a heavy cuticle also can generate strong forces. An especially elaborated use of the lever principle is realized in those land arthropods that jump, which is an especially radical solution to reduce friction with the substratum.

Also, flying in pterygote insects uses the lever principle, which becomes possible because of the stiff cuticle of the wings. The flexibility of movement that is reached consists not only of speed and range but also often the possible maneuvers during flight. Houseflies, for example, are able to hover at one point, to fly backward or upside down, or to turn around within one body length. The lever principle works in such a manner that small movements of muscles generate large movements of the wings.

In summary, there is a broad variance of combinations and types in size and movement dynamics during metazoan evolution, and no singular trend or direction is discernible. However, the overall possibilities for emancipated and free-dwelling movements or strong environmental impacts were essentially enhanced in many groups. The changes that meet the features of Definition 2 are as follows:

- Increases in size with reduction of surface-to-volume ratios and extension of the physiological inertia effect;
- Emancipation from the restrictions of small Reynolds numbers;

(continued)

(continued)

- Generation of instruments for locomotion that widen the capacities of movement and maneuverability or have more physical effects on the environment;
- Generation of a greater range of power, which can be applied to the surroundings; and
- Bilateral organization of the body with a head as a starting point for centralized nervous systems for the generation of larger flexibility of behavior.

## 5.3 Skins

As discussed previously, skins always have to fulfill a double function: They need to protect the animal, and in physical, chemical, and physiological senses, they have to keep the world out. At the same time, however, they often have to admit selected chemicals from that exterior world (water or ions, gases, or nutrients), and they have to admit information about the outside world by giving access to the animal's nervous system. In many cases, skin is also involved in the animal's movement (or stasis). It is virtually impossible to perform all these functions simultaneously with great efficiency. Epithelia and integuments inevitably represent balances between these requirements.

Beginning with the Cambrian radiation of the metazoans, phylogenesis led to diverse reinforcements and differentiations, so that the boundary of the body increasingly supported processes of self-regulation and stabilization toward environmental influences. The extensive structural variations of the integuments within the different groups can be seen more in relation to the phylogenetic position than in relation to the ecology of the respective animal. Thus, the nature of the integument can predispose an animal to a certain environment and exclude it from others (Richards 1984).

Together with the generation of real epithelial tissues, as described in Chap. 4, fibrous materials on their outside may have been an early acquisition in the metazoans, as it would confer potential protection, support, and regulation on the newly multicellular animal. The primitive cuticle may have been no more than a fibrous mass of mucoproteins or mucopolysaccharides, elaborated from the glycocalyx, which protects most animal cells. Simple cuticles of this type can be found in many animal phyla. They may have been a starting point for the extensive variations of integuments (Willmer 1990; Lillywhite and Maderson 1988; Rieger 1984).

The epidermis of invertebrates is virtually always a single layer. Usually, it is assumed that the earliest radiation of the eumetazoans started from such a single-layer epidermis with flagella or cilia. An example for such a simple epidermis is

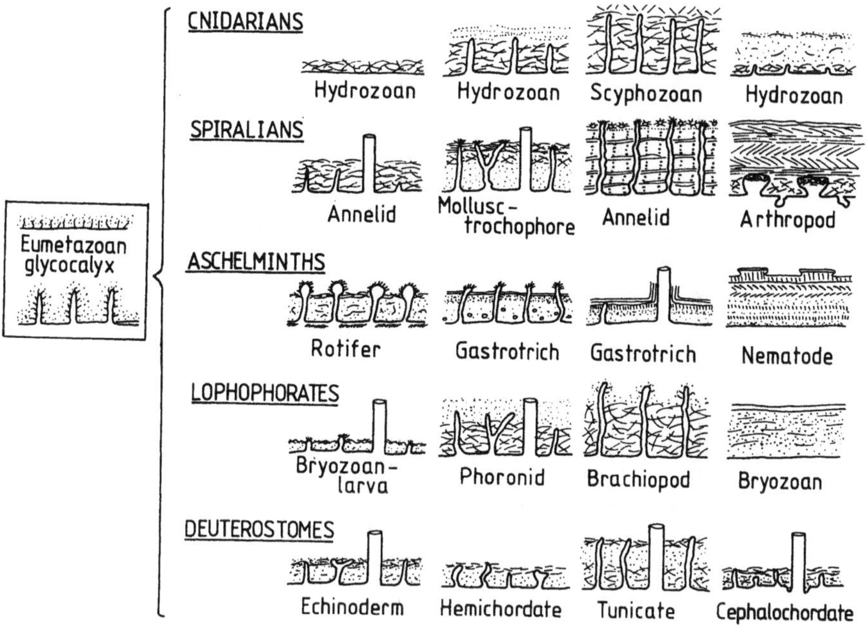


**Fig. 5.4** Structures of the cuticle of invertebrates (From Willmer 1990)

that of the turbellarians. However, some of these examples may also be secondary simplifications. Nonetheless, there are some phyla (ctenophores and nemertins) that hardly have anything beyond such a simple, possibly primitive cuticle, while others generated tremendous specialization. Some general trends are shown in Fig. 5.4. It has been assumed that the tendency to reinforce the cuticle may have been generated about four to five times independently in several groups (Rieger 1984; Willmer 1990). In the platyhelminths and annelids, the cuticles are still variants of the microvillar prototype, but in some animals that are more specialized, the fibrous matrix between the microvilli is strengthened with collagen, chitin, or even calcium carbonate.

Whatever the evolutionary sequence was, in any case there were some trends toward the generation of environmental barriers that were more effective. These can become prominent and can also supply support and stability in face of gravity and other mechanical influences from the environment. To these belong the cuticle of arthropods, consisting of α-chitin and proteins. They seem to be the most important innovation of this group during its evolution (Westheide and Rieger 2006). In its primitive state, it is a thin (just a few micrometers), relatively soft, closed layer such as in onychophorans. Up to the euarthropods, the cuticle evolved into a skeleton of numerous sheets with hard, thick sklerites and segments that were more flexible in between and covered the animal like a knight's armor. Having a low specific weight, it possessed high stability and high resistance with respect to chemical and

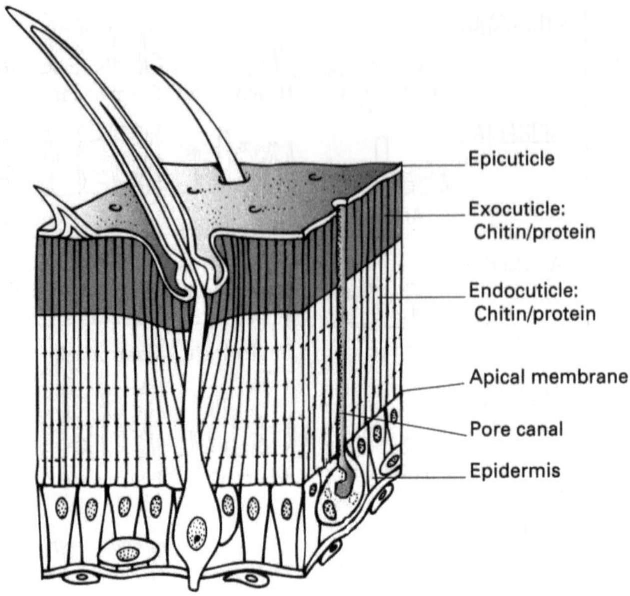


**Fig. 5.5** Structure of a typical insect cuticle (Modified from Willmer et al. 2000, with permission of Blackwell Publishing Ltd.)

mechanical influences. It is also the decisive element during emancipation from water within this group (Jane et al. 2005).

While the sclerotized proteins within the epicuticula especially established mechanical stability, the lipids lowered the permeability for water. Chiefly, the waxy layer reduced evaporation, so that some arthropods were able to reach dry habitats. Desert scorpions, cockroaches (*Periplaneta*), and Tenebrionidae (a large family of beetles) have a cuticle with the lowest permeability for water. With crustaceans, some millipedes, and onychophorans, the waxy layer is lacking. Therefore, they are dependent on biotopes with high humidity.

If one assumes that the onychophorans are the basic group of the arthropods (Westheide and Rieger 2006), it is obvious that an essential part of the evolution of this group consisted of the generation of particularly strong and stiff environmental seclusions (Fig. 5.5). Its diversification included the development of impermeable surface barriers to safeguard against penetration by noxious substances and microbes, to resist mechanical stresses, and to prevent excess fluid loss to the surroundings to sustain life in a terrestrial environment (Jane et al. 2005).

The cuticle is also the basis for the generation of the effective movement apparatus and therewith the high movement capacities in some arthropods. It generates joints and muscular insertions and thus enables the use of the lever principle. Possibly, it was the combination of effective environmental separation with the effective extension of movement capacities that established the prerequisite for the evolutionary success of this group.

However, the chitinous cuticle might also have brought restrictions because it is not suited for larger bodies. It becomes too heavy and stiff with the mass, which has to be built up.

During the evolutionary radiation of the eumetazoans, there might have been conflicts between several features of autonomy. This might be the case for strong environmental seclusion on the one hand and increased movement capacity on the other hand. The respective combination leads to different ecological possibilities and restrictions, which can be seen typically in mollusks. Stiff shells such as in bivalves lead to high mechanical strength (e.g., *Mytilus*, the blue mussel, in the surf) but reduce at the same time the possibilities to move. The other extreme within the mollusks is the cephalopods, among which the coleoids reduced the shell completely and have at most an internal bone or shell, which is used for buoyancy or support. With their jet propulsion, sometimes supported by the arms, some of them are able to swim rapidly. While in *Nautilus* the rigid outer shell prevents fast swimming, *Loligo* reduces the shell to an elastic cartilage and becomes an efficient and fast swimmer of the high sea. However, swimming by jet propulsion is an energy-consuming way to travel compared to the tail propulsion used by fish. The relative efficiency of jet propulsion decreases further as animal size increases.

With the skin of the vertebrates, a completely different principle evolved. Rather than generating a stiff knight's armor, the integument has the potential to combine an effective environmental separation with high flexibility and elasticity, thus synthesizing – again in different combinations – essential features of autonomy. The crucial innovation was that instead of a single-layer epithelium with various external deposits, the epidermis became multilayer and sat on the dermis. This skin, composed of two main layers, is flexibly fixed on the underlying structures by subcutaneous tissues (Fig. 5.6).

The skin of the bony fishes is often thin and rich in unicellular mucilaginous glands. In many species, the whole thickness of the epithelium is built up by live cells, which are involved in secretory functions. The dermis typically contains relatively little of the connective tissue found in tetrapods. The majority of fishes generated a skin with overlapping scales, consisting of hardened substances from the dermis. The principle of building layers, the ability to generate mucus, and the generation of scales brought an effective seclusion toward the watery environment. Fishes living in seawater are protected from a loss of body fluids, while fishes living in freshwater are protected from a too extensive uptake of fluid. Thus, both of them can live as osmoregulators (see Chap. 6). At the same time, the elasticity and flexibility of the body are not disturbed by the skin, but rather supported, so that the typical undulation – as the basis for high movement performance – is possible and supported by the skin.

However, this combination of protection and flexibility was not present in ancestral fishes of the early Paleozoic. The earliest sufficiently known craniates were covered from the head to the base of the tail with heavy bony plates or scales. Also, the early ray fin fishes had extensive and thick protection by means of scales. An environmental separation by such scales may have been favorable for the protection from predators, but the speed of their swimming was constrained by the

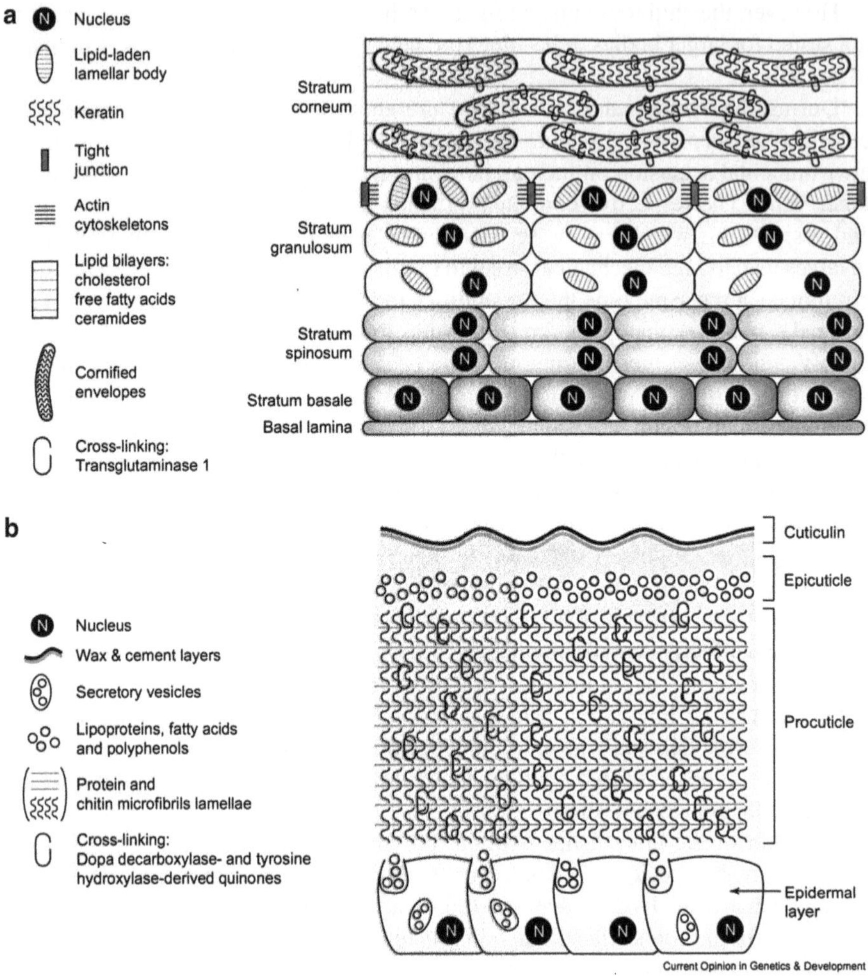


**Fig. 5.6** Schematic diagrams of the mammalian epidermis (**a**) and the fly cuticle (**b**), which, despite depicted structural differences, establish their respective barrier function and resist mechanical stresses to sustain life in a terrestrial environment. In the mammalian epidermis, differentiation of keratinocytes from the basal layer through the spinosum and granulosum layers results in deposition of corneocytes in the functional barrier layer, the stratum corneum. A protein-lipid-reinforced corneocyte and an intercellular lipid matrix provide a durable functional barrier. The essential point is that this principle combines a strong barrier function with high flexibility, elasticity, and low weight. In contrast to the multilayer mammalian epidermis, the fly integument is composed of a single epidermal layer. The fly cuticle comprises protein and chitin fibrils assembled in lamellae that are embedded in a matrix of lipids, which are secreted from vesicles packaged within the epidermal layer. Enzymatic cross-linking of cuticular proteins results in hardening of the cuticle, which further strengthens the barrier function (From Jane et al. 2005, with permission)

great weight of their body and, in many groups, by the close overlap of bony scales. Beginning with the Devonian period in most lines of ray fin fishes, a constant reduction of scales can be observed. The conversion of scales to thin and more flexible elements enabled the animals to bend the body faster and stronger. Thus, the large degree of seclusion was abandoned to gain movement capacity and flexibility, balancing these two requirements.

In contrast to the skin of fishes, that of tetrapods is characterized by general keratinization. The epidermis loses the ability of a mucoid secretion all over the surface, so the skin is dry (with the exception of amphibians). A secluding system is generated from keratinocytes, which grow from a basal layer upward through further layers. They differentiate and die and finally generate a horny layer of relatively stiff and tough cornocytes. They build a unique combination of simultaneous resistance and flexibility.

The skin of extant amphibians possesses some special characteristics, so that it is not clear whether it assumes more transitory features during the evolutionary process to life on land or whether it is a specialization. Its keratinization is thin, and it is active in the exchange of water and gases, thus making it an important organ for osmoregulation and breathing (Alibardi 2003). The accent within the double function of environmental separation and exchange is shifted here toward the exchange functions. This restricts most amphibians to moist habitats, and they must use their movement behavior to avoid drying out. The skin of some terrestrial Paleozoic amphibians (*Stegocephalia*) was probably thicker, dryer, and more similar to that of modern reptilians (Hildebrand and Goslow 2001).

In most reptiles, the skin is so dense that it becomes totally impermeable for water. This is a prerequisite for the full emancipation of reptiles from water environments and moist habitats (Landmann 1986; Chang et al. 2009; Chuong and Homberger 2003; Alibardi 2003, 2009; Wu et al. 2004). The impermeability for water is predominantly reached by the deposition of lipids between the keratinized cells, the reinforcement of the epidermis by hornification with β-keratin. This is more stiff than the more flexible and elastic α-keratin, which is also present in reptiles. Sometimes, the inclusion of β-keratin can be massive, or single scales can ossify, so that they become tough and thick. However, such reinforcement of the skin brings about the restriction typical of all exoskeletons: The animals may lose flexibility of the body during locomotion. Thus, in many reptiles there is balancing by leaving the stiffer scales movable in relation to each other by means of α-keratin between them (Chuong and Homberger 2003; Alibardi 2003). Often, the scales are small (Gekkonidae), which permits more flexibility to the epithelium. In many lizards and snakes both keratins are combined, so that β-keratin is deposited over α-keratin. This composite stratum corneum confers deformable characteristics to the pliable skin, typical of the more agile and light reptiles (Alibardi 2003). Thus, differential distributions of the two types of keratin are possible.

Birds have coverings on their legs similar to reptiles. However, over the rest of the body there is a thin and only weakly keratinized skin, which is only loosely connected to the subcutaneous tissue by way of a network of elastic fibers. Birds have both types of keratin and generate typical features in different body areas according to the local functions. In the scales of the legs, both types are present; in the bill and in claws, β-keratin dominates.

Feathers only contain β-keratin, building a unique combination of a flexible epidermis with stiff and resistant feathers that are highly movable and extremely flexible in relation to each other (Sawyer et al. 1986; Chuong et al. 2003; Sawyer and Knapp 2003). In contrast to scales, feathers can be moved using muscles at their bases. Feathers are essential to the extremely high movement capacity of birds and the regulation of their endothermy, so that birds become "independent of many coincidences and strains, which they encounter within their environment" (Westheide and Rieger 2010, p. 427).

The skin with its appendages is a key feature of mammals. In therapsid reptiles, the progenitors of modern mammals, an integument capable of limiting water loss and protection from the colder environment in which they lived gradually evolved (Alibardi 2003, 2012; Ruben and Jones 2000).

The dermis in mammals can be relatively thick or is quite thin, but in any case keeps its unique properties of strength, extensibility, and elasticity by means of a dense concentration of collagenous and elastic fibers, which weave throughout it. The woven arrangement of the collagenous fibers enables the extension of this network. The epidermis generates a superficial system of seclusion from extracellular substances; this system is produced by dying keratinocytes (Jane et al. 2005). They are comprised of only the more flexible α-keratin. This applies also to scales, claws, hoofs, nails, wool, and much more. Alibardi (2003) sees the soft skin as essential for fine sensitivity and to allow muscles to produce a plastic deformation of the skin, which is a mammalian characteristic.

The essential point for our topic here is that the epidermis is thin and flexible, but yet at the same time highly seclusive (Jane et al. 2005). Evaporation is concentrated on regulated glands. Mammals, as well as birds, have high resistance against the loss of water, which is especially a function of included lipids (Hillman 2001). Finally, hair generated by the skin of mammals also combines extensive protection (thermal, mechanical, ultraviolet [UV]) with high flexibility, as a dense fur can be present without any restriction to body movements (Alibardi 2012).

The reduction of primitive reinforcements of the skin by scales and bony plates, as they are found during the fossil history of the vertebrates, to thin and flexible skin layers relieves the body of overall mass and, especially important, diminishes the torque, which is produced by masses far from the center of gravity when moving on land (Frolich 1997).

This combination of effective environmental seclusion with high elasticity, together with the minimization of weight and destabilizing forces during rapid movement, is a crucial prerequisite for the active lifestyle of mammals and birds (Bereiter-Hahn et al. 1984; Westheide and Rieger 2010). It balances the tasks of the skin to a special extent between environmental seclusion, environmental exchange for homeostasis, and mechanical flexibility, all being features of changes in autonomous capacities.

In summary, different models of skins are able to

- Reinforce the environmental seclusion;
- Support the movement apparatus for maneuverability in the environment;
- Support the flexibility and elasticity in body movements;
- Establish a refined combination of protection and flexibility; and
- Produce feathers or fur, respectively, which besides extensive protection contribute essentially to further functions of autonomy such as movement capacities and temperature insulation.

## 5.4 Respiratory Organs

In the course of metazoan evolution, there were numerous trends to reduce external surfaces for the exchange of respiratory gases and to concentrate them in specialized organs (Maina 2000). In small animals, sufficient gaseous exchange can take place via the body surface. Rates of diffusion are sufficient for animals, whose tissues have a maximum thickness of 1 mm. "Surface animals" can grow much larger, still using direct gas exchange without special organs, as long as they are able to conduct the surrounding water over their surfaces.

There are also larger animals with a "closed" body structure that perform their gas exchange mainly through their skin. They are able to overcome the 1-mm diffusion problem using an efficient internal perfusion system with a flow of body fluid, which passes through an extensive respiratory surface in the skin. Such a combination of direct skin diffusion externally and efficient perfusion internally works in some aquatic invertebrates, including many larval forms and adult aquatic worms, as well as in some semiterrestrial lumbricoids and tropical slugs.

The situation is different in "closed" animals, which use their skin to build a more extensive boundary toward the environment. For a larger body, there would not be enough surface area to supply the $O_2$ needed by the contained volume of metabolizing tissue, and specialized gas exchangers, as either tissues or organs, become necessary (Willmer et al. 2000; Bonner 2004).

Increases in size, organizational closure, and environmental separation via the skin – as features of autonomy – correlate with the emergence of respiratory organs, in which the function of breathing is concentrated. A second tendency, then, is to include the respiratory surfaces into the body, which means that cells and tissues, which perform essential parts of the interaction with the environment, are internalized.

Most animals living in water generate gills as specialized breathing organs. Often, they are protruding evaginated surfaces (Maina 2000), but there are many tendencies to internalize these breathing structures. Many mollusks, for example, internalize their gills into their mantle cavity. In land-dwelling gastropods, the gills are reduced, and the mantle cavity is vascularized and becomes a well-protected (internalized) lung.

In arthropods, different grades and different paths of internalization occur. Small forms, which live in water and have a thin cuticle, are still able to breathe over the whole surface, whereas larger forms generate filament gills. In larger crustaceans, for example, the stiff cuticle makes specialized organs for breathing necessary. External gills occur as lateral protrusions (epipodites) at the base of legs, which can be complex in larger animals. However, a larger breathing surface is often located at the inner side of the carapax, where a current of water is constantly generated for breathing. The gills of the legs can also be covered by the carapax when it laterally reaches down over the legs. In decapods, this principle forms a gill cavity. From a pumping chamber (peribranchial chamber), water for breathing flows through the gill chamber. In crabs, this chamber is closed quite narrowly, having only one inflow opening left. This was favorable for their evolutionary possibilities to change to life on land because the gills became protected against drying out. In some grapsids and in land crabs, the gill cavities are widened, and the inner surfaces are densely supplied with capillaries, so that a kind of lung exists, losing their gills either partially or fully (Westheide and Rieger 2006).

The internalization of gills and their conversion to lungs has been studied in chelicerates (Westheide and Rieger 2006). According to this sequence, the ancestral book gills, as they are present in xiphosurans, are first internalized to book lungs and then to tracheas.

Several times, trachea systems evolved independently in insects and some other terrestrial arthropods (myriapods, onychophorans, some spiders, and many other arachnids). This again involves invagination into the body, but of many separate tubular arrays rather than a single pair of sacs from a single opening as with lungs (Willmer et al. 2000).

In vertebrates also, the skin can serve as a respiratory organ. These include early stages of many fishes, adult eels, some catfish, salamanders, and sea snakes, and especially the more terrestrial anurans. Possibly, the ancestors of the vertebrates relied more on breathing through the skin than through gills, as the extant lancelet still does (Farmer 1999). In many amphibians, the skin is specialized in breathing functions, and some species of fishes take up enough $O_2$ through the skin to cover the requirements of the skin itself (Hildebrand and Goslow 2001).

External gills develop, for example, in some lungfishes and in larvae of some vertebrates. However, the development of internal gills (in bony fishes covered by the operculum) and lungs became more important for the physiology of the vertebrates. If breathing through the skin or through external gills can be regarded as more ancestral, then there is a process of internalization – not of related organs but rather of the function itself.

With the evolution of lungs, the respiratory surfaces are moved completely into the inside of the body (Fronius et al. 2012). Although the air bladder is an old feature of fishes and has in several forms been used for air breathing (Clack 2007; Perry et al. 2001), the full elaboration as lungs took place during the transition to land. In reptiles and then in birds and mammals, respiration is completely internalized, so that gas exchange via the skin plays a minor role. In endotherms, the lungs then must increase their capacity tremendously to satisfy the high $O_2$ demand and thus are in close relation to the emancipation from environmental temperature fluctuations (Chap. 9).

The change to breathing air during the transition to terrestrial animals is also related to the evolutionary increase in metabolic intensiveness (Vermeij 1987; Milewski and Mills 2010) because, by breathing air, much more $O_2$ is available. Fronius et al. (2012) describe that, compared to water, air is a more suitable breathing medium for various reasons: The viscosity of air is much lower than that of water, facilitating ventilation; the concentration of $O_2$ is higher in air than in water, and the diffusion rates of $O_2$ are greater in air than in water. As a result, also the typical metabolic intensiveness of endotherms was possible evolutionally only on land.

The respiratory surface of aquatic as well as terrestrial animals is always wet (Maina 2000; Fronius et al. 2012). The respiratory gases $O_2$ and $CO_2$ diffuse through these surfaces after they have first dissolved in water. Thus, animals with noninternalized breathing organs are always restricted to moist habitats. The emancipation from moist surroundings, which was gained within arthropods, some mollusks, and tetrapods of the vertebrates, requires the internalization of respiratory surfaces.

Internalization of respiratory organs, emancipation from water and from moist habitats, and the increase in metabolic intensiveness with the widening of movement capacities are all features of increased autonomy, which was gained within these groups in many different combinations. These features reached a special distinctness within the functional complex of endothermy, as will be described in Chap. 9.

One might argue that most of these internalizations are related to living on land, so that they might be adaptations in the conventional sense, and to include them into a macroevolutionary tendency such as autonomy is overstated. However, the other perspective is equally reasonable: Acquired possibilities to internalize the respiratory organs (together with other necessary features) generated the capacity of some groups to change to life on land. According to this perspective, the functional change would have come first. The tendencies to internalize breathing organs even in animals that never attempt to leave the water could be a strong argument for this. However, I assume that both views might be wrong. In living systems, we regularly deal with the principle of interdependencies as mentioned in Chap. 3. This might also be the case for changes in the organism-environment relation, including the evolution of breathing systems. Thus, respiratory organs might have evolved in a constant correlation between organismic innovations and the characteristics of the environment.

In summary, there have been multiple concentrations and internalizations of breathing organs or of the function of breathing during animal evolution. These processes stand in relation to

- Increases in size;
- The emancipation from water and moist surroundings;
- The stabilization of body fluids and circulatory systems; and
- The evolution of high metabolic levels in temperature homeostasis.

## 5.5 Circulatory Systems

All living cells need a watery environment. Single-cell organisms directly bathe in water or are at least surrounded by a moist environment. In a certain sense, they have their circulatory system on their outside. Because of their small size, there is a relatively large surface area available for the processes of diffusion and exchange.

When cells aggregate to multicellularity, they become separated from the immediate contact to the external medium in different degrees. Thus, functions must be developed to guarantee the moist surrounding of the cells as well as the necessary exchange processes with the outside world.

Some of the surface animals pump the external medium through their body. Sponges (Fig. 5.7a) drive a current of water through their canals, which are lined with cilia, so that most of the cells still have contact with the water or at most only short distances for the transport of gases and nutrients. Cnidarians (Fig. 5.7b) also use the external medium for this purpose. They not only have water at their external epithelia but also take up water into their gastrovascular cavity and then eject it. They often exhibit complex structures with fine canals for water circulation, as in many medusae, but they do not need an internal circulatory system. With this principle they can reach a considerable size. Even the gastrovascular system of turbellarians is functionally still an external space, and the same holds true for the ambulacral system of echinoderms, which is filled with seawater instead of blood.

On the contrary, organisms that are more secluded circulate their own fluids through their body, which is only indirectly in touch with the surrounding medium (Fig. 5.7). The coelom of some pseudocoelomats (Fig. 5.7c) and coelomats contains a fluid that can perform such circulatory tasks (Ruppert and Carle 1983). The distribution of the fluid can be driven by movements of the body or via cilia. Real circulatory systems, however, have vessels and organs, which drive fluids through them. In open systems (Figs. 5.7d–f) the hemolymph flows through vessels and through intercellular spaces without vessels. The blood of closed circulatory systems (Fig. 5.7g, h) is nearly completely separated from the interstitial fluid. However, the distinction between open and closed circulatory systems is not always clear, and there are intermediary forms. Within mollusks and arthropods, for example, all degrees of both of these forms can be found.

Thus, the function of bathing the cells in a fluid is internalized and emancipated from the surrounding medium with the advantage that the fluid can be regulated. The exchange with the environment takes place more indirectly through specialized surfaces. Therewith, the potential is established to generate a greater distinctness of the internal milieu from the external medium with a stronger self-regulation. Because of their circulatory system, the animals are emancipated from the physical conditions of direct diffusion and can become larger, more complex, and thus more independent (Willmer et al. 2000; Wake 1986; Bonner 2004). With this principle, new possibilities for the generation of larger and more complex body structures opened up (Cowen 1973; Runnegar 1982; Vermeij 1996). In addition, they are a prerequisite for the transition to life on land, making the body independent from the permanent presence of fluids in the environment.

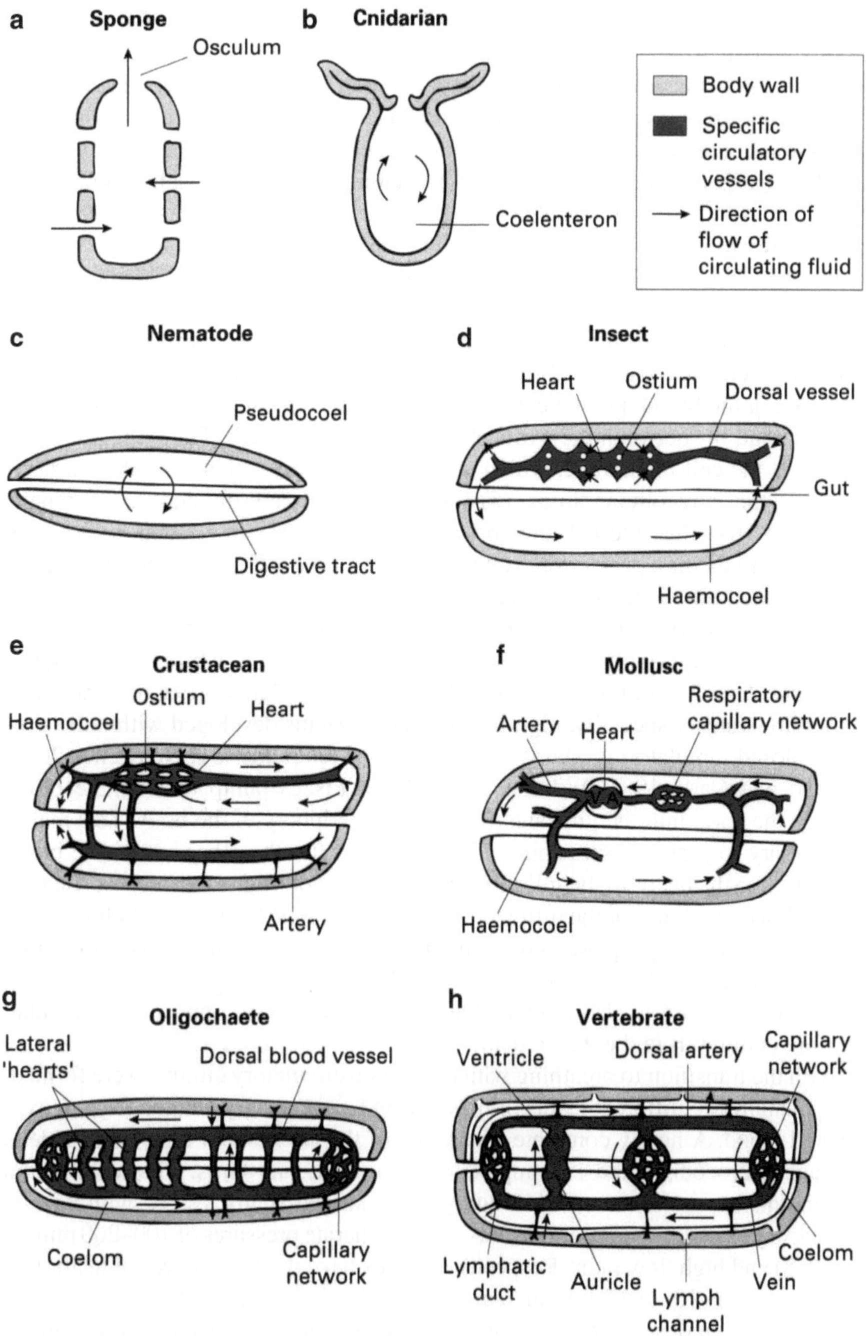


**Fig. 5.7** Diagrammatic views of circulatory patterns in a range of animal taxa (From Willmer et al. 2000, with permission of Blackwell Publishing Ltd.)

Circulatory systems are "key players" for the establishment of homeostasis (Willmer et al. 2000). Besides the exchange of $O_2$ and $CO_2$, water, ions, and nutrients, these systems are central for the regulation of the organism, as they transport signaling molecules such as hormones and thus open new possibilities for integrative system functions. The cellular and humoral components of the defense systems are also distributed via circulation. The circulating fluid can have a hydraulic function, such as in hydrostatic skeletons. Many marine worms use this extensively to generate forces during burrowing through the sediments in which they live.

In organisms with a closed circulatory system, autonomy is generated in several layers of the organism: (1) at external surfaces, where differences between the external world and the circulatory fluid are generated, exchange processes concentrate on specialized organs; (2) in the regulation of circulatory fluids themselves; (3) between the blood and the extracellular fluid, which can have a different composition; and (4) between the extracellular fluid and the cells. Here, the boundary surface is the cell membrane, which controls the exchange between the extracellular and the intracellular fluids. (5) Finally, the cells are divided into compartments, so that the internal membranes direct the processes within the cell. Compared to the first prokaryotic cells, this shows how different layers evolved for self-regulated functions in metazoans.

Open circulatory systems are mostly characterized by large amounts of fluid (at least 30 % of body weight), low pressure, and slow flow rates, while closed systems need considerably less fluid (mammals 8 %), generate higher pressures, and reach higher flow rates. Especially, high-performance systems developed within the generally closed circulatory systems of vertebrates (Heldmaier and Neuweiler 2004; Ruppert and Carle 1983; Willmer et al. 2000). Fishes pump their whole blood through the aorta into the gills and then into the periphery of the body. This makes it necessary to pass several capillary systems one after the other before the blood flows back to the heart. In the gills, high pressure cannot be created because the gills need a thin epithelium for the diffusion of breathing gases. Such an epithelium is not able to withstand higher pressures. This restricts the possibility to generate high blood pressure, and the flow rates are relatively low. For the thin epithelia of lungs, high blood pressure is also unfavorable because there is no opposing hydrostatic pressure coming from the inspired air.

With the transition to breathing with lungs, two circulatory circuits were formed. They contain two different levels of pressure and separate oxygenated from deoxygenated blood. A nearly complete separation of the arterial and the venous side is reached in crocodiles and is completed in mammals and birds. The heart of the endotherms becomes strong, with highly muscularized ventricles and with its own coronary system. In its arterial side, it is able to generate pressures of 100–200 mmHg (Fig. 5.8) and high flow rates. Respectively, birds have the most powerful heart. It is able to meet the 10- to 15-fold increase of $O_2$ requirement during flight.

The evolution of separated pressure systems correlates with two major transitions that are features of an increasing autonomy: the transition to land by the vertebrates (division of pulmonal and body circuits so that the lung can be perfused) and the development of the high movement capacities in connection with the functional complex of endothermy, requiring high capacities for the transportation of $O_2$.

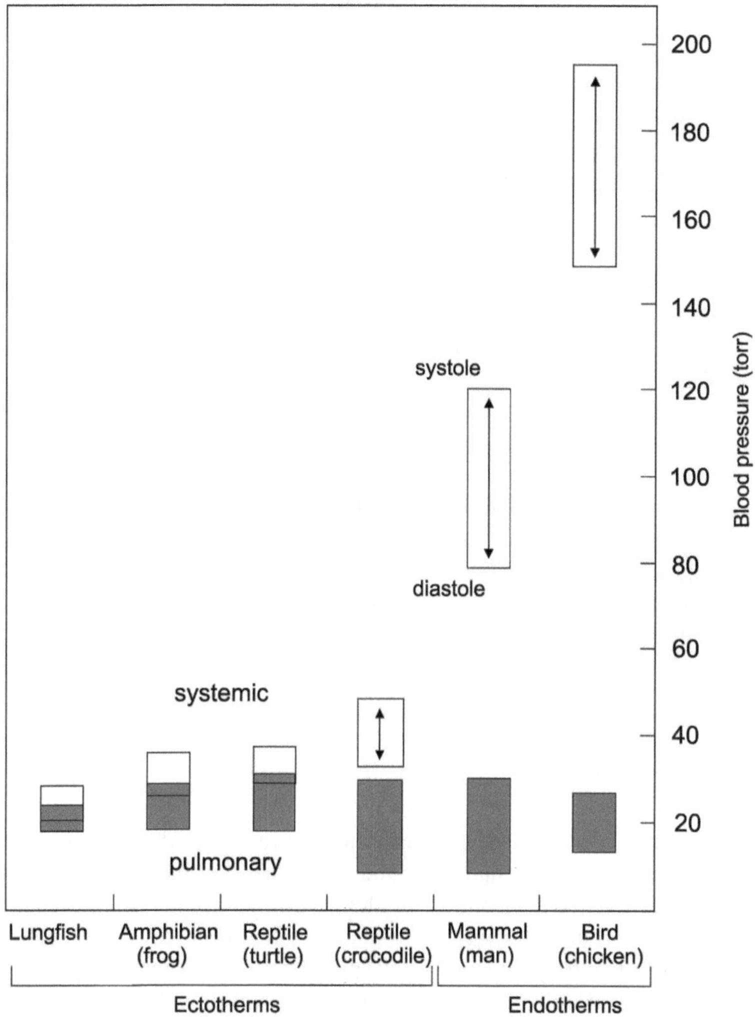


**Fig. 5.8** Comparison of systemic and pulmonal blood pressure systems in vertebrates (From Heldmaier and Neuweiler 2004)

In addition, the composition of blood is controlled closely in higher vertebrates, so that pronounced autonomy is achieved, especially with the help of high-performance kidneys. These developments also correlate with the evolution of physiological regulators (see Chap. 6): "The presence of adequate ventilation and/or transport systems may be the most important determinant of whether an animal is a conformer or a regulator, since conformers seem to be limited mainly by an inability to transport $O_2$ quickly enough to the metabolizing tissues" (Willmer et al. 2000, p. 157).

In summary, the evolution of circulatory systems also introduces further elements that are able to contribute to the functional autonomy of organisms:

- The function to surround the cell with fluid is internalized and thus removed from the outside medium.
- The organism becomes emancipated from the conditions of direct diffusion.
- Internalization is further amplified by closed circulatory systems.
- Larger bodies can be supplied independent of the direct presence of fluids in the surroundings.
- The fluid of circulatory systems can be regulated and thus physiologically stabilized.
- Molecules and signals can be transported to establish enhanced possibilities for regulation and integrative system functions.
- Circulatory systems provide a prerequisite for the emancipation from water and the transition to land in different phyla.
- The generation of separated pressure systems in vertebrates correlates with the emancipation from water (Chap. 6).
- The generation of high-pressure systems correlates with the evolution of endothermy (Chap. 9).

## 5.6 Body Cavities

The occurrence of a body cavity, a fluid-filled space within the framework of cell layers in bilaterian animals, had profound consequences for body size, mobility, and several functions of distribution and homeostasis. Usually, acoelomate, pseudocoelomate, and coelomate forms are distinguished (Fig. 5.9). In acoelomates, there is no uniform extracellular cavity despite the gut. Between the epidermis and the gut, the body is filled with connective tissue and muscles. The organs for excretion, osmoregulation, and reproduction are embedded within the connective tissues. In pseudocoelomates, a fluid-filled cavity exists between the epidermis and gut, but it is not lined by an own epithelium. This cavity derives from the primary body cavity. In the coelomate organization, the coelom (= secondary body cavity) is completely covered by an epithelium, the coelothel. In some groups (e.g., arthropods), primary and secondary body cavities fuse to the mixocoel or the hemocoel, respectively (Westheide and Rieger 2006; Willmer 1990; Dewel 2000).

Functions of body cavities are generation of an efficient hydrostatic skeleton; circulation of gases and metabolites; increase in body size; storage of gonads and gametes; space, within which organs can lie and have independent movements (e.g., heart, gut); deposition area for metabolic wastes, which then are disposed by nephridia; simple distributive system, supplying nutrients and respiratory gases to the tissues; and distribution of signal molecules such as hormones (Clark 1964;

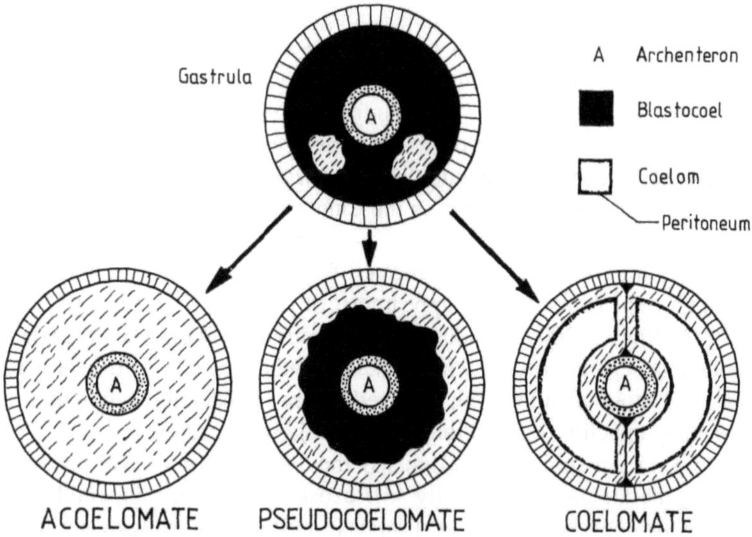


**Fig. 5.9** The three principal types of animal design in relation to body cavities and their derivation from an idealized gastrula (From Willmer 1990)

Willmer 1990). As several of these functions of the body cavities contribute to the autonomy of the organism, they need to be included in the present discussion.

Basically, it can be claimed that the coelom is an internal space, secluded from the environment, which is a typical feature of the animal world beginning with the bilaterians, whereas plants (with a few exceptions in their generative organs) predominantly elaborate the surface principle. Gastrulation, which forms the first body cavity, the blastocoel, has already been mentioned as a typical process of internalization. With a coelom, some functions are internalized that are present in animals without a coelom, but are not included into a fully internalized space. For example, the gastric cavity of polyps also functions as a hydrostatic skeleton and has functions for distribution. However, these are performed with the help of the external medium, the water, as via the mouth the gastric cavity is basically open. The gut of free-living Platyhelminthes also has functions for distribution within the body, but because of the pharynx, it is functionally an external space.

An important function is when the fluid within the coelom, together with surrounding inelastic tissues, can work as an efficient hydrostatic skeleton. It allows, in a huge variety, faster or stronger movements, greater shape changes, and the generation of greater forces to manipulate the environment than the parenchyma tissues of acoelomate worms. Burrowing can become a major way of life, and swimming and crawling movements are facilitated, including swimming in open water (Willmer 1990).

It has been postulated that the earliest bilaterians may have been small organisms in the form of worms, which possibly moved with the help of cilia. Small extant forms of turbellarians and nemertins still employ such a type of movement. However, for larger animals, ciliary propulsion does not generate enough force, so that evolution brought up several innovations to produce more effective forms of mobility, as discussed previously. One basic invention for this was the generation of hydrostatic skeletons based on body cavities. The step from movement by cilia to movement by muscles combined with diverse mechanical devices, including hydrostatic skeletons, was an increase in mobility and flexibility within the environment, which was an essential part of innovations during the Cambrian period and contributed to the increased bioturbation (see Fig. 5.2).

New possibilities of movement by means of hydrostatic skeletons may have been the most important factor for the appearance of coeloms in general (Clark 1964; Gruner 1993; Nielsen 2001). Under this assumption, Siewing (1980) understands the coelom as a separate organ, which, like any other organ, could experience not only manifold reorganizations but also reductions. Especially, segmented organisms, like many annelids, can exhibit effective digging capacities.

Interesting for our subject is particularly the degeneration of coeloms in arthropods. They generate an exceptionally stiff exoskeleton from segmentally ordered plates, while the coelomatic cavities disappear during early ontogenesis. The walls of the coelomatic cavities convert mainly into muscles, which now insert at the exoskeleton. These changes led to the evolution of a completely new type of movement: the movement by levers with the help of extremities, the next stage in flexible movements far beyond the possibilities of hydrostatic skeletons.

The other group that produced a more effective form of movement is the chordates with their – compared to arthropods – reversed anatomical organization with an endoskeleton. Their crucial innovation is the chorda dorsalis. It keeps the body in a constant length and offers resistance to the muscles. Dorsal parts of the coelom wall are converted into muscles, which are segmentally organized as myotomes. Because of the antagonistic contraction of the myotomes, the body is able to bend sideward, producing effective propulsion. The location of the chorda causes the propulsion to come from the dorsal body parts; the ventral areas of the body do not make an active contribution. Thus, ventrally the coelomatic segmentation is given up, and the coelom mainly serves the flexible fixing of internal organs. Thus, in chordates there also is an obvious relation between the generation of a stiff skeleton for the next dimension of flexibility and the reduction of the coelom as a hydrostatic skeleton.

Besides these functions of mobility, coeloms serve just as means to reach a larger body size. With the generation of coeloms, it was possible to increase body size without filling the whole body with metabolically active and thus costly tissues. Willmer (1990, p. 24) thus summarizes: "In an important sense, then, a body cavity is primarily beneficial by virtue of the size increase it allows, giving less risk of predation by other small creatures, a greater size range of foodstuffs that can be handled, and increased possibilities of homeostatic control and independence of the environment."

The functional separation of digestion, distribution, and hydrostatic skeleton that becomes possible by means of the coelom makes each of these functions more effective. While, for example, in a polyp all three functions are performed by the same organ, in a coelomat they are separated into different systems. The gut is kept separate from the body wall, so that the two are able to undergo peristaltic waves independently and often in different directions. The gut can be closed and thus internalize digestion and generate higher enzymatic concentrations to perform extracellular digestion, while in primitive guts (polyps, platyhelminths), intracellular digestion dominates. Animals that generate a secondary body cavity can transform the residues of the primary body cavity into a vascular system for circulation and distribution, while the secondary cavity can be specialized into a hydrostatic system. In segmented animals such as the annelids, the vascular system can run through the whole body as a continuous system, while the coelom is separated into isolated segments. Thus, the coelom of any type can contribute to differentiations in the sense of division of labor. The related systems can then work much more effectively.

In bilaterians, the occurrence of organs for excretion and osmoregulation is in close relation to the organization of the body cavity and of body size (Dewel 2000). Metanephridia, which are characteristic for coelomates and related animals with a vascular system, are formed in the coelom or its walls. The ancestral pattern separates the location of ultrafiltration of the primary urine from the organs of excretion. Urine is directly filtrated into the fluid of the coelom; from the coelom, the metanephridium absorbs it and processes the fluid further before it is excreted. In the nephridial organs of vertebrates and of arthropods, the location of ultrafiltration and excretion then move closer together but are still in relation to the coelom. In this sense, the coelom is also involved in the regulation of body fluids and the removal of metabolic wastes. Thus, it is an essential organ for homeostasis.

Finally, the coelom is involved in the generation of germ cells. In bilaterians, the germ cells are rarely included within the gastrodermis, as it is the rule in coelenterates, but rather is more internalized under the coelothel and released into the coelom. Within the invertebrates, it is remarkable that especially the germ cells are already deeply internalized to a special degree.

In summary, body cavities contribute to an increase in autonomy with the following functions:

- Generation of an inner space with internalization of functions;
- Generation of a hydrostatic skeleton, which can increase movement capacities and mechanical influences on the environment;
- Generation of distribution systems that internalize circulation and distribution of fluids and substances such that larger bodies can be supplied and functions of coordination, regulation, and homeostasis can be supported;
- Increase in body size without filling the whole body with metabolically active and thus costly tissues;
- Gain in relative independence of organs and thus increased performance abilities;
- Increased specializations because of functional separation in digestion, distribution, and hydrostatic skeleton;
- Support of excretion and osmoregulation as functions of homeostasis; and
- Internalization of germ cells.

## 5.7 What Is Special About Chordates?

Carroll (1988) describes a possible ancestral vertebrate as a small, fishlike animal in the form of a spindle, basically constructed for an active swimming way of life, with a head, which was integrated into the body form (Fig. 5.10). A series of segmented muscles, the myomeres, was arranged along both sides of the body. Their fibers ran lengthwise between transversally oriented myosepts. Medially, these myosepts were anchored at the notochord (chorda dorsalis), the stiffening axial rod, which at the same time was considerably elastic for sideward bendings. Swimming took place by means of muscular contractions, which in short intervals ran alternating over both sides of the body, thus generating a lateral undulatory movement. Typical for the ancestral organization of a vertebrate is also the branchial gut.

Single elements of these characteristic structures are already present in invertebrate chordates, to which the extant cephalochordates (with *Branchiostoma*), the urochordates (tunicates), and the hemichordates (acorn worms and *Pterobranchia*) belong. *Branchiostoma* (Acrania) may come near such an ancestral form and has been discussed since the nineteenth century as a forerunner of all vertebrates. Carroll (1988) assumes that this organization must have evolved as an adaptation to an active lifestyle, including swimming in open water. Filtering detritus from seawater and spending most of its life buried in sand as the extant *Branchiostoma* does might have evolved secondarily. However, if forced, *Branchiostoma* is a forceful and effective swimmer (Lacalli 2012).

Some modern phylogenetic analyses using genomic data even see *Branchiostoma* near the basis of the deuterostomes (Delsuc et al. 2006; Putnam et al. 2008). According to Gee (2006), the last common ancestor of deuterostomes may have

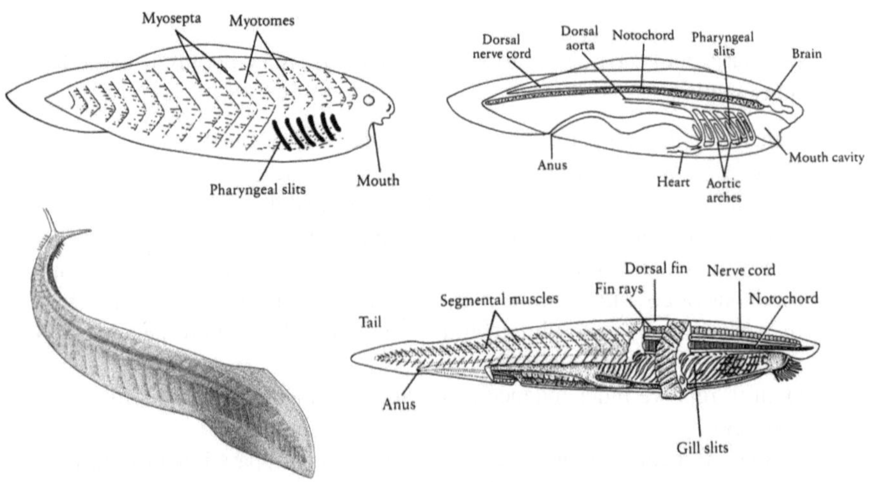


**Fig. 5.10** *Above*: Hypothetical ancestral vertebrate (From Carroll 1988, modified). *Bottom left*: Pikia (From Briggs et al. 1994, modified). *Bottom right*: Branchiostoma (From Carroll 1988, modified)

been a free-living, bilaterally symmetrical creature with segmented body wall musculature and possibly a reasonably sophisticated central nervous system.

The fossil Acrania also have signs of a segmental structure of the muscles, as *Pikaia* from the Burgess Shale fauna of Canada and *Yunnanozoon* from China show, both remains from Middle Cambrian times (Shu et al. 1996; Chen et al. 1995; Conway Morris and Caron 2012). Also, *Myllokunmingia* and *Haikouichthys* from the Lower Cambrian of South China share some characteristics with the chordates (Shu et al. 1999).

Whether such early chordates were rarely moving filter feeders or already moved more actively is a matter of discussion (Lacalli 2012). In any case, the construction of a central axis with muscles that insert on it – something unparalleled among invertebrates – is the basis for the evolution of great movement capacities, including fast swimming, large maneuverability, and forceful sprints, whenever this potential was elaborated and used in the manner of the fishes we have today. Generally, the evolution of this capacity was a central element in the history of chordates (Koob and Long 2000; Gans 1989). A crucial innovation during this transition was the chorda-myomere system. The stiffening through the notochord – and later through the vertebral column of the fishes – offers an elastic and strong insertion to the muscles. It avoids changes in body shape or, as in hydrostatic skeletons, keeps the body form constant only through energy-consuming muscular forces (Gutmann 1981, 1985; Satoh et al. 2012; Lacalli 2012). The antagonism of the muscles to the axis, which releases elastic forces; the antagonism of the muscles on both sides of the axis; and the concentrated transmission of forces on just one axis make the system effective. The elasticity of the axis enables continuous propulsion, and the concentration of forces permits high acceleration. This is the basis for directed and fast movements with stamina. Thus, high autonomy of movements was at the evolutionary basis of the chordates and vertebrates (Reichholf 1992b).

In fishes, this principle became more elaborated. The packages of myomeres became thicker, and an advantageous relation of tension and force was generated. The muscles of a modern bony fish (Fig. 5.11) spiral along the body from the vertebral column to the skin and back again. The fibers are oriented in such a way that their contractions are optimal in all areas, so that within the available amount of contraction high effectiveness on the vertebral column is possible. The muscular force inserts with a fourfold transmission ratio at the vertebral column.

With the evolution of the vertebral column, further skeletal elements were generated, which stiffen and stabilize the axis; at the same time, the system of joints keeps it flexible. This widens frequency and forces to bend the body. A three-dimensional system of ligaments between the vertebrae saves and releases elastic energy, which can quickly be used during sprints. During slow movements, it supports the active work of the muscles by passive counterforces. In addition, the skin, which has several layers now, can work as a sort of external ligament and thus complements the elasticity of the dorsal column. With further stiffening of the tail fin and the development of stabilizing unpaired fins, the force that can be transmitted to the water is further increased (Koob and Long 2000; Lauder 2000).

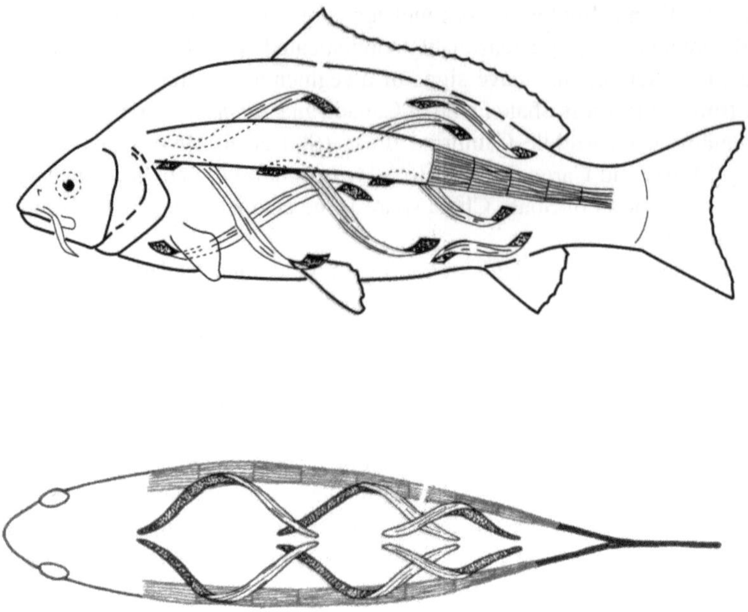

Fig. 5.11 The muscular system of a carp (From Heldmaier and Neuweiler 2003)

However, undulation also needs neuronal coordination of the alternating contractions, so that the evolution of central pattern generators in the spinal cord (units of motoric excitation; see Chap. 8) must have been in correlation to the origin of this principle of movement. This movement capacity in water is unattained within invertebrates. Perhaps only the jet propulsion of some cephalopods may be equally effective.

The reachable dimensions become obvious if one looks at the best swimmers of the open sea. Thus, the Atlantic bluefin tuna (*Thunnus thynnus*) reaches a maximal speed of 80 km/h; its average speed is about 40 km/h. These tunas belong to the most widely migrating fishes, with distances of 5,000 km per year. These tunas are able to keep their body temperature at about 31 °C by way of endogenous heat production, which again is related to their movement capacity. Within the bony fishes, this endothermy is a specialization. Similarly fast migrating fishes are the marlins (*Makaira*). In large jumps, these fishes, which can be several meters long, spring from time to time high out of the water. They also reach speeds of 80 km/h (Scheiba 1990).

Sharks also are fast hunters and good endurance swimmers, with the bigger ones faster than the smaller ones (Scheiba 1990). A blue shark (*Prionace*), which had been marked at the northern coast of Africa, was captured in Australia 2 months later. Thus, it must have traveled around 300 km per day. However, the largest sharks, such as the basking shark (*Cetorhinus maximus*) and the whale shark (*Rhincodon typus*), are slow and lethargic, although they also migrate over long distances.

Besides these pronounced swimmers, there are many special adaptations within the world of fishes. To this bewildering diversity belong those that lie on the ground and wait for prey, those that dig themselves into the ground, others that attach themselves to objects or other animals, those that are highly maneuverable within narrow spaces, and many more. The basic potential of movement, which the vertebral column offered, was elaborated differently during the adaptive radiations.

If the theory advocated here is correct, it might be possible to reconstruct the evolution of vertebrates as a differential elaboration on features of autonomy combined with environmental adaptations. However, as mentioned, some of these features might also come into conflict with each other. Thus, a domination of environmental separation may restrict mobility as in early fishes with stiff exoskeletons. On the other hand, a thin epidermis would offer only weak seclusion to a large body (*Branchiostoma* has, as most invertebrates, only a single-layer epidermis). A balancing of extremes may have led to the generation of modern fishes with a multilayer epidermis, scales movable in relation to each other, and a mucous cover giving sufficient environmental seclusion and simultaneous flexibility. The principle of balancing may also apply to other characteristics, such as size, for example. An extensive increase in size may restrict flexibility during movement in many cases because of the mass, which has to be moved. On the other hand, a certain size seems to be necessary for high-performance swimmers.

Griffith (1994) also sees this extension of flexibility within the environment as an essential characteristic of the vertebrates, which was elaborated on the way from an early chordate to the real vertebrates.

In summary, a specialty of the basic organization of chordates or vertebrates is the construction of a central axis with muscles that insert at it. In the basic form, this was the chorda-myomere system, which was further elaborated by the complex vertebral column and changes of the muscular arrangements, together with specializations of the skin and the central nervous systems. All this together delivers the possibility for directed and fast movements with stamina. Increases in flexibility within the environment belong to the set of resources that can be involved to change autonomous capacities (Definition 2). Thus, the basis of vertebrates was a mechanic principle that from the outset had the potential to develop constructions for high autonomy of movements. This potential was elaborated in different degrees in water. Then, especially during the transition to land, it was also the decisive feature to stiffen the whole body to counteract gravity, another prerequisite for emancipation from environmental conditions.

(continued)

(continued)

With the formation of the vertebral column, a whole series of features developed, which allowed for the active lifestyle we usually associate with vertebrates and which are discussed in other chapters. Pertinent to these features are specialized sensory organs and a brain, which integrates the extensive sensory information coming from them; an extended metabolic capacity; an elaborated circulatory system; breathing organs with a large capacity; organs for digestion, which deliver energy for high activity levels; a glomerular kidney; a complex endocrine and immunological system; as well as cartilage and bones as mechanically strong building materials.

## 5.8 Adjustments for Movement on Land

The transition from water to land in vertebrates brought a series of features of autonomy, such as independent management of liquids, environmental seclusion to prevent dehydration, internalization of gas exchange, and many more.

Most of these features did not evolve all at once but were developed during the further history of tetrapods. In the following paragraphs, I look at the radically new way of movement on land with the help of legs, which led to a tremendous increase in the potential to move freely within the environment.

It does not seem to be obligatory to come up with such a radical innovation as the tetrapods did. There are invertebrate groups that went on land only employing the same principle of movement as in water, such as the gastropods and animals, which use hydrostatic skeletons. However, today we can hardly imagine how a vertebrate should have moved on land using fishlike undulation. Partially, the way salamanders, for example, still use undulation might give a clue. The interesting point, however, is that from the beginning the evolution of limbs with an internal skeleton was involved, and that these limbs literally step by step increased the autonomy of movement in tetrapods, leading to high-performance runners and flyers, which we can observe within modern fauna.

A series of fossils from the late Devonian period show how evolution might have taken place from lobe-finned fish, which already had bones being homologous to those of later tetrapods within their fleshy fins, to early tetrapods, which increasingly used these extremities to push themselves up from the ground (Fig. 5.12). Presumably, such early tetrapods lived mainly in shallow waters or in borderline areas between water and land (Clack 2002, 2007; Ahlberg and Clak 2006; Shubin 2008).

Some of the oldest known fossils that illustrate the transition to land are those of *Ichthyostega*, a 360-million-year-old amphibian (Upper Devonian), which resembled fishes in many features of its skeleton but also possessed pelvic (hip) and pectoral (shoulder) girdles and limbs, allowing movement on land (Pierce et al. 2012). Ahlberg et al. (2005) provide a reconstruction and functional analysis, which show that

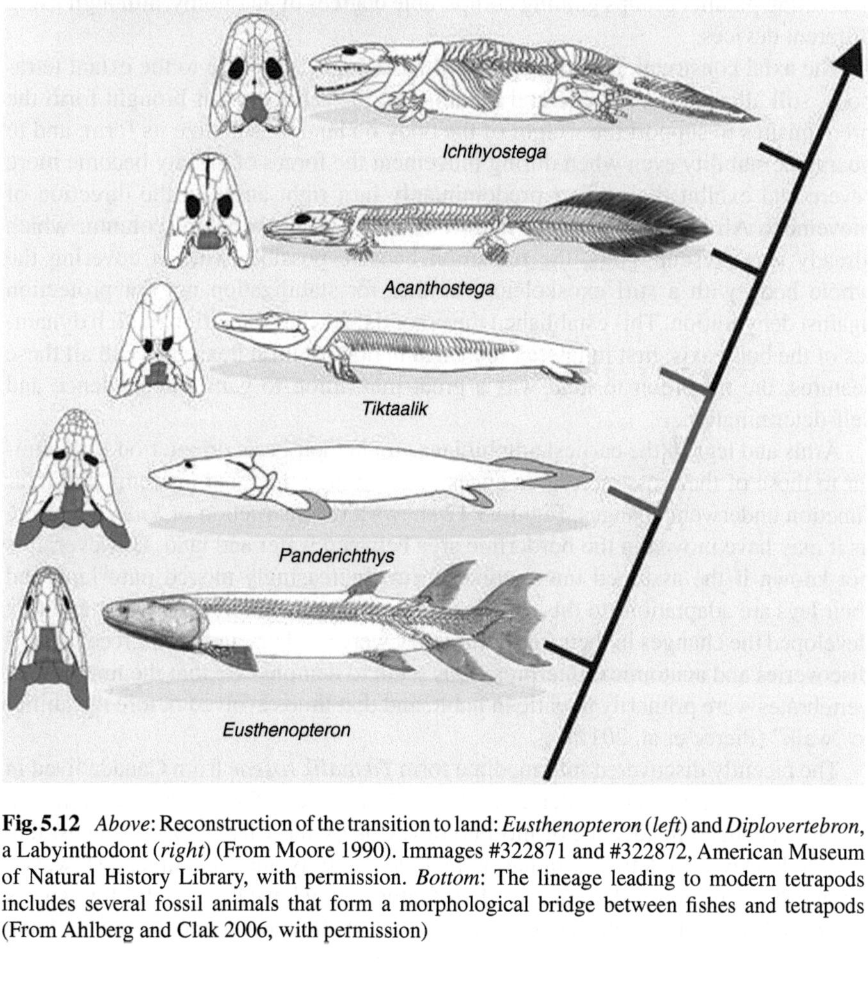


**Fig. 5.12** *Above*: Reconstruction of the transition to land: *Eusthenopteron* (*left*) and *Diplovertebron*, a Labyinthodont (*right*) (From Moore 1990). Immages #322871 and #322872, American Museum of Natural History Library, with permission. *Bottom*: The lineage leading to modern tetrapods includes several fossil animals that form a morphological bridge between fishes and tetrapods (From Ahlberg and Clak 2006, with permission)

*Ichthyostega* had a unique structure in vertebrae and ribs that enabled the trunk to be supported above the ground. However, they assume that the trunk must have been too rigid to undulate during swimming in water, which was compensated by a tail that could be moved forcefully from side to side. This combination is unique, as modern amphibians also retain to a large extent fish-like undulations to move on land, only gradually using their legs to propel the body forward. Therefore, Carroll (2005a) assumes that this might have been a single, but in the long run unsuccessful, way for the transition to land. He sees at least 11 lineages of advanced lobe-finned fishes and early amphibians from the Upper Devonian, only 1 of which is a plausible close relative of later land vertebrates, and interprets this as a 15-million-year history of "experimentation" among the descendants of fishes, which had already developed fins with a central bony axis, a swim bladder (the fish's equivalent of lungs), and paired nostrils opening into the mouth.

Thus, here again we encounter the principle of diversity of attempts, just as it is present in other major transformations, either generating unsuccessful lineages or generating parallel groups gaining comparable degrees of autonomy, although using different devices.

The axial construction of the group, which actually gave rise to the extant tetrapods, still allowed dynamic lateral flexions. At the same time, it brought forth the prerequisites to support the weight of the body on land, to stabilize its form, and to guarantee stability even when during movement the forces of gravity become more severe and exhibit their effect predominantly in a right angle to the direction of movement. All this was possible mainly because of the vertebral column, which already was present. Thus, the transition became possible without covering the whole body with a stiff exoskeleton, neither for stabilization nor for protection against dehydration. This established the potential for the generation of high dynamics of the body axis, first in lateral, then also in dorsoventral flexions. With all these features, the transition to land was a great innovation to gain independence and self-determination.

Arms and legs of the earliest amphibians are, in their basic orientation, still similar to those of the crossopterygian ancestors, although the inner anatomy and their function underwent changes. Figure 5.12 shows a reconstruction of *Eustenopteron*, as it may have moved in the borderline area between water and land. However, it is not known if the assumed intermediate forms increasingly moved onto land and their legs are adaptations to this, or whether these forms still lived in water and first developed the changes in their fins before they were used on land. More recent fossil discoveries and anatomical interpretations seem to demonstrate that the first limbed vertebrates were primarily aquatic in habit, and that limbs evolved before the ability to "walk" (Pierce et al. 2012).

The recently discovered intermediate form *Tiktaalik roseae* from Canada lived in flat freshwater areas (Daeschler et al. 2006). Ahlberg and Clak (2006) assume that function changed in advance of morphology. Coates and Clack (1995) suggest that the whole Devonian tetrapod radiation included entirely aquatic animals. However, Kemp (2005) proposes that most of their features may have been useful for aspects of the environment, which were common to both shallow water and the muddy bank

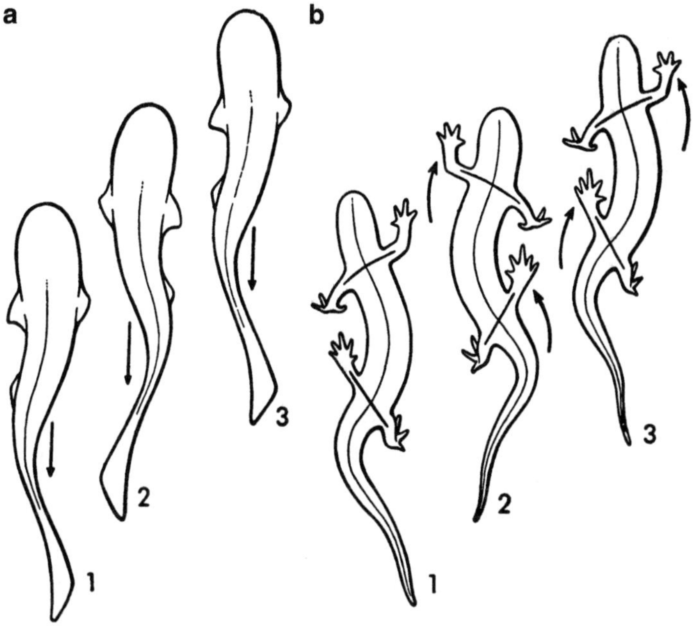


**Fig. 5.13** Undulation of the body axis in fish and urodels (Modified from Romer and Parsons 1977)

alongside. In his view, it is more plausible that tetrapods originated as a result of a transition along an ecological gradient with no abrupt barriers. To be adapted to shallow, low-oxygen water, a vertebrate requires the capacities for substratum locomotion, air breathing, and aerial sense organs. Even radical differences between the two habitats, such as loss of buoyancy, increased dryness, and exposure to heat, are in effect gradual across the boundary insofar as an organism can ameliorate them by temporarily returning to the protection of the water. From this perspective, the radiation of early tetrapods consists of lineages, which might have been emancipated in different degrees along the water-to-land gradient (Clack et al. 2012).

Once into the Carboniferous, a substantial radiation of tetrapods commenced, and there were lineages showing a wide range of respective degrees of terrestrial emancipations (Clack 2007). Some also reduced or even lost the limbs and developed streamlined bodies, indicating that the divergence of forms between increasing movement capacities and pure adaptations continued. Presumably, they mostly retained the anamniotic egg laid in water, which hatched into a gilled, fully aquatic larval form, thus still not completely emancipating themselves from water.

The predominant means of locomotion in ancestral tetrapods may still be exhibited in the extant urodels (Westheide and Rieger 2006; Moore 1990; Romer and Parsons 1977; Carroll 1988). The main propulsion is still generated by lateral undulation of trunk and tail, similar to the axial propulsion in fishes (Fig. 5.13). The limbs serve for anchoring on the ground, and in fast movements, they are

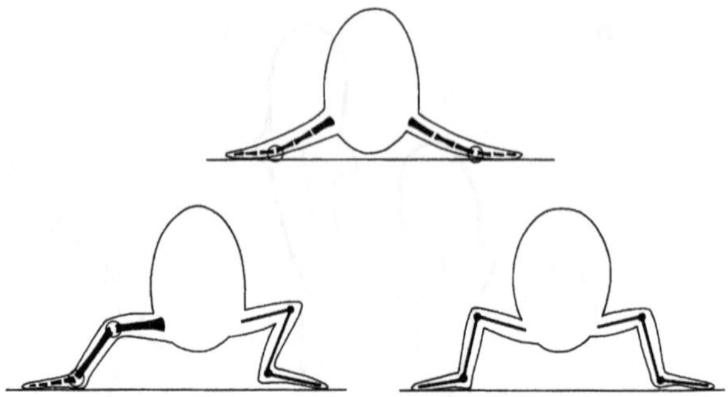

**Fig. 5.14** The generation of joints in the legs during the transition from water to land. *Above*: A fish, moving along the soil in water. *Bottom*: Amphib with undulatory movement (*left*), lifting the body from the soil (e.g., in reptiles) (Modified from Starck 1982)

passively protracted and retracted. Only during slow walking are the extremities used in a diagonal alternating manner. However, here also the lateral flexion of the trunk is responsible for the main stride length. Generally, there is a sprawling posture of the limbs, in which the animal lies with its belly on the ground. Especially, the early labyrinthodont amphibians still used this fish-like movement.

Then, within reptilians and with great diversity, the extremities increasingly gained a more active and independent role during movement. In some groups, evolutionary changes moved the limbs further under the body and enabled the animals to elevate themselves from the ground. To move and support the body, shoulder and pelvic girdles had to be strengthened and stabilized. The ancestral association of the shoulder with the head is loosened, so that the latter increasingly becomes independent from the trunk, and the shoulder blade is fixed at the chest. To be able to carry the weight of the body, especially ventral parts at the thorax, the coracoid, are strengthened, and the shoulder girdle is closed by an interclavicula in the center line with the breastbone to form a U-shaped arch around the thorax. Two strong transversal muscles carry the main load and prevent the forelegs from spreading (Figs. 5.14 and 5.15). The pelvic girdle, which was not fixed to the vertebral column in fishes being nearly weightless in water, forms the three parts of the pelvis and articulates in tetrapods with the ileum at a pair of fused vertebrae in the sacral region. The pelvis forms the joint for the femur and attachments for the musculature of the hind legs. The pelvic girdle is fixed by some muscles that run to the belly and others that connect to the ischium and to the lower leg.

With these transformations, the lifting of the trunk above the ground, against the forces of gravity, becomes possible, so that only distal parts of the legs touch the ground. In this way, many geckos and chameleons within the extant reptiles, for example, are well emancipated from touching the ground.

In sideward-oriented limbs, the body must be elevated with the help of muscular force, and the limbs must be short. By getting the limbs more underneath the body, the main part of the weight is shifted onto the bone system, and the muscles are released from carrying the body to a large extent. Because of this, they can be used more economically, and the levers of the legs can now become longer. Within lizards, for example, such tendencies can be observed. Some have longer hind legs and

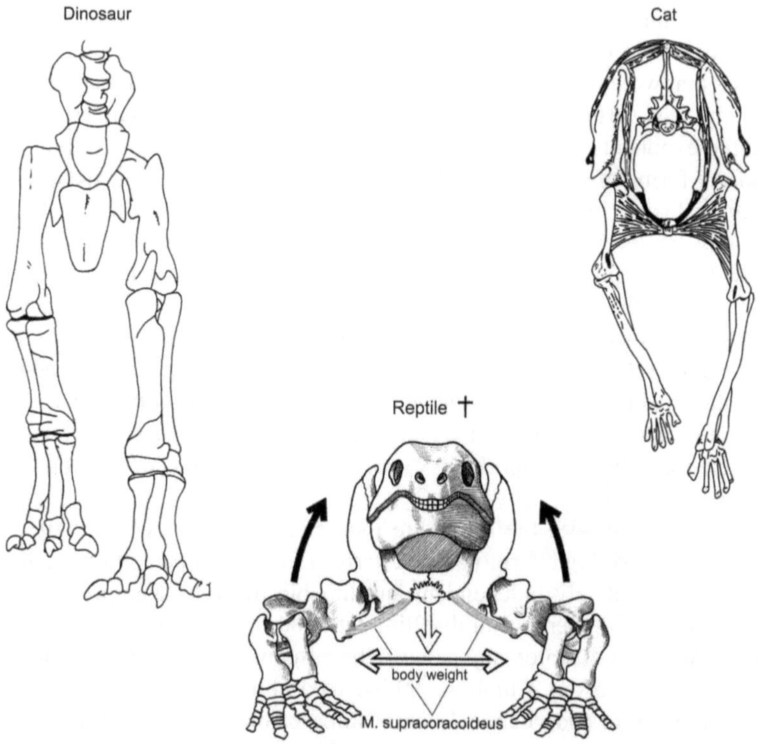


**Fig. 5.15** Position of legs: *Above left*: Dinosaur. *Above right*: Mammals. *Bottom*: Ancestral reptile (Modified from Heldmaier and Neuweiler 2003; Carroll 1988)

are able to elevate the forelimbs during escape behavior and sprint bipedally. The zebra-tailed lizard (*Callisaurus draconoides*) thus reaches speeds up to 4–5 m/s.

However, this principle was elaborated more extensively in three other groups: dinosaurs, birds (which built on the emancipation of movement already gained in theropods), and mammals.

The group of Thecodontia includes some relatively ancestral archosaurians. They already had some changes in their skeleton, which are related to the upright posture and to a more effective forward and backward stroke of the limbs compared to the more primitive diapsids and lepidosaurians (Carroll 1988). In the succeeding eldest groups of dinosaurs, evolutionary changes arose, by which means a nearly vertical leg position became possible (Fig. 5.15).

Many dinosaurs, such as the theropods, walked bipedally. Among them, there were relatively small forms, which probably were considerably quick because of this type of movement. The sauropods were quadruped, and also within ornithishians, both types of locomotion were represented.

It is difficult to reconstruct how fast dinosaurs could run. They may have been better runners than previously thought (Alexander 1991; Lockley 1991, 1999). Probably, small and medium-size biped dinosaurs (theropods and ornithopods) were good runners, while larger biped dinosaurs (larger carnosaurs and large ornithopods) were only moderately good runners. Quadruped ornithishians and sauropods that were more massive may have been comparatively lethargic.

With the legs vertically under the body, the widespread bipedal forms, and the often agile exponents, the dinosaurs may have been a group that exhibits its own radiation of forms of movement. They emancipated themselves far from ancestral crawling, and some of them obviously reached a considerable degree of flexibility. Presumably, however, they did not reach the speed and the maneuverability, which we know from extant mammals. Hotton (1980) showed that the large cartilages in the joints of the dinosaurs excluded such acceleration as is possible in many mammals. Also, the form of the joints in many dinosaurs indicates rather short steps in slow gaits rather than the speed of gazelles or the jaguar. It is not known how much stamina they had to maintain fast movements. However, a predominantly reptilian-like metabolism may have set limits (see Chap. 9). The pterosaurians, whose phylogenetic relation to the dinosaurs is not clear, were even able to widen their movement capacities into the air.

Mammals, along with the birds, are the vertebrates most completely able to cope with the physiological rigors of the terrestrial environment, gaining maximal potential for a relatively independent and flexible lifestyle. Kemp (2005, p. 14) writes: "Whilst all the terrestrial dwelling tetrapods can operate in the absence of the buoyancy effect of water, and can use the gaseous oxygen available, mammals have in addition evolved a highly sophisticated ability to regulate precisely the internal temperature and chemical composition of their bodies in the face of the extremes of fluctuating temperature and the dehydrating conditions of dry land."

To this spectrum of emancipation belongs the strong tendency in mammals to transfer the limbs vertically underneath the body, so that not the transverse muscles, but rather the bony pillars of the limbs and the muscle sling of the shoulder blade support the body (exceptions within extant mammals are monotremata, chiropterans, pinnipedia, cetaceans; incomplete in many small mammals). Smaller muscles stabilize the joints. The limbs move vertically backward and forward and thus can reach a large step length. The muscles, relieved from supporting functions, can be optimized for dynamics during fast walking and running. The strong ventral part of the bones of the shoulder girdle in amphibians and primitive reptiles is reduced. The breast muscle keeps the upper arm underneath the body. The shoulder blade is integrated into the movement of the limb and rotates in its sling, elongating the lever function.

Figure 5.16 displays this reorganization of the supporting apparatus (Starck 1982). Humerus and femur are narrowed to the trunk, and hand and foot are moved under the body. The rotation in forelimbs and hind limbs took place in the opposite direction. As hand and foot retain their orientation to the front, radius and ulna cross over each other. In many forms, the ability to rotate the forearm is retained, which often enables greater flexibility of the hand concerning pronation and supination. With these changes, especially the forelegs gained flexibility and lost static stability, which was replaced by complex neuronal control generating dynamic stability (Kemp 2005). In many movements, the contact to the ground can be

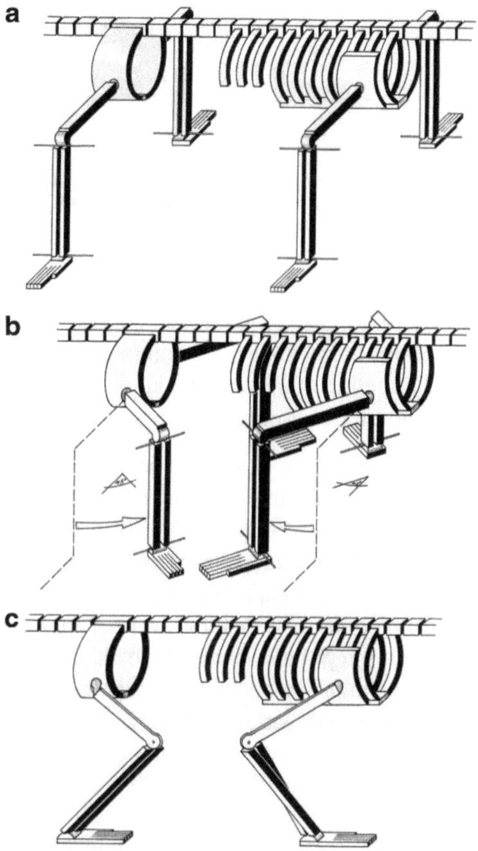


**Fig. 5.16** Schematic representation of the changes in the extremities from reptiles to mammals. (**a**) Primitive land tetrapodes; (**b**) hypothetical transition; (**c**) mammal (From Starck 1982)

reduced. In some phases of running, dogs and galloping horses can even have all their legs in the air simultaneously.

The described reorganization took place during the evolution of the mammalian organization, beginning with the early Synapsida ("mammal-like reptiles": Pelycosauria, Therapsida) in the late Carboniferous and Permian periods and mainly continuing throughout the Mesozoic. Because of the rich fossil findings, it is possible to follow these changes in some detail, comparing and reconstructing their skeletal organization (Kemp 2005; Kümmell 2009, 2012).

There are four aspects of an increase in flexibility, which emerged during the evolution of the Synapsids (Kümmell 2009):

1. In some groups, the reachable speed increased. This also includes the evolution of different gaits.
2. The multitude of reachable habitats increased.

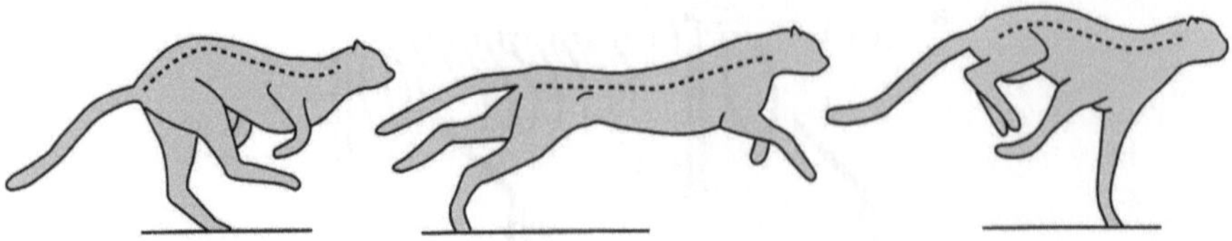

**Fig. 5.17** Elasticity of the vertebral column (From Heldmaier and Neuweiler 2003)

3. The diversity of forms of adaptation increased.
4. The agility of locomotion was enhanced. Mammals are more capable of rapid changes in speed and direction of movement and of coping more effectively with irregular terrains, including tree climbing.

However, in mammals again there is a broad spectrum of specializations. With the mammalian radiation in Tertiary times leading to ungulates, carnivores, rodents, and primates, the potential for movement flexibility was specialized in different ways. According to the respective adaptations to the environment, the animals reached high dexterity or high speed capacities, became able to dig or to climb, and more. Their variety and flexibility exceeds that of their reptilian-like ancestors by far. Even compared to modern reptilians, some of which gained flexibility compared to their ancestors, many mammals are obviously high performers (Pflumm 1996; Lockley 1991, 1999; Bauwens and Garland 1995).

Within extant mammals, a summit of speed and maneuverability is found in cats (Fig. 5.17). Kinetic energy of the accelerated body is stored in elastic structures when the legs come down to the ground and is released in the following cycle. The cheetah thus can make 3.5 jumps in a second, each jump 7 m long, reaching speeds of 100–120 km/h (Heldmaier and Neuweiler 2003).

Flying is a particularly strong overcoming of gravity, which is only reached in two groups: insects and vertebrates (pterosaurians, birds, bats). In both groups, the lever principle is involved. Because of the lower density of the medium, high speeds and long ranges can be reached by many animals. Including the respective specializations, flying is one of the apexes in the emancipation of movements. Some known top speeds are achieved as follows: carrier pigeon with tail wind, 177 km/h; peregrine falcon (*Falco peregrinus*) in a nosedive, 290 km/h; and Asiatic needle-tailed swift, *Hirundapus caudacutus*, one of the fastest flying birds in flapping flight, speeds up to 170 km/h.

However, these specializations for high movability can at the same time bring about some one-sidedness. For example, the forelegs of birds are specialized for the task of flying; thus, they are not available for object manipulation, which has been elaborated in some mammals. To some extent, this is compensated by corresponding abilities of head and bill, which needed a different neurological solution (see Chap. 8). The tendency to flexibility of manipulation is nonetheless also discernible (Fig. 5.17).

It is especially interesting that there seem to be strong correlations between the features discussed in the present chapter, pointing at transitions between systems rather than at the generation of single characters. One such correlation is emphasized by McCarthy and Enquist (2005), who studied the increases in body size, complexity, and metabolic intensity throughout macroevolution. They found a correlation between complexity (as measured by the number of cell types) and metabolic intensity on a macroevolutionary level. According to their model, an increase in biological complexity leads to an increase in metabolic intensity as additional intercellular networks are involved and require increased amounts of energy per unit of body mass. Each new cell type requires cellular and tissue integration, such as hormones and cell-to-cell signals. These new cell types must interact and communicate with the other cell types. Basically, clades derived more recently have more cell types and increased metabolic intensity than clades that are more ancient. Increased complexity and metabolic intensity require additional energy input. They further suggest that these processes are in some relation to increases in body size, although not in a linear and direct manner, but constrained by conditions of the body plan and the environment. They state that an increase in cell types allows the organism to perform new functions. Also, Milewski and Mills (2010) propose that a basic biological imperative of all organisms is to maximize energy intensity, defined as the average rate of energy used per unit area of the earth's surface.

Thus, an aggregate of body size, complexity, and metabolic intensity might contribute to open new possibilities for organisms: Increased size generally stabilizes physiological robustness and can increase movement capacities, and increased complexity together with increased metabolism allow for more functions and functional flexibility. However, these relations need more research than is available so far.

The relationship between the increase in metabolic intensity and extended functional possibilities of the organism becomes especially clear in the evolution of endothermy, which is discussed in detail in Chap. 9.

Perry and Carrier (2006) describe a coupled evolution of breathing and locomotion in chordates because the increase in metabolic rate related to locomotor activity places demands on the cardiorespiratory apparatus. As the respiratory faculty becomes more refined, increasingly aerobic life strategies can be explored, and this activity is in turn expedited by a higher-performance respiratory apparatus. The authors call this a leapfrogging of respiratory and locomotor faculties, which begins in noncraniate chordates and continues in water-breathing and air-breathing vertebrates. This phenomenon is described in some physiological detail in three related works (Taylor et al. 2006; Vasilakos et al. 2006; Klein and Owerkowicz 2006).

In my terms developed here, these are all elements for opening new levels of flexibility and possibilities within chordates, as described in the present chapter, and are strongly interdependent with each other, thus pointing to a system-level evolution of autonomic functions.

In summary, the transition from water to land by vertebrates brought a spectrum of flexibility and independency of movement, which according to Definition 2 are features of increasing autonomy:

- The axial construction of vertebrates brought forth the prerequisites to support the weight of the body on land and to generate high dynamics of the body axis.
- Lifting the trunk above the ground, against the forces of gravity, generated new possibilities for the use of the lever principle of legs.
- Further tendencies to have the limbs increasingly underneath the body made the lever principle more effective and economical, again enhancing movement possibilities.
- Especially in mammals and birds, increases in the flexibility and efficiency of movement were reached by reorganizations of the limbs and the vertebral column.

Compared to early crawling by undulation, a tremendous potential for mobility and maneuverability within the environment was generated. However, a spectrum of types of movement was developed on each level, including secondary returns to aquatic lifestyles with the related adaptations.

The lever system of the limbs with its differentiated muscles not only enabled the development of speed and maneuverability during locomotion, but also in some mammals the system reached further flexibility in such a way that limbs became able to manipulate the environment. Examples are digging, scraping, kicking, grabbing of prey, or moving objects. Extensive flexibility was reached with the ability to grasp, such as with raccoons, marsupials, and squirrels (using both paws working together) and especially within the primates (also with just one hand). Primates use their hands as sensorimotor organs to explore the environment and to manipulate it in a literal sense. Together with the related precision, a new quality of flexibility with high degrees of freedom evolved. This capacity, moreover, requires extensive differentiation of neuronal control; thus, this topic is discussed further in Chap. 8.

# Chapter 6
# Fluid Management in Animals

Since Claude Bernard's formulation of the "milieu intérieur," the functions of homeostasis have been studied extensively. Not all animals, however, run sophisticated homeostasis, which consumes a large amount of metabolic energy. Most invertebrates of the sea live in a milieu with basically constant osmotic concentrations, oxygen levels, temperature, pH, and other parameters. These organisms adopt their homeostasis predominantly from their environment.

It is assumed that earliest organisms of Precambrian times developed in marginal areas of shallow seas. It is further assumed that the water of those seas was not too different from today's seawater, with an average salt concentration. Thus, single-cell organisms as well as metazoans went through their early evolution in marine conditions. This origin seems to be reflected in the intracellular concentration of salt in most modern cells. Most groups of metazoans stayed marine anyway and have body fluids that are roughly iso-osmotic with seawater. Groups that inhabit freshwater or land deviate from this composition. Their cells as well as extracellular fluids have about one fourth to one third of the concentration of salt in seawater.

Organisms living in environments such as estuaries, freshwater, and land often underlie major changes of external factors. Physiology distinguishes different types of answers to changing environments (Fig. 6.1). Avoiders try to move away from environmental problems either in space or in time. Conformers tolerate the more or less changing conditions by letting their own internal conditions follow those of the outside world. They do not attempt to maintain a homeostatic condition for the whole body. Regulators maintain some or all of the components of their internal functions close to a "normal" level, irrespective of external variations. These three types can be present in different combinations concerning factors such as temperature, osmotic concentration, oxygen concentration, or pH (Willmer et al. 2000).

Conformers need functions on the physiological and biochemical levels in the form of enzymes and stabilized membranes to continue life processes even under larger environmental variations. Energetically conforming is less costly, but reduced possibilities of activities, growth, and reproduction must be accepted if larger changes of the external milieu occur. On the contrary, regulation requires an

B. Rosslenbroich, *On the Origin of Autonomy: A New Look at the Major Transitions in Evolution*, History, Philosophy and Theory of the Life Sciences 5,
DOI 10.1007/978-3-319-04141-4_6, © Springer International Publishing Switzerland 2014

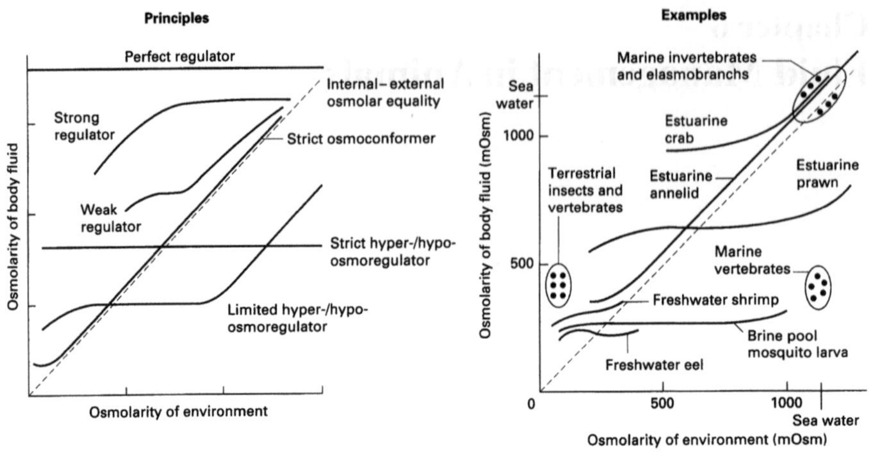


**Fig. 6.1** Categories of responses of animal body fluids to variations in external concentrations (From Willmer et al. 2000, with permission of Blackwell Publishing Ltd.)

extensive amount of energy. Extracellular fluids and the extracellular matrix are involved to buffer the cells against excessive environmental stress. Controlling this cellular and extracellular balance requires energy and is therefore a resource-intensive process (Natochin and Chernigovskaya 1997). Commonly, there are limits to both regulation and conforming, and there is no such thing as perfect regulators or perfect conformers. According to Willmer et al. (2000), smaller and soft-bodied animals are more likely to be avoiders and conformers. They can use microhabitats with concealment in protected crevices or burrows or on and in other organisms. Because they have a high surface-to-volume ratio, they will experience relatively rapid fluxes across their surfaces, and working to restore a status quo against these fluxes would be metabolically expensive. They often have little built-in protection against swelling and shrinking and lack the complex outer layers that can be modified to give some insulation or impermeability.

Large animals are more likely to be regulators in all environments: "Animals with hard outer layers (exoskeletons), of small and medium size, may have better options for some regulation and a greater independence of their environments" (Willmer et al. 2000, p. 15). They have lower surface-to-volume ratios so that rates of change of state internally are much slower, giving them an "inertia" effect that smoothes the fluctuations and gives time for regulatory functions to operate. They have better options for energy storage and for other resources, such as water and thermal energy. In terrestrial habitats, where environmental changes are faster, all of these factors may work together to make regulation the only real option for a large animal.

Most invertebrates tend to be conformers concerning several factors (Willmer et al. 2000; Heldmaier and Neuweiler 2004; Hill et al. 2008). However, many mollusks and many arthropods, such as crustaceans, spiders, scorpions, millipedes, and most insects, are regulators for several factors. Vertebrates generally tend to be regulators. Only birds and mammals are thermoregulators.

Osmoregulation and maintenance of body fluids can serve as an example for the evolution of functions for homeostatic stabilization of the internal milieu (constant volume, osmolarity, ionic concentration). During evolution, this occurred several times independently from each other. Figure 6.1 (right side) gives examples of different possible answers from body fluids to changes in the osmotic composition of the environment.

The intracellular milieu of protists is sheltered from the external environment only by their cellular membrane and some production of a glycocalyx. There is no equivalent to the extracellular buffer zone of metazoans. Protists are small and thus have a high surface-to-volume ratio, which leads to high exchange rates of water and ions via their cell membranes. Nevertheless, protists can be found in a broad spectrum of different aquatic habitats, such as freshwater, brackish water, seawater, and even hypersaline water.

The essential point is that the basis of their adaptation is an adjustment of their osmolarity to the environment. Freshwater has an osmolarity of 10–40 mOsm, while seawater has an osmolarity of approximately 1,100 mOsm. Species of protists living in seawater have an internal osmolarity within the range of the surrounding medium. Freshwater protists, such as many ciliates, flagellates, and amoeba, must iono- and osmoregulate, but their intracellular osmotic concentration is greatly reduced (50–150 mOsm) and thus adjusted to the freshwater, possibly as far as their intracellular composition allows. *Paramecium* in freshwater have an intracellular osmolarity of 111 mOsm (Withers 1992). The resulting influx of water is compensated with the help of their contractile vacuole.

Osmotic adjustment is achieved through changes in the concentration of osmotically relevant substances. *Dunaliella*, a halophilic green alga, and some yeasts and fungi use glycerol for balancing the osmolarity (Madigan et al. 2000; Brown and Borowitzka 1979). Species of *Dunaliella* grow in a broad spectrum of salinities up to salt lakes. Some protists are euryhaline and survive in freshwater as well as in seawater. For example, *Paramecium calkinsi* can live in a medium from 10 to 2,000 mOsm by adjusting its cellular osmolarity through changes in the concentration of amino acids.

Comparable adjustments are found in lower metazoans. Most marine invertebrates are osmoconformers. Mostly, they are stenohaline and isoosmotic with the seawater and need only small regulative corrections (Hill et al. 2008; Willmer et al. 2000). In freshwater, however, sponges, for example, reduce their osmolarity to a large extent. Thus, *Spongilla* has an osmolarity in its extracellular fluid of 55 mOsm. The necessary osmoregulation is performed on the cellular level, using contractile vacuoles. The freshwater polyp *Hydra* has an osmolarity of 45 mOsm in its extracellular fluid. In contrast, *Aurelia*, a jellyfish from the North Atlantic, has an osmolarity of 1,050 mOsm, living as a typical osmoconformer. Thus, although lower freshwater animals are osmoregulators, the basis of their osmotic stability is an adjustment to the surrounding milieu.

In contrast to these adaptations of the internal osmolarity to outside conditions, several groups of metazoans evolved that have stabilization of body fluids, which leads to an osmolarity essentially deviating from that of their outer medium.

These animals are always osmoregulators. The ability to (1) maintain the internal fluids on an own, species typical level, also if this is very different from the osmolarity of the environment and/or (2) to stabilize this osmolarity on a constant level against the fluctuations in the environment is an element of emancipation from environmental influences and thus the establishment of physiological autonomy. The basic principle is present in every osmoregulation, which always contributes to cellular autonomy of some extent. Widening of this autonomy is generated, however, where larger differences are established and maintained through regulation.

Within invertebrates, there are several marine crabs that are able to maintain their own deviating osmotic level. These include, for example, *Pachygrapsis*, *Uca*, *Leptograpsis*, *Eriocheir*, and *Sesarma*, which are hypoosmotic. *Pachygrapsis marmoratus*, for example, has an osmotic concentration of 86 % seawater. Some branchiopod crustaceans (brine shrimp; *Artemia*, *Parartemia*) are strong hypoosmotic regulators in seawater as well as in hypersaline water. *Octopus vulgaris* is slightly hyperosmotic to seawater at about 1,300 mOsm (Withers 1992; Hill et al. 2008).

In freshwater, there are again some crustaceans that have an osmolarity that differs from that of the medium. Also, *Lumbricus* (300 mOsm) and *Hirudo* (200 mOsm) are able to regulate (Willmer et al. 2000). Typical is high osmolarity in larvae of insects (250–499 mOsm) as well as adult insects. Figure 6.2 shows some features of freshwater animals.

All animals that started to invade terrestrial habitats had to establish sturdy management of their body fluids. Many of them stayed dependent on moist micro-habitats (platyhelminths, annelids, most mollusks, onychophorans), while others generated functions to survive in dry environments. They had to develop effective osmoregulation, autonomous fluid management, and tough boundaries through enclosing integuments. Their independence from a moist surrounding allowed them to spread into nearly all available habitats on land and to go back into water secondarily, whereby they mostly maintained the once-acquired level of homeostasis (see crocodiles, penguins, seals, whales).

The transition to land has occasionally been mentioned as an example of emancipation in earlier literature (Kuhn-Schnyder 1967; Lange 1976). It was criticized that the dependence on water was only exchanged for dependence on land (Dobzhansky et al. 1977; Simpson 1971). However, the essential point is that emancipation during this transition consists of the capability to live within a milieu that is fundamentally different from the milieu the cells and tissues need as their direct environment. This is possible through the self-generation of their own stabilized milieu intérieur. The fluid environment that all cells need can also be maintained within dry habitats by means of extension and regulation of the extracellular matrix and effective seclusion through integuments. What animals in the sea obtain from their marine habitats is now functionally internalized. The animals carry their own "pond" with them, in which their cells bathe and stabilize this against adverse circumstances and fluctuations.

Increased autonomy in this sense was developed by many groups of arthropods. Their tough and secluding integuments gave them a built-in resistance against

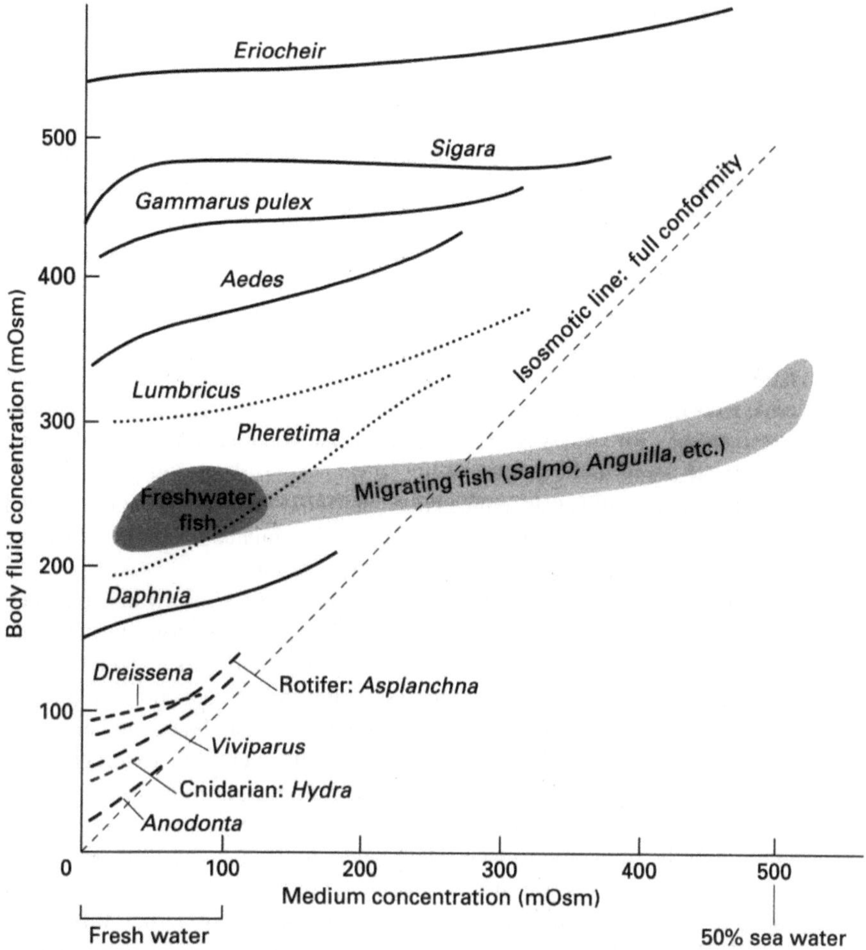


**Fig. 6.2** The relationship between internal body fluid concentration and external medium for various freshwater animals. Arthropods are shown by *solid lines*, mollusks by *dashed lines*, and worms by *dotted lines*, with teleost fish represented by the *tinted area* (From Willmer et al. 2000, with permission of Blackwell Publishing Ltd.)

swelling and shrinking through fluid uptake and loss. The external surface reduces permeability and can be partly isolated through fine cuticular hair. Thus, a high degree of regulation is energetically possible in spite of the small size. On the other hand, avoiding behavior supports this and is still an essential part of the stabilization of the whole physiology. By means of their effective integuments as well as their efficient organs for excretion, arachnids and insects in particular are strong regulators. This is the basis for their capacity to settle within all the different habitats on land even up to extreme terrestrial environments.

## 6.1 Vertebrates

Also, vertebrates generated an increasing osmotic autonomy. The osmolarity of body fluids of teleosts is between 250 and 500 mOsm, whether they live in freshwater, brackish water, or sea water. With this, the teleosts distinguish their internal milieu from the environment and are osmoregulators, as shown in Figs. 6.1 and 6.2. Main organs of osmoregulation are the intestinal epithelium, gills with their chloride cells, and the kidneys (Whittamore 2012). Species that migrate between seawater and freshwater (salmon, eel) are able to keep their internal milieu constant through changes in their osmoregulatory organs even against such massive environmental changes. The reduced osmolarity is assumed to be related to the evolutionary history of the taxon in freshwater. Here, the interesting point is that those species that went back to seawater not only adapted (or readapted) to the osmolarity of seawater but also maintained the level that was once achieved. One could speculate whether this is a model of how organisms gain autonomy: During some adaptation to freshwater, the osmolarity of body fluids was reduced to some extent and then maintained when these organisms went back to seawater, now on a self-regulated level, for which the necessary physiological processes first had to be generated.

The blood of primitive jawless hagfishes is isoosmotic to seawater, but they regulate the concentration of single ions (Ca, Mg, $SO_4^{2-}$ are less concentrated than in seawater, Na and Cl are more concentrated). Cartilaginous fishes and the extant coelacanth (*Latimeria*) are also isoosmotic to seawater. A low concentration of electrolytes is balanced through organic osmolytes such as urea and trimethylaminoxid (TMAO). Thus, most cartilaginous fish are osmoconformers, but ionoregulators. The water balance is maintained by avoiding significant osmotic influxes or effluxes.

A more extensive stabilization of body fluids by some groups of vertebrates occurred through the transition to land (Heldmaier and Neuweiler 2004; Hill et al. 2008). A prerequisite for this was strict osmoregulation and ionoregulation, as well as the capacity to keep the fluid within the body living in an external milieu that fundamentally differed from the internal milieu. Amphibians have an intermediate organization and are still dependent on moist surroundings. However, it is controversial whether this also was an intermediate stage within the evolutionary process or a specialized side group (Feder 1992).

In particular, the amphibian lifestyle carries a large osmoregulatory burden. The skin and bladder have functions to balance water movements. The skin of amphibians must be seen as specialized for water transport instead of an underdeveloped barrier against water loss (Frolich 1997). Another function is the high tolerance of many amphibians for water loss. *Bufo terrestris* tolerates a water loss of 43 % of its body weight, including the resulting increase in the osmolarity of the body fluid. No other vertebrate tolerates water loss of this dimension. Also within amphibians, there are some species that stabilize their fluid balance to such an extent that they can live in dry habitats (Shoemaker and Nagy 1977; Schmuck and Linsenmair 1997).

The accumulation of nitrogenous wastes is an essential limitation for terrestrial life in extant amphibians. Species that live mainly in water surroundings produce much urine with a low concentration as water is freely available. They excrete nitrogenous wastes partly as urea through their urine and partly as ammonia by diffusion through their skin into the water. The diffusion of ammonia to the air, however, is less. As ammonia is toxic, these species change, when on land, to increased production of urea. However, many aquatic and terrestrial amphibians excrete no urine whatsoever on land. Here, the bladder functions as a fluid repository, from which water can be reabsorbed gradually, while urea accumulates in the body fluids. Thus, most amphibians are limited in the duration of their terrestrial excursions by the concentration of urea within their blood plasma. They must return to water to rehydrate, and then urine is produced and excreted (Martin and Nagy 1997; Shoemaker et al. 1992).

Some few species of frogs living in arid areas are able to produce uric acid, thus reducing their need for water. The same species can have low rates of evaporative water loss by producing a waxy waterproofing secretion that is spread over the skin, as in the case of *Phyllomedusa*, and by several layers of dead skin, as in the case of *Chromatins* (Martin and Nagy 1997; Shoemaker et al. 1992).

Generally, most amphibians have a rapid turnover of water. This behavior, especially the choice of appropriate microhabitats, is essential to maintain the water balance. Species that are more terrestrially adapted and those in arid areas are only able to maintain water balance if they stay inactive in lairs (Martin and Nagy 1997).

Pough (1983) describes amphibians as well known for their inconspicuous and hidden lifestyle and for their low annual need for energy. The advantages of such a low metabolic rate are low need for food and high tolerance for cold conditions. Amphibians can be active at lower body temperatures than reptiles. Frogs that swim under the ice and salamanders that walk over the snow can regularly be observed. Several species of frogs let themselves freeze in the ice during the winter.

Because of their high flux rates, amphibians have, as Feder (1992, p. 5) summarizes, only limited independence from the direct environment: "In many cases these fluxes seemingly permit only limited independence from the immediate environment. Accordingly, the internal milieu of amphibians may be far less fixed than that of many other vertebrates." And, as Spotila et al. (1992, p. 59) formulate: "The usually moist skin of amphibians intimately couples them to their physical environment by the exchange of energy, water, and respiratory gases. … Nearly all active amphibians are tied to wet habitats in both their larval and adult stages."

## 6.2 Amniotes

Fluid balance stabilization was a central innovation in the evolution from anamniotes to amniotes. However, the physiological details of paleozoic amphibians, from which the amniotes took their origin, can hardly be reconstructed. Thus, it is not clear whether the permeable skin was a character of primitive amphibians or a derived specialization (Carroll 1988; Feder 1992; Martin and Nagy 1997). The basic features of their physiology, however, should have been to some extent comparable to that of today's amphibians. During the transition to amniotes, there must have been an emancipation from moist microhabitats, thus generating new ecological possibilities.

Reptiles generally have a requirement for water, which amounts to only 1–5 % of that of amphibians, as well as widely reduced water flux. This stabilization was possible by means of some special characteristics: The skin of reptiles has high resistance to water loss caused by evaporation. It has a high concentration of phospholipids in the stratum corneum and is covered by horny layers and scales. However, vertebrates do not reach the level of closure seen in some arthropods: The lower surface-to-volume ratio compared to the smaller insects has the effect that, in relation to their body weight, water loss is reduced to a comparable extent, so that some species can even live without access to drinking water (e.g., desert reptiles). The skin of reptiles has no respiratory function that could disturb the osmoregulatory functions (Hillman 2001; Martin and Nagy 1997; Willmer et al. 2000; Heldmaier and Neuweiler 2004).

The slightly higher metabolic rate of reptiles compared to amphibians delivers more metabolic water, which can be around 10–20 % of the total water needed by some reptiles of arid habitats. They produce relatively dry feces, and some are able to excrete excessive salt from nutrients through salt glands with low loss of water. Reptiles adapted to dry desert environments can store water in their bladder and can use small amounts of water in their food effectively. In addition, there are special adaptations of some species, such as the use of drops on the vegetation or the use of rain or condensation through furrows in the skin to lead drops to the mouth.

With reptiles, evaporation is also often the main way to lose water. However, the rate can be largely reduced by inhabitants of arid areas, whereas reptiles of humid habitats show higher rates. Some reptiles of moist tropical soils can have water exchange that is nearly the same as in amphibians, but the quantity of their body fluids has nearly no fluctuations. To the basically more stabilized physiology of reptiles, obviously a larger adaptive component is added. It might be species specific (Shoemaker and Nagy 1977; Willmer et al. 2000).

In animals with highly regulated body fluids and a strong homeostatic system, especially the terrestrial vertebrates and insects, the excretory system is responsible for regulating water and certain ion levels. This includes the regulatory capacity of excretory organs such as the kidneys in vertebrates as well as the way to release nitrogenous wastes. Most reptiles excrete the largest amount of their nitrogenous wastes as uric acid (uricothely), which precipitates at low concentrations because of its low solubility. It can be excreted as a nearly dry paste without toxic effects, thus saving water that otherwise would be needed to dissolve and excrete urea or ammonia. Together with uric acid, salts can be excreted, also saving water that otherwise would be needed for their excretion. With respect to the production of urea or uric acid, there is a strong adaptive component involved along with the phylogenetic one. Therefore, a few amphibians of arid areas (xeric tree-living frogs of the genera *Chiromantis* and *Phyllomedusa*) are also able to use uric acid. Turtles are adapted to their respective habitat: Sea turtles excrete liquid urine with urea into the water, and terrestric turtles generate at least 50 % uric acid. In semiaquatic turtles, there are intermediate conditions.

In the transition from amphibians to reptiles, the respiratory surfaces are finally internalized, which reduces the diffusive exchange of water with the environment to a large extent (Feder 1992). The amniotic egg, which defines the group of amniotes, made it possible to remove the reproductive cycle from the aquatic environment as well (Stewart 1997; Packard and Seymour 1997; Westheide and Rieger 2010). The inclusion of the embryo into the amnion and the chorion is a typical internalization (see Chap. 7). The embryo is protected within the amniotic fluid, which replaces an external aquatic milieu. Likewise, land vertebrates also develop within a fluid, analogous to their evolutionary history, but this milieu is now self-generated. The extraembryonal membranes take over the gas exchange, the storage of waste materials, and the transfer of nutrients to the embryo. The amniotic egg has been characterized as a "terrestric" egg, which emancipates the development of the embryo from water (Needham 1931; Romer 1957, 1967).

It can be assumed that the transition to a reduced need for water and dry, impermeable skin was connected to changes in temperature regulation (Martin and Nagy 1997). Wet skin leads to high heat loss. Exposed to full sunlight, the body temperature of a frog with wet skin is only a few degrees higher than that of a frog in a shadowy place. For two ectotherm animals of the same size, if one has moist skin and the other dry skin, the first will be cooler through evaporation, even if both are exposed to the sun and at the same external temperature. Elevated surrounding temperature will lead to rising body temperatures in animals with moist as well as with dry skins, but in different amounts. In a model developed by Spotila et al. (1992), under identical conditions the temperature in animals with dry skin will rise, for example, from 26 to 42.4 °C, while in animals with moist skin it will increase from 23.4 to 28.3 °C. Under natural conditions, the body temperatures of amphibians are usually lower than those of reptiles (Hutchison and Dupre 1992). However, reptiles have a preferred range of temperature for their activity. Although they are ectotherms, many regulate their body temperature through their behavior to some extent. The significance of this regulation is shown by the different temperatures for activity that different species of reptiles can have within the same habitat. Vice versa, reptiles of the same species living in different habitats can be active at the same temperature (Bartholomew 1982). A higher body temperature raises the metabolic rate and thus also the possibilities of activity and movement within the environment. Together with the more autonomous fluid balance, this opens access to a more extensive radius of action (Bennett 1991; Martin and Nagy 1997). On the other hand, dry skin leads to higher sensibility for some environmental influences, such as sunlight, wind, and thermal convection, without the evaporative effect through the skin (Spotila et al. 1992). With this, a larger range of body temperatures is achievable, together with higher sensitivity for many environmental factors. This enables and requires some regulation of temperature through behavior.

Overall, reptiles need more energy compared to amphibians. At the same temperature, the standard metabolic rate of amphibians is on average about two thirds that of reptiles of comparable mass (Pough 1983). A semiterrestrial animal, which stays colder by means of permanent evaporative loss of water, is restricted to a niche of low metabolic turnover.

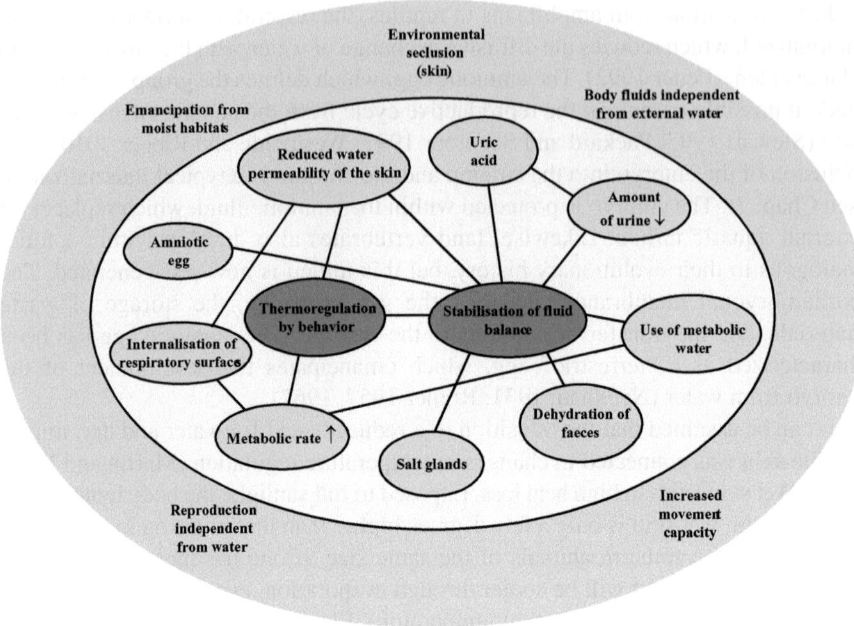


**Fig. 6.3** Functional complex of increased autonomy in the transition to the reptilian organization

These physiological components are a clear extension of autonomy during the transition from amphibians to reptiles (Fig. 6.3). Generally, terrestrial surroundings – as compared to water – are more demanding because of larger and less-foreseeable fluctuations. Reptiles confront this with their increased emancipation from environmental influences. This is in accordance with the formulation of Martin and Nagy (1997, p. 403): "Reptiles completed the transition to land by truly exploiting the land's resources as a permanent habitat. … To permit full exploitation of a more rapidly fluctuating external environment, we suggest that this evolutionary step hinges on a dramatic increase in homeostatic regulation of the internal environment. … In this manner, reptiles have obtained greater independence from the hydric environment by means of a more fixed internal milieu than was the case in anamniote tetrapods."

## 6.3 The Kidney as an Organ of Autonomy

Land animals are confronted with two opposing problems: to retain body fluids and at the same time excrete metabolic waste. The amount of excreted water together with waste must be restricted in most terrestrial animals. Thus, the terrestrial environment is basically an unfavorable milieu for organisms with cells, which not only permanently need a watery surrounding but also have to excrete their waste

**Table 6.1** Comparison of capability of concentrating urine in some vertebrates (maximum values) (Data from Withers 1992; Heldmaier and Neuweiler 2004)

| Reptiles | mOsm/l | Mammals | mOsm/l |
|---|---|---|---|
| Gecko | 325 | Canadian beaver | 537 |
| Crocodile | 267 | Domestic cow | 1,160 |
| Iguanid | 362 | Human being | 1,430 |
| Horned lizard | 327 | Horse | 1,545 |
| Desert tortoise | 337 | Domestic dog | 2,087 |
| **Birds** | | Domestic cat | 3,250 |
| Emu | 459 | Hamster | 5,340 |
| Chicken | 538 | Spiny mouse | 6,039 |
| Senegal dove | 661 | Jerboa | 6,500 |
| House sparrow | 826 | House mouse | 7,000 |
| House finch | 850 | Chinchilla | 7,599 |
| Honeyeater | 925 | Australian jumping mouse | 9,370 |
| Zebra finch | 1,005 | | |
| Savannah sparrow | 2,000 | | |

(Hillman 2001). The spectrum of functions land vertebrates developed to cope with these contradictory demands is large and reaches from the use of salt glands to the excretion of uric acid, reabsorption of water in the intestine, highly concentrating nephridia, and many others.

Whereas reptiles and birds predominantly rely on their uricothely and the role of their cloaca for the reabsorption of water, mammals mostly rely on the generation of urea (Heldmaier and Neuweiler 2004). However, they developed a kidney, which is able to produce highly concentrated urine. Most terrestrial vertebrates have an osmotic concentration of blood plasma of about 200–300 mOsm. Only the kidneys of mammals and to a lesser extent those of birds concentrate urine well above the osmolarity of blood plasma (Table 6.1). However, here great variations exist between the species, which depends on the favored habitats. Most mammals of temperate climates reach a urine osmolarity of about 1,000–2,000 mOsm, whereas small mammals of arid habitats are able to concentrate urine up to 5,000 or 9,000 mOsm (Hillman 2001; Withers 1992; Sinke and Deen 2011). Thus, the release of nitrogen can be high; at the same time, much water is saved. This is crucial for endotherm animals with a metabolic rate about ten times higher than that of ectotherms, taking up more nutrients and thus having to cope with a high amount of nitrogenous waste. For this concentration, the loop of Henle within the kidney is mainly responsible. The mammalian nephron also has a remarkable capacity for altering the clearance and fractional reabsorption of ions, osmolytes, and water because it is under hormonal control. Human urine, for example, can vary in osmotic concentration from about 60 mOsm to about 1,200 mOsm. Thus, it is possible to keep the fluid balance of the organism constant. Other mammals can produce even more dilute – or more concentrated – urine, depending on their iono-osmoregulatory demands. In general, mammals have a high capacity of systemic osmoregulation (Heldmaier and Neuweiler 2004; Sinke and Deen 2011).

Marine mammals maintained their high regulation capacities when they secondarily populated the sea (Ortiz 2001; Wang et al. 2012). In face of the seawater, they retained a physiological independence and robustness that fall within the range of that of land mammals. Seals and whales have highly efficient kidneys that can remove excess salt from the body fluids and at the same time retain water combined with an exact regulation of fluid balance. This establishes their autonomy, their distinction from the environment, and the assertion of their physiological independence. This is a fundamental difference compared to conformers in seas. However, the special situation required many adaptations of the system. Thus, it should be possible to analyze in some detail which of the functions can be interpreted as features of autonomy and which are adaptations to seawater.

Natochin and Chernigovskaya (1997) see two essential trends in the evolution of functions for water and salt homeostasis: (1) increased precision to keep the physicochemical properties of body fluids constant and (2) increased number of regulative factors and complexity of the controlling systems.

In summary, it can be maintained that organisms that are more primitive live in a water milieu, which the cells need as their environment. They only have basic regulation of their ion composition and receive their homeostasis predominantly from their environment. This is also true for a large amount of evolutionarily advanced invertebrates, which live as conformers.

In contrast to this, other animals stabilize their fluid management and thereby fulfill several of the principles of increases in autonomy, according to Definition 2 (Chap. 3): They internalize their homeostasis and maintain it actively. They emancipate their fluid composition from outside factors, and they maintain robustness toward changes in their environment.

Several times and with different solutions, this self-regulation was extended so far that life in terrestrial habitats became possible. Kidneys of mammals and birds have a high capacity for regulation, leading to an extensively controlled body fluid composition. This physiological autonomy is even maintained if animals secondarily go back into the water.

Regulators generally have more options on flexibility within the environment and for the enhancement of self-referential, intrinsic functions and thus for the evolution of interactive as well as constitutive autonomy of the respective animals.

This view is identical with the description of Smith (1953, p. 4), who points out the importance of physiological independence for humans: "It is no exaggeration to say that the composition of the blood is determined not by what the mouth takes in but by what the kidneys keep: they are the master chemists of our internal environment, which, so to speak, they manufacture in reverse by working it over completely some fifteen times a day. … We must acknowledge that our kidneys constitute the major foundation of our physiological freedom."

# Chapter 7
# Reproduction

In correlation with the evolution of morphological and physiological features of autonomy within the vertebrates described so far, there are also different levels of protection and internalization of embryonic development (Schad 1977, 1992; Simpson 1971; Grant 1985). Although there are many variants, the most frequent and most fundamental mode of embryonic development within bony fishes is external fertilization and the development within the external environment, the water. With amphibians also, embryonic development most often takes place within the external medium, but the eggs are often surrounded by an egg jelly of several layers, thus providing some additional external protection. The generation of a fluid space, the amnion, which surrounds the embryo, is the definitive innovation of amniotes (reptiles, birds, and mammals). The origin of the amniotic egg is most often seen to be related to the emancipation of the reptiles from water (Needham 1931; Romer 1957; Stewart 1997; Ferner and Mess 2011; Westheide and Rieger 2010). Embryonic membranes and a shell protect the eggs from some environmental influences so that embryonic development can take place independently from water as a medium.

However, the shells of reptiles are mostly as thin as parchment. Only some species (e.g., land turtles inhabiting arid areas) have eggs with shells that are more calcified. Birds' eggs are regularly more calcified, providing a rigid outer shell for protection. In addition, brooding by birds yields relative emancipation from the external temperature for embryonic development. Higher birds build complex nests, which represent an additional shelter. In some species, the nest is completely closed, so that just a small opening remains as an entrance (Fig. 7.1).

Within mammals, there is an obligatory transition of development into the uterus, which means that embryonic development is internalized into the body of the mother, nearly completely independent from environmental influences. These circumstances are extensively described and well known in many of their variations (Siewing 1969; Westheide and Rieger 2010; Packard and Seymour 1997; Romer and Parsons 1977; Ferner and Mess 2011). However, it has only rarely been noticed that the increasing enclosure and internalization of development in face of the environment is the continuation of other processes of evolutionary internalizations and

B. Rosslenbroich, *On the Origin of Autonomy: A New Look at the Major Transitions in Evolution*, History, Philosophy and Theory of the Life Sciences 5,
DOI 10.1007/978-3-319-04141-4_7, © Springer International Publishing Switzerland 2014

**Fig. 7.1** Weaver bird at its nest

correlates with further features of autonomy in the adult organisms. Also, parental care, which is regularly present in mammals and birds and occasionally occurs in other vertebrate groups, is another related evolutionary pattern.

As examples, the generation of the amnion and the generation of viviparity are described in the following paragraphs in more detail.

## 7.1 The Origin of the Amnion

A main difference between extant amphibians and other tetrapods is the type of egg they produce: Amphibians generate anamniotic eggs; all other tetrapods generate amniotic eggs (Fig. 7.2). The amniotic egg has four extraembryonal membranes: the amnion, the chorion, the allantois, and the yolk sac. These provide protection, fluid retention, and large surfaces with extensive vascularization for gaseous exchange. By means of the yolk, the embryo is supplied with nutrients.

The evolutionary origin of these membranes and especially of the amnion is still unclear (Elinson and Beckham 2002; Laurin 2004; Mess et al. 2003). Most hypotheses connect their appearance with the emancipation of reproduction from water (Romer 1957; Martin and Nagy 1997; Mess et al. 2003; Blackburn and Flemming 2009; Ferner and Mess 2011). Thus, Stewart (1997, p. 291) formulates, for example: "Conceptually, the amniotic egg is a 'terrestrial' egg; an achievement of embryonic emancipation from the aquatic realm, replete with specializations for an independent existence." And Westheide and Rieger (2010, p. 360) maintain: "The crucial autapomorphy of the amniotes is an embryonic membrane serving as a repository

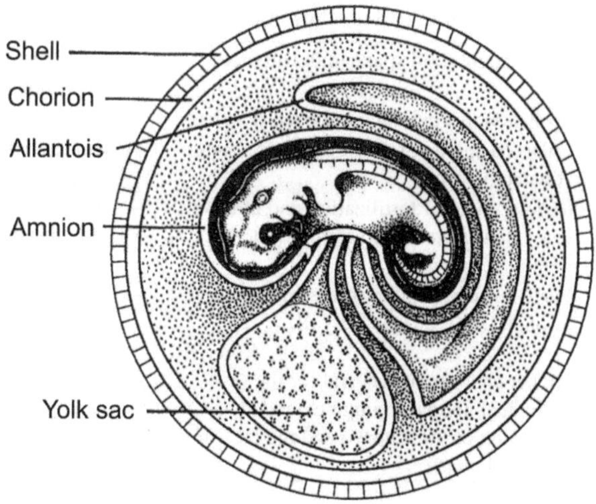


**Fig. 7.2** Schematic representation of an amniotic egg (Modified from Romer and Parsons 1977)

for the amniotic fluid within an egg, which has a rigid calcified shell. It rendered its bearer for the first time irrevocably independent from a water-bound development via a larva. This was the prerequisite for the evolution of the vertebrates on land, which was so successful."

Romer (1957) holds the opinion that the emancipation of reproduction from the watery environment by means of the amnion preceded the emancipation of the adult animals from water. While the earliest reptiles still had an aquatic lifestyle, they possibly already had an amniotic egg. In this sense, the generation of the amnion might have been the key innovation for the full transition to land, thus preparing the tremendous evolution of the amniotes to modern birds and mammals. "It was the egg which came ashore first; the adult followed later" (Romer 1957, p. 61). For this hypothesis, Romer outlined a scenario with small stretches of water that might have dried out periodically, thus favoring this emancipation.

Laurin and Reisz (1997) propose an alternative explanation. According to this hypothesis, the development of some of the extraembryonal membranes might originally have been related to a temporary retention of the embryo within the mother after internal fertilization. This would result directly in the possibility of more extensive interaction between embryos and mothers, which is also supported by the relatively frequent viviparity among reptiles.

Carroll (1970, 1991, 1997) proposes that the modification leading to the amniotic egg also made possible the generation of larger eggs and consequently larger body sizes. Without the generation of membranes for gaseous exchange, the size of an egg is limited because of diffusion problems. Also, a small egg can contain only a limited stock of nutrient material. Thus, for larger animals, which hatch as young that are more developed, larger eggs with the supporting membranes would become necessary (see also Packard and Seymour 1997). Finally, the animals hatch with a completely developed organization as miniature adults. The larval phase, during which huge losses often occur and which thus makes a high number of larvae (r-strategy) necessary, can be omitted. Therefore, there is a change of

**Table 7.1** Features of autonomy of the amniotic egg

| | |
|---|---|
| – Independent fluid space in the amnion | → Emancipation of reproduction from a watery environment |
| – Seclusion through membranes and a shell | → Environmental separation |
| – Enlarged egg and relatively larger young at hatching time | → Homeostatic stabilization through reduction of relative surface |
| – Regular internal fertilization and more or less extended retention of the embryo within the female | → Internalization of fertilization and a part of embryonic maturation |
| – Early nutrient supply of the embryo through larger amount of yolk | → Supply of the young becomes more independent from the environment; no larval phase that depends on external nutrients |
| – Higher chances of survival for the individual animal through extended protection of the embryo and the lack of the larval phase | → Increased robustness of the individual |

reproduction to higher parental investment into the care of their young with lower numbers of offspring, whereby the individuals have a higher likelihood of survival (k-strategy).

Whatever the essential innovations might have been, in any case these different scenarios indicate functional complex of the amniotic egg, which again contains features of increasing autonomy. These are listed in Table 7.1.

## 7.2 Evolution of Viviparity

The predominant and most ancestral mode of reproduction in fishes, amphibians, and reptiles is oviparity, the production of eggs that are laid in the outside world. Diverging from this general picture, in all three groups a wide variety of viviparity, the birth of living young, evolved within certain lineages or certain species.

Lecithotrophe viviparity, in which the egg yolk provides the nutrients for development, can be distinguished from matrotrophe viviparity, in which the female provides nutrition to her embryos through the uterine wall. Lecithotrophy and matrotrophy represent extremes of a continuum, as viviparous species tend to exhibit both nutritional modes and species are characterized according to the predominant source of developmental nutrients (Blackburn 1995; Blackburn and Flemming 2009; Westheide and Rieger 2010; Murphy and Thompson 2011).

From the known data, Blackburn (1992) assembled an overview of the generation of viviparity within the vertebrates. He distinguished phylogenetically related developments from convergences in different lineages and found that viviparity was generated at least 132 times independently from each other, with 98 of these cases found within reptiles. He also quantified the convergent generation of a substantial matrotrophy and found at least 24 independent origins. The majority of these were found within fishes.

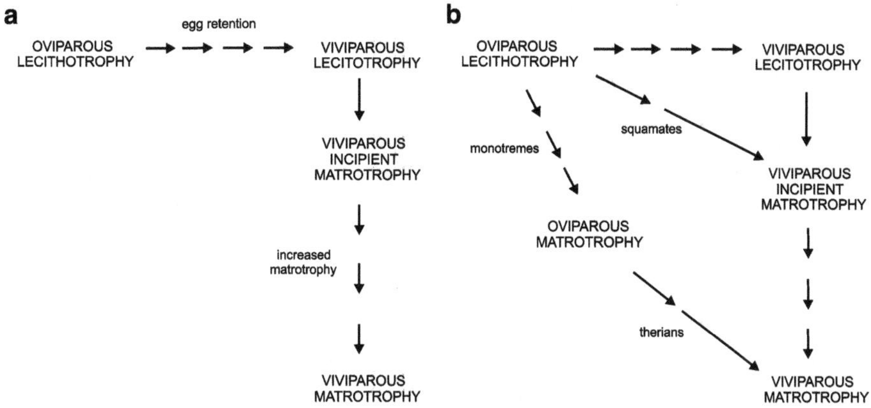


**Fig. 7.3** *Left*: Traditional scenario for the evolution of viviparity and matrotrophy. *Right*: Multiple hypothetical scenarios for reproductive evolution (Redrawn from Blackburn 1992)

For the evolution of viviparity, often three successive hypothetical stages are assumed (Angelini and Ghiara 1991; Packard et al. 1977; Blackburn 1992) (Fig. 7.3a): First, there might be some retention of the fertilized eggs, leading to viviparity. As a next stage in such viviparous forms, the amount of substances supplied from the female may develop from a rudimentary matrotrophy to a more extensive matrotrophy, which in a third step can lead to the generation of a placenta. The generation of more elaborate placental structures with close associations of fetal and maternal tissues is usually seen as highly derived. It is seldom present outside the mammalian class. However, there are some examples for this.

The traditional scenario corresponds to observations in many groups of vertebrates (e.g., Chondrichthyans). However, according to Blackburn (1992), it has to be expanded. Thus, it has been demonstrated that in all viviparous Squamata (lizards and snakes), at least some transition of nutrients from the mother to the embryo takes place. He proposes that beginning matrotrophy could also have been generated simultaneously with viviparity rather than in separate modifications (Fig. 7.3b). The evolution of viviparity could also have followed a "punctuated equilibrium" model rather than a gradualistic mode of evolution.

The generation of a placenta originated several times convergently within the vertebrates (Blackburn 1992; Rothchild 2003; Ferner and Mess 2011; Sheng and Foley 2012; Blackburn and Flemming 2010; Fioroni 1992). Under conditions of viviparity, the membranes that surround the embryo lie apposed to tissues of the uterine oviduct, such that physiological exchange occurs with maternal tissues. Any organ that is formed through apposition of fetal and maternal tissues and that functions in physiological exchange qualifies as a placentation. The nutrition of the embryo then takes place either partly or completely by way of the blood of the female. In some single cases, placenta-like formations are already present in bony fishes. Thus, the million fish (*Lebistes* sp.) generates a vascularized placenta involving the yolk sac. Similar placentations have been found in several groups of cartilaginous fishes and some lizards and snakes, also involving other embryonic membranes. These examples of nonmammalian placentas even involve intensive and close contacts between the maternal tissue and the embryo.

Although the evolutionary sequences are only rudimentarily known, it can be stated that during evolution the internalization of embryonic growth has been reached several times independently to different degrees. However, this mode of reproduction did not become the rule previous to the therian mammals. Internalized fertilization is the prerequisite for all forms of viviparity, either by means of copulation or by way of absorbed sperm.

According to Blackburn (1992), even the maternal supply of the young with nutrients after birth or after hatching is a form of matrotrophy. For this purpose, mammals use a secretion from the body, milk, which allows the young to remain independent from the ingestion of external nutrients during nursing. For the young, suckling means a form of buffer against fluctuations in nutrition from the environment concerning quality, quantity, and toxicity. In addition, the high-energy content of milk allows even small animals to stabilize their homeothermy early in their life.

The reproduction of recent mammals, compared to reptiles, is characterized by the combination of parental care, lactation, yolk reduction, and placentation. Interesting stages preceding the full internalization of the embryonic development are found for primitive mammals. Again, no linear process is to be expected, but ancestral patterns form both the evolutionary substrate for the origin of the diversity seen in living mammals and the baseline for understanding what sort of changes occurred in these primitive patterns during evolution (Renfree 1991; Freyer and Renfree 2009; Mess and Carter 2006).

Two groups are important for understanding this baseline: the monotremes, which are mammals that provide their young with milk but still lay eggs (platypus and echidna, restricted to Australia and New Guinea), and the marsupials (including kangaroos, koalas, wombats, and many others, with most living in Central America, South America, and Australasia). They stand in contrast to the typical mammals, which have full placentation and thus are called "Placentalia" (= eutherians, including rodents, chiropterans, Carnivora, Primates, and many more).

The eggs of monotremes are smaller relative to the body weight and have less yolk than those of reptiles. However, they still have a yolk, so that the type of cleavage is similar to that of reptiles. In monotremes, the eggs take in maternal nutrients from uterine glands through the parchment-like shell membrane and grow about fourfold in size before being laid. Westheide and Rieger (2010) see this as an ancient form of placentation. The young animals that hatch from the laid eggs are still premature and are fed with milk for several months. Generally, the ontogenesis is already more united with the adult female than in any reptile.

Marsupials show further elements of internalization of ontogenesis, including viviparity. In many ways, the ontogenesis resembles that of monotremes, but there is no maturation of the egg outside the female. Instead, there is a sort of "intrauterine hatching" by opening the shell membrane. The persistence of a shell coating during most of the development time is seen as an ancestral character that prevailed during the transition from oviparity to viviparity. No eutherian mammal has such a shell coating (Menkhorst et al. 2009). After birth, the young are still closely united with the female. In some species, this takes place without the inclusion into a pouch, as the young only hang on the mother's nipples. The more derived inclusion of the young in a pouch, as in most marsupials, is a secondary process of internalization. The gestation time in marsupials is usually shorter than in eutherians, and the newborn animals are extremely premature, which means that the internalization of embryogenesis is still not as highly developed as in eutherians. It is complemented by the long milk feeding time and a close connection to the mother during this time, in most species including protection in a pouch. The exchange of nutrients and gases between embryo and mother animal depends more on a placenta than in monotremes. However, there is seldom an invasive form of the placenta in marsupials, although such forms also exist for the last days of gestation (Carter 2001; Westheide and Rieger 2010). Usually, embryonic membranes only attach closely to the uterine wall without any penetration into its tissues, but the intensity of the contact between fetal and maternal membranes often increases during ontogenesis.

The prolonged development within the uterus is a main characteristic of placental mammals. Here, the placenta performs the complete supply of the embryo until birth of offspring that are more developed than in monotremes and marsupials. Within the different groups and species, there is a considerable variety of types of placentation, which is difficult to systematize (Mess and Carter 2006; Lewitus and Soligo 2011). However, for the topic discussed here, it is remarkable that there are different grades of internalization of embryonic tissues into the surface of the uterus. This concerns the degree of degradation of maternal tissues during implantation as well as different ways in which the embryo is attached to the uterine wall.

In some forms of placentation, the adjoining embryonic and maternal tissues remain intact, and there is only a relatively loose connection (epitheliochorial type of placenta, which predominates, for example, in odd-toed ungulates, lemurs, whales, pangolins). All other placentas, in which the tissue of the uterus is degraded to different degrees, are summarized as invasive placentas: If just the epithelium at the surface of the uterus is degraded, the embryonic tissue reaches the uterine connective tissue (syndesmochorial type of placenta; many ruminants, sloths). Deeper penetration into the uterus brings the embryonic tissue into close contact with the blood vessels (endotheliochorial type of placenta; in most carnivores, elephants, and some bats). The closest connection is generated when the inner surface of the uterine blood vessels is also degraded and the blood of the mother directly washes around the embryonic membrane (hemochorial type of placenta; rodents, many insectivores, some bats, hyrax, sirens, and most primates, including humans). However, there are often mixed forms of these variations of placentation within different areas of contact.

There are also differences in how much of the embryo is enclosed by the uterine tissue. In some animals, the embryo lies centrally in the cavity of the uterus (central implantation; e.g., rabbits) or is more or less superficially in a side cavity of the uterus (eccentric implantation; e.g., rats, mice). In other forms, the whole embryo penetrates into the tissue of the uterus and is finally completely enclosed by the maternal material (interstitial implantation; e.g., hedgehogs, some bats, monkeys, humans).

These forms of implantation do not necessarily correlate with the degree of degradation of maternal tissues. However, often in interstitial implantation the tissue degradation is more extensive (endotheliochorial or hemochorial). In central implantation, the uterine tissue also can be degraded in certain areas more or less extensively, and the fetal membranes can even reach the blood vessels. In other cases, the contact can be superficial (e.g., horses).

There is no clear sequence of these types of placentation, which would allow reconstruction of chronological order. Rather, different combinations are found in mosaic fashion. Recent studies revealed that the ancestral placental mammal may have had an invasive placenta (Elliot and Crespi 2009). This means that less-invasive modes of placentation, as found in cows and horses, for instance, are derived rather than primitive modes as previously thought. They possibly originated several times independently (Luckett 1993; Carter 2001; Mess and Carter 2006; Enders and Carter 2012). However, they all exhibit different types of internalization of the growing embryo.

Characteristically, humans, who also exhibit far-reaching autonomization in some other features, generate a particularly deep internalization of the embryo during pregnancy (Carter and Pijnenborg 2011). The embryo is implanted relatively early (around the sixth day after ovulation). The implantation is interstitial and the embryo is so completely integrated into the tissue that it loses the contact to the uterine cavity, and the placenta is hemochorial.

In summary, comparing different forms of reproduction in vertebrates, there are processes of internalization, separation from environmental influences, and support for the regulation of body functions of the embryo. These are all features that are characteristic for increases in autonomy according to Definition 2. The discussed examples are as follows:

- The evolution of the amniotic egg, which protects the embryo from environmental influences by membranes and a shell. This is presumably related to the emancipation of reptiles from water. Possibly, it is also related to increases in body size.
- The internalization of fertilization.
- Bird eggs are more calcified, so that environmental seclusion is enhanced.
- The brooding by birds within a nest, yielding a shelter of varying degrees and regulation of embryonal body temperature by the parents.
- The evolution of viviparity, which exhibits different grades of internalization of the embryo into the body of the mother and various intensities of contact with her tissues.
- Different grades of internalization of the embryo into the surface of the uterus within placental mammals.

# Chapter 8
# Nervous Systems and the Flexibility of Movements

Bacteria have a perception of some factors in their environment that are relevant to them. They use membrane receptors, for example, to detect gradients of nutrient particles or toxins. They can, to a certain extent, integrate this information and "decide" whether they will continue swimming in the same direction or whether they turn around and swim in the opposite direction. The successive concentrations must briefly be memorized and compared to each other to detect a gradient. Such processes have been studied thoroughly in *Escherichia coli*. The bacterium perceives its environment through its receptors, which are protein molecules embedded in the cell wall. *Escherichia coli* has more than a dozen different types of receptors on its surface. The receptors bind to specific chemicals outside the cell and communicate with other chemical processes inside the cell, which may directly influence the flagellar beat. Such a direct influence on the beating pattern of flagella is also found in eukaryotic single cells, as in, for example, *Chlamydomonas*. In *Paramecium*, the mechanical stimulus bumping against an obstacle activates an influx of $Ca^{2+}$ at the membrane, which leads to a reversal of beats of the cilia. In single cells, mainly $Ca^{2+}$ and $K^{+}$ channels can be found, which mediate changes of the membrane potential, whereas $Na^{+}$ channels are found in metazoans – in cnidarians for the first time (Allman 1999; Heldmaier and Neuweiler 2003; Hill et al. 2008).

Comparable to single cells, neurons of metazoans have receptors on their surface. Here, however, incoming signals are transmitted into action potentials, not into the direct change of movement. The action potentials influence other neurons, through which only indirectly is a reaction (e.g., of contractile organs) initiated. Therewith, an additional level of integration is present: First, the sensible neuron integrates incoming stimuli and activates action potentials or not. If an excitation occurs, the signals are processed in one or several postconnected neurons, and depending on the result, a reaction is initiated or not. Both principles show basic elements of independence regarding the physical environment and a certain amount of self-determination. Even single cells do not behave like billiard balls, which when hit only move according to the physical principles of cause-effect relations, but rather they calculate the incoming signals. The advantages of a nutrient, for

B. Rosslenbroich, *On the Origin of Autonomy: A New Look at the Major Transitions in Evolution*, History, Philosophy and Theory of the Life Sciences 5,
DOI 10.1007/978-3-319-04141-4_8, © Springer International Publishing Switzerland 2014

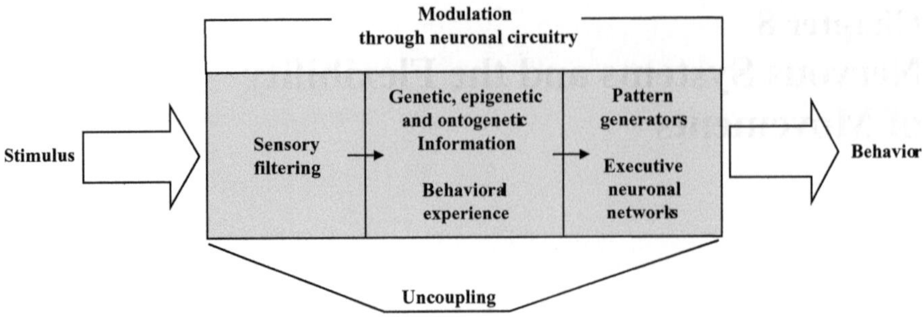


**Fig. 8.1** The general principle of modulation by the central nervous system

example, may be weighed against a concentration of toxins. However, with the generation of nervous systems, the capacity for this indirect and self-determined reaction is largely expanded. An increasing *uncoupling* or detachment of signals and reactions is introduced, so that the signals are processed in neuronal intermediate steps before a reaction follows.

This is what Allman (1999, p. 9) expresses: "Cells, like people, are immersed in a flood of information, which they must evaluate in order to generate an adaptive response. Even *E. coli* must integrate information from more than a dozen different receptor types to make the simple binary decision as to whether to rotate its flagellar motors clockwise or counterclockwise. … One advantage possessed by multicellular organisms is that they can channel this flood of information by creating a dam between the external world and the interior of the organism."

The simplest neuronal circuit is the reflex arc. It is assumed that a basic reflex arc may have consisted of just one single receptor cell, which would have had both receptor and motor functions. However, monosynaptic and polysynaptic circuits essentially widen the possibilities of modulation of the reaction. As soon as interneurons and synapses are integrated, many more excitatory or inhibiting influences can be involved. The reaction is not directly initiated by the sensory input, but uncoupled from it and prepared by way of more or less complex processing (Fig. 8.1). This modulation makes possible more flexible, self-determined reactions of different degrees to environmental stimuli and thus is a feature of autonomy, according to the definition in Chap. 3.

The principle of uncoupling is central for my further comparisons of neuronal and behavioral abilities of animals. It is a crucial feature for the phylogenetic generation of a constitutive autonomy in nervous systems as processing that is more internal becomes possible. However, interactive autonomy also is enhanced as the variation and possibilities of reactions toward the environment are expanded.

The generation of complex centralized brains is seen as one of the major evolutionary patterns in the history of animals. Monophyly (i.e., presence of a centralized nervous system in urbilateria) versus polyphyly (i.e., multiple origins by parallel centralization of nervous systems within several lineages) are two conflicting scenarios to explain these patterns (Northcutt 2012; Moroz 2012). In any case, there has been an evolution from diffuse nerve nets to concentrations of nervous tissues in ganglia and brains with an increase in the number of interneurons. Typically, the simpler circuits, such as the monosynaptic arc, are retained during evolution, and circuits that are more complex are added to them. The larger number of interneurons between input and output correlates with widened behavioral flexibility and is the basis of an increased learning potential (Neuweiler 2008).

The concentration of neurons within central nervous systems provided the potential for many neurons to be located near each other, so that the possibilities for generating connections and networks were increased. In addition, more neurons are present only for the integration and modulation of incoming and outgoing signals. This widens the possibilities for modulation to an even larger extent. In densely packed areas of the brain, as in cortices of vertebrates and optical lobes of arthropods, the neurons are organized in layers. This optimizes the possibilities for building up connections.

The widening of central nervous modulations with increasingly flexible behaviors is observed in highly cephalized groups (Neuweiler 2008). Within invertebrates, the cephalopods are most prominent. Their nervous system builds up a relatively large brain, which is enclosed by a cartilage capsule (internalization). Especially, the octopus shows impressive capacities to learn and to memorize (Fiorito and Scotto 1992; Boal et al. 2000), on which Conway Morris (2003, p. 215) comments: "Evidently the octopus, far from being a rule-bound machine, is capable of acting in an autonomous fashion."

Within arthropods, centralizations occurred in several lineages. Some insects, for example, show impressive learning capacities, which has been especially demonstrated in bees (Menzel and Giufra 2001; Giufra et al. 2001). Within vertebrates, centralization is pronounced, which is discussed in some detail in Chap. 10.

Enhanced behavioral flexibility based on complex neuronal systems developed several times convergently, starting with different body plans within the various phyla. Superimposed systems increasingly dominated the function of local motor circuits and simple reflex chains. Insects with an experimentally removed brain are still able to carry out coordinated leg movements after stimulation. However, they need the superimposed subesophageal ganglion to initiate such movements. Spinal circuits of a brainless frog still perform some reflexes, whereas in mammals without a brain, no movements are possible even with an intact spinal cord. Here, the brain's capacity of modulation and flexibility dominates the nervous system.

Generally, the more complex the neuronal networks become, the larger the possibilities for modulation of their functions will be and the more flexibly the organisms not only can react within their environment but also can act actively. This is a possible answer to the question of Bullock (1995) regarding the qualitative differences in the function of differently complex nervous systems: Flexibility of behavior emancipates organisms from simple stimulus-response relationships to stimuli from the environment and introduces increasing degrees of flexibility within the behavioral possibilities.

Characteristically, the neuronal networks themselves are exceptionally plastic. As opposed to technical electrical circuits, their connections are not static but rather highly plastic. In adaptation to experiences, they can change themselves functionally and in many cases also structurally.

The more a physiological and morphological interior is established, the more organisms are also able to perceive the external world via the sensorineuronal communication system. The more an organism establishes autonomy, the more it seems to be able to generate a bridge between the inside and the outside. It is interesting to observe that nerve nets were not present before there was an effective seclusion of tissue from the environment through intercellular structures (basiepithelial and subepithelial nerve nets) (see Chap. 4).

Nervous systems that are more complex and larger are especially built in organisms with high movement capacity. This is physiologically understandable; during higher mobility, more extensive neuronal performance is needed because organisms face a fast series of unforeseeable environmental situations. In this sense, Allman (1999, p. 2) also formulates that brains enlarge the independence from the environment: "The essential role of brains is to serve as a buffer against environmental variation."

Both widened movement capacity and central nervous systems, which enable flexible behavior, are features of autonomy and correlate with each other. The indirectness of reactions and the possibility of their modulation in a self-determined way are also present in other physiological functions, which Lewontin (2000, p. 63) describes as follows:

> Finally, organisms determine by their biology the actual physical nature of signals from the outside. They transduce one physical signal into quite a different one, and it is the result of the transduction that is perceived by the organism's functions as an environmental variable. For a mammal, when the temperature of the air rises, the increased thermal agitation of the molecules does not result in a matched increase in thermal agitation of molecules inside the animal. The smallest initial change in internal temperature is converted by the hypothalamus to an endocrine signal which results in a large number of internal chemical and neural and anatomical changes such as concentrations of hormones, of blood sugar, of breathing rate, of the chemical activity of sweat glands, of the contraction of muscle fibers in the skin. … It is the biology, indeed the genes, of an organism that determines its effective environment, by establishing the way in which external physical signals become incorporated into its reactions. The common external phenomena of the physical and biotic world pass through a transforming filter created by the peculiar biology of each species, and it is the output of this transformation that reaches the organism and is relevant to it.

Textbooks of comparative neurology have often described the increasing complexity and differentiation of nervous systems within vertebrates as related to increased flexibility of neuronal control of movements and to behavioral repertoires that are more complex. It has been overseen, however, that this is part of a more universal process of autonomization of organisms. To point out the evidence for this, an overview of some functions of neuronal motoric flexibilization is provided next. The first part follows a hypothesis of Dubbeldam (2001) on the evolution of central pattern generators (CPGs). The second part follows well-known principles of general neurophysiology.

## 8.1 The Uncoupling of Central Pattern Generators

The neural system for locomotion control in vertebrates consists of two main components: the CPGs in the spinal cord and the superimposed systems in the centers of the brain stem and the forebrain. The CPGs generate autogenously the rhythmic patterns of muscle activity that are characteristic for locomotion. Each segment of the spinal cord contains two pairs of CPGs, one for each side. The left and the right CPGs of a segment can interact through reciprocal inhibitory connections. Further interactions occur between CPGs of successive spinal segments. The superimposed systems in the brain modulate the activity of the CPGs (Heldmaier and Neuweiler 2003; Kiehn 2006; Gordon and Whelan 2006; Tresch et al. 2002; Dickinson 2006; Grillner et al. 2008).

Many details of the basic organization of CPGs are known from studies of lampreys (Petromyzontida). Their locomotion results from undulatory movements of the whole body, generated by coordinated activity of the CPGs in the spinal cord. The activation signals move on their own rostrocaudally from segment to segment with a phase delay of 1 % between successive segments. This type of undulation can also be found in other groups of fishes, but in many species, undulation is restricted to only part of the body and is more or less supported by movements of fins. In these cases, CPGs can be situated at different places depending on which fins are involved and the location of these fins on the body.

Within amphibians, the Gymnophiona have no limbs and use an undulatory type of locomotion. Urodels move through a combination of undulatory locomotion and limb activity. The situation in urodels may be seen as a transition from the undulatory movement to locomotion mainly with limbs. Within the reptilians, the snakes use this type of locomotion, which may have been reacquired secondarily. In lizards, body movements and limb movements both contribute to locomotion, but limb movements seem to be more independent, in some species completely independent, from the body movements. In walking crocodiles, locomotion depends primarily on the movements of the limbs, whereas axial movements, including the tail, are important during swimming.

The basic principle of motor control by means of CPGs has changed little during evolution, although there have been refinements. These consist of a richer assortment of neuroactive substances within the CPGs and superimposed systems, a richer repertoire of sensible influences allowing more flexibility of the pattern produced by a specific CPG, and changed patterns of interactions between CPGs. The most prominent changes of this system occurred during the transition from the undulatory swimming movements to locomotion by limbs in land vertebrates. In both ways of locomotion, the rhythmic pattern of movements is generated by the same type of CPGs. Dubbeldam (2001) assumes that the main change was an uncoupling of the CPGs of successive segments. He describes this by a comparison of the locomotor patterns in different groups of vertebrates, which enables him to specify what kind of changes may have occurred in the CPGs. He emphasizes that the transition was not a simple linear process, but that rather many branches exist (Fig. 8.2).

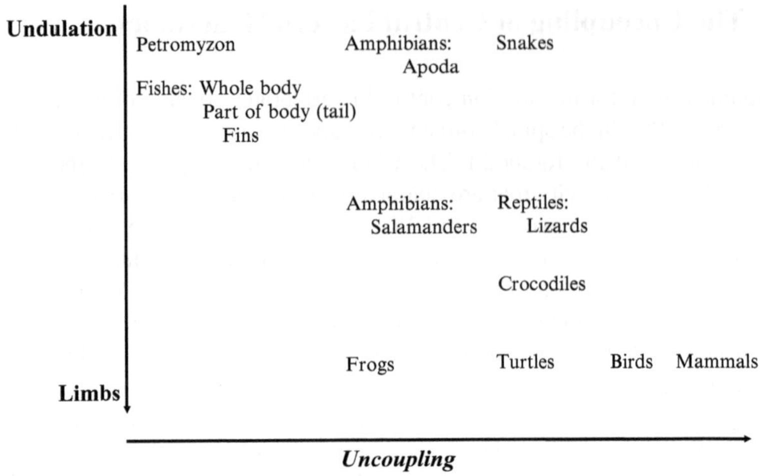


**Fig. 8.2** Two trends in the evolution of locomotion: the transition from undulatory movements of the body to locomotion by limbs (*vertical axis*) and the transition from strictly coupled to uncoupled activity of central pattern generators (*horizontal axis*) (Modified from Dubbeldam 2001)

To control movements of the body as well as those of limbs in land vertebrates, the CPGs must provide activation cycles that are more differentiated and that are less dependent on each other. In locomotion exclusively by limbs, only a small number of CPGs may be involved, uncoupled from those of other areas of the trunk.

Coupling between groups of CPGs of left and right fore- and hind legs might be responsible for the alternating patterns of movements of the respective legs. Derived from reptilian progenitors, birds and mammals developed different types of locomotion. In birds, the movements of forelimbs and hind limbs have been uncoupled, and the locomotion pattern of the forelimbs has changed from alternating to synchronous movements. In many birds, the hind limbs are also synchronized when hopping is favored over walking.

In most mammals, locomotion is performed with legs, which are placed under the body, resulting in a rich variety of different gaits. Most common is the gait with alternating use of the legs, but also nonalternating gaits are used in jumping and galloping. Movements of the body axis are dorsoventrally involved. In those mammals that mainly move by means of their four legs, the CPGs for locomotion are restricted basically to the lumbar and cervicothoracic segments. In this case, further uncoupling of the activity of different groups of muscles has occurred, allowing more differential actions of the limbs (Lacquaniti et al. 2012; Tresch et al. 2002). Each unit in a CPG network may control the activity of single muscles or of a group of muscles controlling the movements of one joint.

In many species, each limb can be used independently, for walking as well as for grasping, which some mammals develop. Whereas grasping is an integrated movement of a whole limb, further uncoupling of the activity of groups of muscles has occurred in primates, allowing finely tuned finger movements. These differentiated finger movements are under direct control of the motor cortex, overruling the CPGs.

Yet, the role of CPGs in locomotion of higher vertebrates is not clear in all details. However, for the present discussion, it is only important that the principle of evolutionary uncoupling of the system of CPGs exists. Dubbeldam (2001) assumes that such an uncoupling is not restricted to locomotion. Further rhythmical movements, such as jaw movements during feeding or vocalization in some species and breathing, are also under the control of CPGs and need to be uncoupled so finely tuned movements can be performed. The uncoupling of the CPGs enables greater flexibility and a richer repertoire of movement patterns. It introduces degrees of freedom for movement first by ensuring the basic rhythmic pattern of movement so that it is available without too much neuronal occupation of higher-level brain centers and second by generating the possibility to vary the movement pattern, thus becoming a neurophysiological basis for the widening of the behavioral repertoire, including play in some species.

## 8.2 Increasing Flexibility via Differentiation of the Motor System

Because fossilized animals provide only limited morphological data and unfortunately no details of neural organization and connectivity, many of the inferences about the pattern of motor organization in primitive vertebrates must be based on the comparative study of living vertebrates and their close living relatives. Knowledge of the spinal and supraspinal networks of extant anamniotic animals such as lampreys allows some speculation about the properties of the motor system of early vertebrates. According to Fetcho (1992), early vertebrates almost certainly had a myomeric axial musculature consisting of a series of muscle blocks (myomeres) lying along each side of the vertebral column. This arrangement is also evident in *Pikaia*, a fossilized animal from the Cambrian, which belongs to the earliest known chordates (Fig. 5.10). A fossil with comparable features is *Yunnanozoon* from even older layers of the middle Cambrian of China. Also, *Myllokunmingia* and *Haikouichthys* from the lower Cambrian of south China have some features of chordates (Conway Morris and Caron 2012; Shu et al. 1999; Chen et al. 1995). This myomeric musculature is also present in extant cartilaginous and bony fishes, salamanders, and anuran tadpoles.

Early vertebrates might have used this axial musculature and the motor units for bending the body during swimming. The spinal network, which initiated lateral undulations, may already have had excitatory and inhibitory interneurons, which were connected to each other. Mainly, two supraspinal systems might have activated the spinal networks: Reticulospinal neurons initiated the rhythmical movements for swimming, while Mauthner cells initiated fast movements during fleeing reactions (Table 8.1).

During further evolution, when the tetrapods also appeared, the myomeres were broken into a large number of different muscles that can be separately controlled by the nervous system (Fig. 8.3). Nearly all of the muscles in living vertebrates arose from the segmental axial musculature in early vertebrates. This includes the

**Table 8.1** Evolutionary changes in the ancestral motor system (From ten Donkelaar 2001, after Fetcho 1992)

| |
|---|
| Features of the ancestral motor system |
| Myomeric axial musculature |
| Two types of muscle fibers: nonspiking slow muscle fibers and spiking fast muscle fibers |
| Two types of motoneurons: small and large motoneurons innervating the slow and fast muscle fibers, respectively |
| A spinal network composed of at least two types of interneurons: an excitatory interneuron and an inhibitory commissural interneuron |
| Two major descending supraspinal systems to activate the spinal networks: reticulospinal neurons and, for rapid escape movements, Mauthner cells |
| Evolutionary changes in the motor apparatus |
| Substantial differentiation of the ancestral myomeric musculature into many individual axial muscles |
| Development of paired appendages and their associated musculature |
| Loss of segregation of muscle fiber types in many muscles |
| Development of a topographic map of the motor column onto the embryonic myotome |
| Development of muscle spindles and, in mammals, separate gamma motor innervation of muscle spindles |
| Development of an adequate neural control system for the steering of limb movements |
| Development of the corticospinal tract in mammals |

muscles of the fins and limbs, as well as diverse arrangements of axial muscles (Fetcho 1992). A topographic arrangement of the motoneurons in the spinal cord arose according to the location of the innervated muscles. There are no hints of such an arrangement in lampreys, bony fishes, or urodels. Thus, they were probably not present in early vertebrates and were generated during the transition to amniotes. Fetcho (1987) assumes that the generation of such a topographic map and of separate groups of motoneurons was important for the ability to activate the many different muscles, which originated from the respective myotome independently from each other, thus generating increasing degrees of freedom during movement.

It is likely that the progenitors of land vertebrates within the crossopterygian rhipidistians used their paired fins to "walk" on the ground, as can be observed in extant lungfishes today. The only living crossopterygian, *Latimeria chalumnae*, uses coordinated alternating movement of its paired fins for slow swimming without walking on the sea ground. The pattern of movement is comparable to that of tetrapods (Fricke et al. 1987).

Urodels move by a combination of lateral undulation of the trunk with protraction and retraction of the limbs, which sprawl to the sides of the body. The segments of the spinal cord are not yet able to coordinate the alternating movement of the pairs of limbs. The spinal segments of the forelegs and the hind legs are still in close connection to their respective neighboring segments and do not work independently from the undulation of the body axis. When urodels run, they change from a standing wave of undulation to a moving wave, which generates the main momentum, during which the extremities only serve for anchoring on the ground and are passively moved (Romer and Parsons 1977; ten Donkelaar 2001; Dubbeldam 2001).

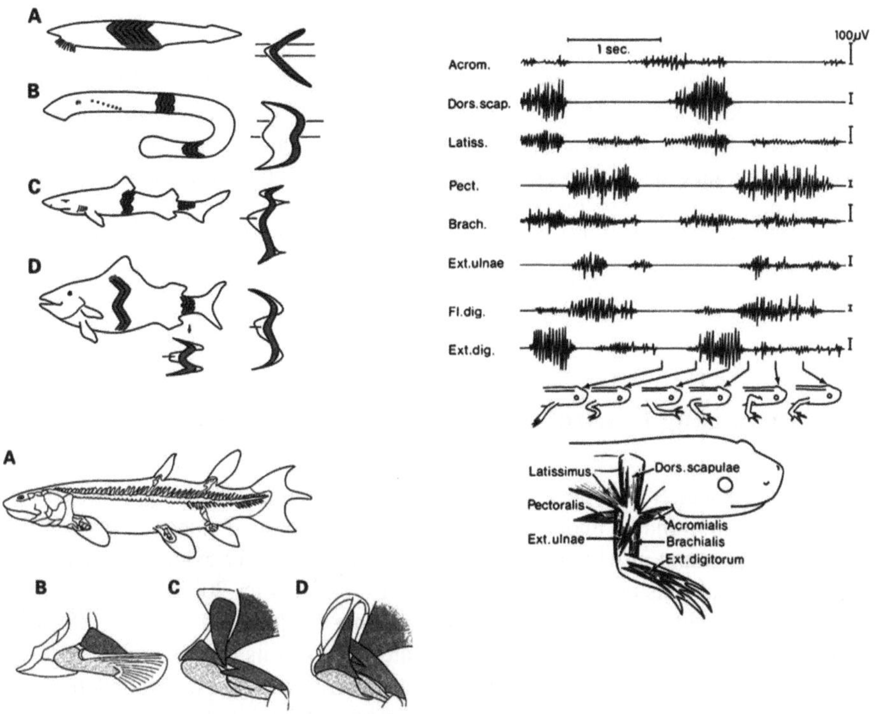


**Fig. 8.3** *Upper left*: The myomere pattern of fishlike chordates. *A* Amphioxus, *B Petromyzon*, *C* a cartilaginous fish, *D* a bony fish. *Bottom left*: *A* lobe-finned fish, representing a possible ancestor of the land vertebrates. *B–D* Dorsal (*dark gray*) and ventral (*light gray*) muscle masses and their derivatives in *B* the pectoral fin of a teleost; *C* the forelimb of a lizard; and *D* the forelimb of an opossum. *Right*: Activity pattern of eight forelimb muscles recorded electromyographically in the freely moving newt. The position of the limb is shown in different phases of the step, corresponding with the electromyographic data from muscles (Modified from ten Donkelaar 2001)

Nevertheless, the limb movement of urodels is complicated. Figure 8.3 shows the electromyographic recording of the patterns of activity of the forelimb of a newt. Antagonistic muscle pairs contract in both protraction and retraction. This contraction of antagonistic muscles is an important function to stabilize the joints and to regulate the movement. Similar antagonistic contractions were registered in *Varanus exanthematicus* and in the North American opossum *Didelphis virginiana* (Jenkins and Goslow 1983). A number of shoulder muscles of these two species are inserted in a way comparable to amphibians, and the latter demonstrate similar patterns of activity. Jenkins and Goslow (1983) interpret this similarity as a primitive pattern that was taken over from an early common ancestor of amniotes, which would already have had this general basis for the potentially flexible limb movements. With the generation of limbs, the neuronal control networks also developed, which made the highly differentiated coordination of the different muscles possible. Groups of interneurons are essential for the coordination between CPGs, central control, and peripheral feedback. They integrate locomotion into the behavioral context and modulate the coordination and length of stride.

The breakup of the myomeres into a large number of discrete axial muscles, as well as the development of paired fins and limbs and the associated limb muscles with their differentiated innervation in tetrapods, is not only the prerequisite for movement on land but also becomes the basis for an extension of the flexibility of movement, although not so much elaborated in amphibians and reptiles, but particularly in mammals and birds. This includes the use of limbs for other tasks than locomotion, such as reaching, grasping, and manipulation of food as in carnivores within mammals and many others.

## 8.3 Increasing Control of Movements via the Brain

The extent to which the spinal cord operates as an independent organ or, on the contrary, is subordinated under the instructions of the brain depends in part on the type of environment in which the animals live, but to a larger extent it depends on their evolutionary history (Butler and Hodos 1996; Striedter 2005). Generally, the spinal cord in nontetrapods (agnathans, cartilaginous fishes, and ray-finned fishes), which inhabit an aquatic environment, operates more on its own than in tetrapods. It tends to carry out many postural and locomotor activities more or less independent of the brain. Even if the spinal cord has been surgically isolated from the brain, the CPGs are still able to produce rhythmic locomotor movements of the body and of the fins. The coordination with sensory inputs of the brain also takes place in part on the level of the spinal cord, supported by structures of the brain stem as in the tectum, the cerebellum, and the reticular formation.

The superimposed systems in centers of the brain stem and the forebrain influence the activity of the CPGs and the motoneurons of various species in different amounts (Heldmaier and Neuweiler 2003; Grillner et al. 2008; Dickinson 2006). The most direct system in the brain stem, which is superimposed on the spinal cord, is the reticular formation, which is the most ancestral descendant motor system by which the vertebrate brain exerts control over movements in all classes from cyclostomes to mammals. In lampreys, the posterior rhombencephalic reticular nucleus is the major descending system of the spinal locomotor circuitry. Although the initiation of movements is increasingly transferred into the forebrain during evolution, the generation of locomotor patterns as well as the control of their execution remain tasks for the brain stem. Here, complete sequences of movement can be generated. In anamniotes as well as in amniotes, including mammals, reticulospinal cells exert their influence by direct contacts with motoneurons as well as indirect effects via contacts with promotor interneurons. The evolutionary changes introduced increases in the motoric flexibility and the variety of movement possibilities.

The reticular formation itself can be activated by rostral brain stem structures and by a variety of sensory stimuli. The "mesencephalic locomotor region" (MLR) is the next-higher control unit rostral from the reticular formation. It was first identified in decerebrated cats and was probably present in the earliest vertebrates, in which it turned on the swimming networks in the spinal cord via activation of reticulospinal pathways. An MLR has been described in representatives of many groups of vertebrates, including most anamniotes as well as mammals. However, it is not clear whether the MLR exists in all vertebrates. Together with the reticular formation, it forms the brain stem's locomotor region.

For the generation of differentiated control of movements, the rubrospinal tract, which originates from the nucleus ruber in the midbrain, and the cerebellum were of increasing importance (ten Donkelaar 2001). It is not present in agnathans and is also absent in many bony and cartilaginous fishes or is only small (Butler and Hodos 1996). Elements of a rubrospinal tract were not identified, for example, in the small-spotted catshark (*Scyliorhinus canicula*), but in the thornback ray (*Raja clavata*), in the common goldfish (*Carassius auratus*), in the African lungfish (*Protopterus amphibians*), in the European fire salamander (*Salamandra salamandra*), and in anurans (ten Donkelaar 2001). In dogfish sharks, there is a nucleus ruber, but it has no projections to the spinal cord (Butler and Hodos 1996). In contrast, no rubrospinal tract was found in some examined amphibians without legs. ten Donkelaar (2001) conjectures that the rubrospinal tract might be related to the evolution of paired limbs. Accordingly, for example, the thornback ray uses its enlarged pectoral fins for locomotion. Presumably, snakes have secondarily lost their rubrospinal tract.

Only in birds and mammals does the rubrospinal tract become prominent and thus essentially widens the capacity for voluntary movements. Their connections are basically the same in both classes (Butler and Hodos 1996). The tract begins with a large somatotopically organized nucleus ruber, which receives influences from the cortex and corrections from the cerebellum. The tract protrudes to areas of the spinal cord especially responsible for the distal musculature of the limbs and thus for goal-directed limb movements and grasping. Therefore, the rubrospinal tract influences the muscles of distal joints by way of direct motoneural connections. In primates and especially in human beings, the rubrospinal tract is reduced in favor of the pyramidal tract.

In mammals and in human beings, the basal ganglia are located beneath the cortex, lateral from the thalamus, and consist functionally of several nuclei. They contribute essentially to the generation of voluntary movement programs, the generation of time structures of movements and of their fine control. The cerebellum and the basal ganglia are involved in the fine-tuning of movements in mammals. These two centers of motoric programs are not connected to each other and work in parallel, which obviously is an effective system for dealing with the wide variety of motoric repertoires and for choosing one motor action among many possibilities. The cerebellum influences motor centers predominantly through excitation.

Several models exist for basal ganglia functions. One such model describes their functions predominantly in the connection and coordination of cortically induced movements. An alternative view sees their task more in filtering of the appropriate action from general motoric activity. General inhibition through the basal ganglia would thus only be released for motoric activities, which are initiated by the cortex. Accordingly, intentional movements would be promoted, whereas unintended and antagonistic muscular activities would be suppressed.

The largest input station of the basal ganglia, the striatum, collects inputs from the entire neocortex and sends processed information through other parts of the basal ganglia to areas of the frontal cortex, which are especially implicated in motoric planning and execution. In addition, the basal ganglia have a unique and highly complex circuit arrangement with multiple internal loops within themselves. By means of repetitive feedback cycles between basal ganglia and neocortex and the processing in the internal circuits, movement programs can be filtered and finely tuned. The basal ganglia are essentially responsible for the activation of only those muscles (of more than 600 in mammals) needed for highly differentiated intentional movements. The large number of movements possible for a limb has been called the "degrees-of-freedom problem in motor planning." The neural loops within the basal ganglia contribute essentially to selecting movements of appropriate muscles to perform a movement initiated by cortical areas. However, many details of this system are still not understood.

This refinement of motoric coordination evolved stepwise. Basal ganglia are found in all vertebrates, but in fishes and amphibians they are small, and cortical influences are sparse. In amniotes, the more numerous basal ganglia project increasingly to the cortex. In reptiles and birds, the control of the CPGs and the motoneurons is mainly organized via the striatum and connections to a number of descending spinal tracts. In mammals, the connections of the basal ganglia to the cortex increase. In primates and in human beings, the neocortex dominates the entrance and exit relationships of the basal ganglia (Dudel et al. 2001; Heldmaier and Neuweiler 2003; Neuweiler 2008).

In mammals, all these centers of movement control are increasingly subordinated to the influence of the neocortex (Fig. 8.4). Each consciously and intentionally initiated movement has its origin here. For this reason, the motoneurons are arranged in the neocortex according to the body parts they innervate. Basal ganglia and the cerebellum are connected with the neocortex by extensive nerve tracts and build separate feedback loops for control and estimation of the outcome of the planned actions from the neocortex. During brain evolution in primates, both parts grew simultaneously with the neocortex and are especially large in human beings (Neuweiler 2008).

Also in mammals, most standard movements can be carried out without the influence of the cortex, but they would be slower and automaton-like. However, voluntary movements and goal-directed movements, which need sensory guidance, are not possible without cortical control. This exemplifies that, with the enlargement of the cortex, especially the more flexible, goal-directed, and – in the meaning supported here – more autonomous movements become possible.

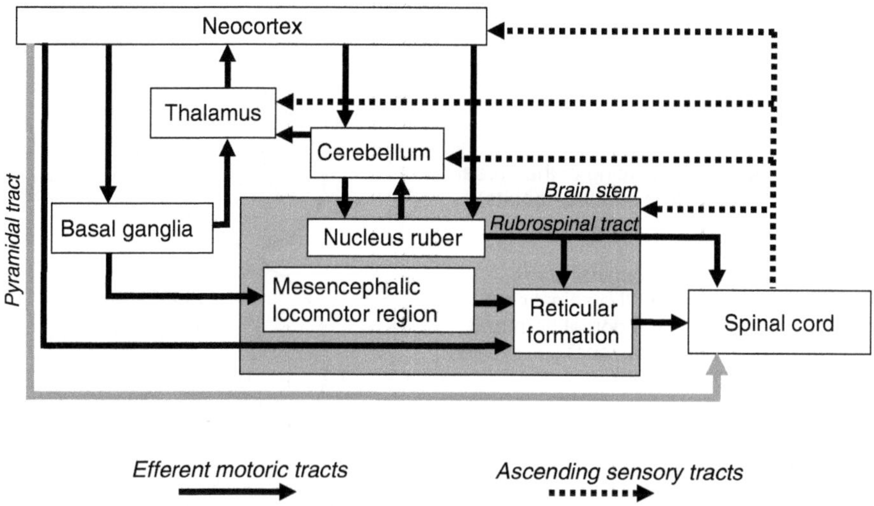


**Fig. 8.4** The most important motoric tracts and some of the sensory nerve tracts important during locomotion in the human being

## 8.4 The Pyramidal Tract

In mammals, the pyramidal tract is gradually added to the descending pathways (Heldmaier and Neuweiler 2003; Neuweiler 2008; Lemon 2008). It constitutes a direct connection from the cortex to the motoneurons, bypassing the motoric centers of the brain stem and the forebrain. Some fibers are added to the motoric brain nerves, which innervate the facial muscles and thus bring them increasingly under voluntary and flexible influences. Fibers of the long pyramidal tract control the motoneurons in the spinal cord, which innervate muscles of the trunk and limbs. In monotremes and marsupials, relatively primitive mammals, the tract is still small. Within the placental mammals, it is generally more developed, but it gains importance especially in primates. The evolution of the pyramidal tract is related to increased precision in the use of fingers, especially in apes, but also in raccoons and in many rodents, such as squirrels and hamsters.

Many diapsid reptiles and birds are able to use their toes in a goal-directed way, as seen in grasping by birds of prey. However, these capabilities are limited compared to those of a squirrel, which turns a nut to gnaw on it; of an ape, which turns a piece of food between two fingers; or of a human being, who plays the piano. This precision is mediated by the pyramidal tract and requires cooperation with an elaborated somatosensory system for the fine-tuning of movements to the immediate consequences of the actions (Butler and Hodos 1996; Heldmaier and Neuweiler 2003).

Mammals can be divided into four groups according to the terminal distribution of pyramidal fibers, which display an increasingly wider distribution area of cortical fibers in the spinal gray matter (Table 8.2). This corresponds to different degrees of

**Table 8.2** The corticospinal tract: patterns of spinal termination (Modified from Kuypers 1981; ten Donkelaar 2001)

Group 1
Mammals with corticospinal fibers extending only to cervical or midthoracic segments and terminating in the dorsal horn (opossum, tree shrews, rabbits)

Group 2
Mammals with corticospinal fibers extending throughout the spinal cord and terminating in the dorsal horn and intermediate zone (rat, most carnivores)

Group 3
Mammals with corticospinal fibers extending throughout the spinal cord and terminating in the dorsal horn, intermediate zone, and dorsolateral parts of the lateral motoneuronal cell groups (raccoon, primates including Saimiri, Galago, and rhesus monkeys)

Group 4
Mammals with corticospinal fibers extending throughout the spinal cord and terminating in the dorsal horn, intermdiate zone, and both dorsolateral and ventromedial parts of the lateral motoneuronal cell groups (Cebus, apes such as chimpanzee, humans)

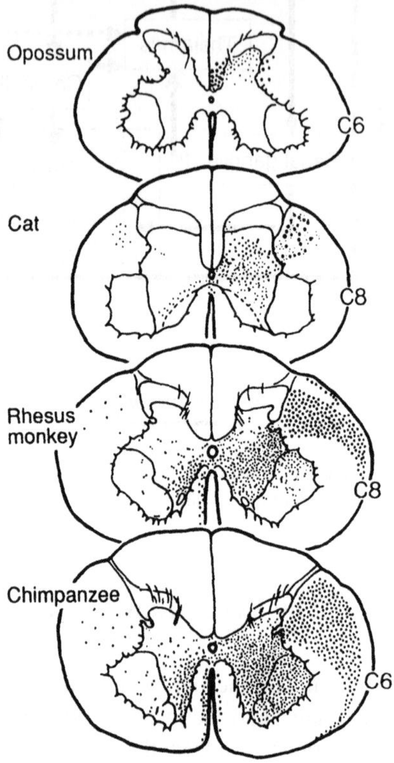


dexterity and independent voluntary limb movements. Direct corticomotoneuronal connections to motoneurons innervating hand and finger muscles are only found in primates and a few carnivores, such as raccoons and kinkajous. Heffner and Masterton (1983) developed an index for the dexterity of the forelimb and studied 69 species of mammals with it (Table 8.3). They demonstrated that the size and the range of the pyramidal tract in the spinal cord correlate with the dexterity of the species.

In primates and in some few carnivores, as for example in raccoons, such connections directly end on the motoneurons, which innervate muscles of the hands and fingers. It is assumed that this derived characteristic appeared independently in primates and carnivores in relation to the formation of their manual dexterity. However, direct corticomotoneuronal projections are not restricted to motoneurons of the forelimbs. In many apes, the motocortex also projects to motoneurons of the hind legs, whereas in spider monkeys (*Ateles*) and woolly monkeys (*Lagothrix*), it also

**Table 8.3** Index of dexterity in different mammals (After Heffner and Masterton 1983, from ten Donkelaar 2001)

| Function | Dexterity index | Digit type | Species |
|---|---|---|---|
| Specialization for locomotion | 1 | Fused or restrained digits | Ungulates |
| | 2 | Separate digits that do not converge when flexed | Hedgehogs, most carnivores |
| Simple hand | 3 | Convergent but not prehensile (not capable of holding an object in one hand) | Opossums, tree shrews |
| Specialization for manipulation | 4 | Prehensile digits, nonopposable thumb | Tarsiers |
| | 5 | Prehensile digits, pseudo-opposable thumb | Lemurs |
| | 6 | Opposable thumb, capable of power and limited precision grips | Old World monkeys, apes |
| | 7 | Opposable thumb, capable of precision grip in opposition to each finger | Humans |

extends to motoneurons that innervate the muscles of the prehensile tail (Kuypers 1981; Petras 1968; Heffner and Masterton 1983).

In birds, a different neuronal solution than in mammals developed for a differentiated manipulation of objects. Here, the beak and the hind limbs take over some functions that are comparable to those of the forelimbs in some mammals and the hands of primates, although they do not reach the same precision and versatility. Birds of prey, ravens, and parrots are well known for fine manipulation of objects in their environment. This is typical parallel evolution (convergence) of a feature with not only a different phylogenetic history but also increased degrees of autonomy generated.

In contrast to the elaborate somatosensory and motoric systems of mammals, those of other amniotes seem to be developed only modestly (Butler and Hodos 1996). Because of the influence of the cortex, a larger scope for the initiation of voluntary movements arises. Thus, birds and especially mammals developed versatile use of their extremities, which in many cases goes well beyond the pure functions of movement. In human beings, the forelimbs are completely emancipated from locomotion.

However, with the growing number and complexity of behavioral patterns, many movements are programmed as retrievable patterns, which run largely on their own. This applies for innate as well as for learned movements. The basal ganglia in the forebrain and the cerebellum are responsible for the generation of such programs and the supervision of their realization. This frees the higher centers of the brain from the burden of working out all movements in every detail and preserves their capacities for further flexible tasks. The cortex enables mainly the conscious initiation of movements, the generation of new combinations of movements, and the solution of special tasks such as walking on a ladder or a beam, which require sight control as well.

Human beings have a large range of goal-directed actions, which reflects not only our cognitive abilities but also the high plasticity of movements. This includes, for example, the performance of completely new movements. It is enabled through the hierarchical system of reflexes, rhythmical patterns, and voluntary movements. Combing one's hair, driving a car, or preparing a gourmet dinner are voluntary movements that are complicated and predominantly learned. The ability to learn movement patterns increased during evolution.

In particular, the more precise movements must be learned under continuous control and correction. Even walking and precise grasping must laboriously be learned by children. To the extent that actions are permanently mastered and do not need further corrections, the programs run unconsciously, so that consciousness can be directed intentionally toward different things. Thus, we have no difficulty in having a philosophical discussion while we are walking and at the same time smoking a cigarette.

Compared to cognitive abilities, motoric intelligence is usually underestimated. In this area, considerable differences exist between human beings and animals (Heldmaier and Neuweiler 2003; Neuweiler 2008; Gibson 2002). Human beings, for example, are able to throw a ball precisely into a net. Here, the complicated time structure of a multitude of exactly processed muscle actions needs to be planned in advance by the brain. Not even apes possess such capacities for this type of precise and "intelligent" behavior. They hit a nail with a hammer only after many attempts, and they can throw a stone or a branch only in a vague direction. Beyond this, flexible linkage of complex sequences of muscle activities is also the prerequisite for speech. Thus, the evolution of complex somatosensory and motoric systems exhibits a tremendous increase in flexibility, self-determination, and degrees of freedom and thus meets the criteria of increasing autonomy.

In summary, it can be stated that the evolution of nervous systems introduced the neuronal prerequisites for increasing possibilities of flexibility and self-determination and thus matches features of Definition 2. This concerns interactive autonomy, as the possibilities to act and react in the environment are elaborated, as well as constitutive autonomy, as the amount of internal neuronal processing is increased. The principles discussed are as follows:

- The formation of nervous systems in general, which expand the capacities for indirect and modulated reactions to environmental situations. An uncoupling of signals and reactions is introduced so that there are increasing abilities to process information before reactions will follow.
- The generation of complex centralized ganglia and brains, which enlarge the capacities of neuronal processing and correlate with widened behavioral flexibilities.

(continued)

(continued)

- The correlation of nervous systems that are more complex and with sophisticated movement capacities.
- The uncoupling of CPGs in the spinal cord, enabling greater flexibility and a richer repertoire of movement patterns.
- The increasing possibilities for movement through differentiations of the motor system.
- The evolution of sophisticated neuronal control systems for motoric actions.
- The evolution of the pyramidal tract for precise limb and hand movements and its correspondence to different degrees of independent voluntary limb movements and dexterity.

As long as locomotory movements are initiated by independently working motoric units of the spinal cord with some control from the brain stem, these movements are relatively stereotypic and less flexible. With the transition to locomotion on land and with control through higher brain centers, increasing degrees of freedom have been achieved. The modular structure of the control systems consisting of partly independent circuits as well as the control and modulation by superimposed brain centers gave the system even further degrees of flexibility.

# Chapter 9
# Endothermy

Most birds and mammals continuously maintain high rates of endogenous heat production (endothermy), combined with the regulative capability of keeping their body temperature within narrow limits (homeothermy) even when ambient temperatures are fluctuating. These homeothermic endotherms have high basal metabolic rates and levels of daily energy expenditure, which require a high food acquisition rate (Table 9.1). In contrast, vertebrates such as amphibians and reptiles are predominantly dependent on ambient temperature (ectothermic) and have significantly lower energy needs. Partial endothermy has arisen in a number of species, among which are some sharks, tunas, reptiles, and even some insects.

Origin and generation of endothermy in mammals and birds are a challenge for evolutionary biology, as it is still not clear which forces initiated and selected this "wasteful strategy of energy use" (Pough 1980; Koteja 2000; Clavijo-Baque and Bozinovic 2012; Clarke and Pörtner 2010). In terms of adaptation and efficiency considerations, it even becomes a paradox that such an energetically expensive lifestyle became so successful.

The extensive modifications that are needed to reach endothermic physiology took place independently in the avian and mammalian lineages. However, it has remained difficult to determine when each of these lineages attained endothermic status (Nespolo et al. 2011; Clarke and Pörtner 2010).

Hillenius and Ruben (2004) state that there is some evidence coming from the presence of respiratory turbinates within the nasal cavity of mammalian precursors that sheds light on the history of endothermy. These respiratory turbinates are important in endotherms for the retention of heat and moisture from the permanently high breathing volumes and thus are indicative of elevated resting metabolic rates. They occurred in early synapsids during Late Permian times. Attachment scars of respiratory turbinates in fossils of two lineages of Late Permian/Early Triassic synapsids, the therocephalians and the cynodonts, suggest that resting metabolic rates were gradually expanding in these taxa and that endotherm-like metabolic rates may have been attained some 30–40 million years before the appearance of the earliest "true" mammals.

B. Rosslenbroich, *On the Origin of Autonomy: A New Look at the Major Transitions in Evolution*, History, Philosophy and Theory of the Life Sciences 5,
DOI 10.1007/978-3-319-04141-4_9, © Springer International Publishing Switzerland 2014

**Table 9.1** Aspects characterizing typical endothermic temperature physiology

| |
|---|
| The basal or resting metabolic rate is high. It varies typically from five to ten times that of an ectotherm of similar body size |
| The body temperature is higher than the animal's normal ambient temperature and lies between about 28 and 40 °C in mammals and between 38 and 42 °C in birds, depending on species |
| The core body temperature is maintained at a constant value. In mammals, it normally varies not more than 1–2 °C during the daily cycle, whereas birds can be a bit more flexible, depending on species and physiological status. Exceptions are hibernation and torpor |
| Main organs of internalized heat production are brain, liver, heart, kidneys, and gut |
| The maximum aerobic metabolic rate (ability to generate energy aerobically, mainly during movement) that the organism is capable of sustaining is greatly elevated over that of ectotherms. There is an approximately constant ratio of 5–10 between basal rate and maximum aerobic rate in ectotherms as well as in endotherms. As the basal rate in endotherms is generally higher, the maximum aerobic rate greatly exceeds that of ectotherms |

Within the bird lineage, although less clearly detectable, turbinals seem to appear first in mesozoic ornithurine birds, implying that increased ventilation rates were reached relatively late and were not present in the earliest known groups of birds. Also, some postcranial features that are associated with the avian lung/air sac system appear late. Hillenius and Ruben (2004) assume that avian endothermy evolved well after the initial appearance of flight. Radiation in the Early Cretaceous period thus would have been primarily birds that did not yet have the full metabolic capacity and endurance for long-distance flight associated with modern birds (Table 9.1).

The following chapter includes a brief overview in evolutionary biology on how and why the complex of endothermy might have evolved. The overview exemplifies that often single causes are sought for evolutionary events, but that evidently a more synthetic systems view is needed to develop appropriate theories. This is apparent for the transition from ectothermy to endothermy. Then, it is demonstrated that this functional complex has a common denominator, which is the creation of an increased autonomy and environmental independence.

## 9.1 Theories of the Evolution of Endothermy

Endothermy is thought to have developed independently in birds and mammals, even though similar factors for its generation have been considered. An older theory centers on selection for enhanced thermoregulatory capacities per se. It is argued that a high and stable body temperature increases metabolic efficiency through thermal effects on enzymatic processes (Cowels 1958; McNab 1978; Hopson 1973). Another theory, the "thermal niche expansion model" (Hayes and Garland 1995; Kemp 2006), connects homeothermy with the capability of animals to be active in thermal niches from which they had previously been excluded. Such a thermal niche could be nocturnal activity. Heath (1968) proposes a model for the evolution of endothermy in mammals in connection with the development of an erect, not sprawling, stance. This erect stance would require greater muscle tone, which would result in increased metabolic rates and consequently endothermy.

Another scenario, which is preferred by most authors today, suggests that the initial factor in this transition was a selection in favor of increased capacities for sustainable, aerobically supported movements. This "aerobic capacity model" consists of two main parts: (1) The increased activity capacity is gained through higher rates of aerobiosis during movement, as present in modern mammals and birds; and (2) maximal and resting levels of metabolism have a constant linkage, so that they may not change independently from one another. As a high resting metabolic rate is itself connected to endothermy, the endothermic condition might have developed together with the changes in movement capacities (Hayes 2010; Bennett 1991; Ruben 1996; Hillenius and Ruben 2004; Clarke and Pörtner 2010; Nespolo et al. 2011; Clavijo-Baque and Bozinovic 2012).

Alternative models propose a primary role for parental care in mammalian and avian evolution. Farmer (2000) expects endothermy to have evolved as a means to provide parents with the ability to control incubation temperatures for their offspring. Koteja (2000) saw the starting point in the selection in favor of increased parental effort, in particular feeding offspring, generating an increased total energy expenditure and requirement. He describes this as a complex of enhanced parental care with the necessary increased locomotor capacity, rates of food consumption, energy expenditure, and increased basal metabolic rates as a self-reinforcing evolutionary process, which he calls the "assimilation capacity model."

McNab (1978) postulates a two-step scenario in which large carnivorous reptilian ancestors of mammals first gained inertial endothermy because of their mass, thereby developing additional features to stabilize body temperature. During a decrease in size, the endothermic capacity would then have been preserved by an increase in the mass-specific resting metabolic rate. Schweitzer and Marshall (2001) present a model for the evolution of endothermy in birds, which starts with a mutation leading to increased oxygen affinity of the hemoglobin molecule in a common ancestor. This is followed by a cascade of changes, such as multiplication of mitochondria in the tissues, increased adenosine triphosphate (ATP) production, generation of waste heat, and several others, including increased capabilities for sustained activity.

Each of these scenarios has various problems and shortcomings. For example, it has been argued that small increases in resting metabolic rate, as they might have occurred in the beginning of a shift to endothermy, would not enable animals to elevate their body temperature significantly above that of the environment, and the additional energy expenditure would outweigh the benefits to be gained. This applies to the model of enhanced thermoregulatory capacities as well as to the thermal niche expansion model (Hayes 2010; Pough 1980; Bennett and Ruben 1979). The postural change model is problematic because there is no evidence for greater energy costs for erect limb posture compared to sprawling posture (Hayes and Garland 1995).

The aerobic capacity model has the problem that no necessary physiological link is known between resting and activity metabolism. The relation of a five- to tenfold increase in aerobic capacity above resting levels during movement represents a mean value, which hides considerable variation in different species. Thus, the related thermal features also have only an indirect connection to activity capacities (Hayes and Garland 1995; Hayes 2010; Swanson et al. 2012). Considerable deviations were found in some varanids as well as in some snakes, which exhibited enhanced oxygen capacities during exercise without elevated resting metabolism (Bennett 1972; Gleeson et al. 1980; Ruben 1976). Mammals have a sixfold increase in mean capacity, but in different species, it can range between less than twofold in sloths and elevenfold or more in shrews (Taigen 1983). The naked mole rat (*Heterocephalus glaber*) has a low basal metabolic rate and normal aerobic capacity during movement (McNab 1966). Anuran amphibians exhibit even larger variations, with values between sixfold and twentysevenfold or more (Taigen 1983; Taigen et al. 1982). These exceptions do not question some general linkage of resting and maximal rates of metabolism, but they do show that there seems to be no strong cause-and-effect relation. In addition, the maximum aerobic metabolism depends on muscular work, whereas the basal metabolic rate results primarily from the metabolism of some organs (Koteja 2000).

If increased aerobic capacity during activity was the single main factor preceding others, a more efficient solution would have been to expand the aerobic capacity during movement without the costs of increased basic metabolism. Thus, the animals could benefit from both increased endurance and low maintenance costs. From an energetic point of view, such a hybrid type of physiology could have been superior to both ectothermic and endothermic strategies of energy use, so that selection in favor of high aerobic capacity alone would be unlikely to support wasteful endothermic energetics (Wuethrich 1999; Koteja 2000; Bennett 1991). The physiologic possibility of such a solution cannot be excluded because the variations in some species show that there seem to be different degrees of correlations. Likewise, behavioral thermoregulation of high body temperatures, as in many extant reptiles, could have led to the benefits of a high and relatively constant temperature without increased metabolic costs (Bennett 1991).

In addition, the generation of external insulatory layers, which are necessary especially for small endotherms, cannot be inferred as a single initial factor because insulation alone does not result in significant homeothermy and there is no physiologic reason why it would have been followed by internal heat production (Ruben 1995; Bennett 1991). Some data suggest that the development of fur or feathers was independent from the development of endothermy in either group (Ruben et al. 2003). The parental care models (Farmer 2000; Koteja 2000) as well as the model of increased oxygen affinity of hemoglobin (Schweitzer and Marshall 2001) are speculative. Both features seem rather to be linked to the whole endothermy complex. However, there is no current way to decide which of these factors was the initial one, which followed, and what evolutionary role they played within the emergent properties of the whole physiological process.

## 9.2 The Functional Complex

Although some authors see various factors in connection with each other, the main problem with these different scenarios is that they focus on single factors, isolated from the whole complex, that set off the shift to endothermy. Sometimes, it seems to

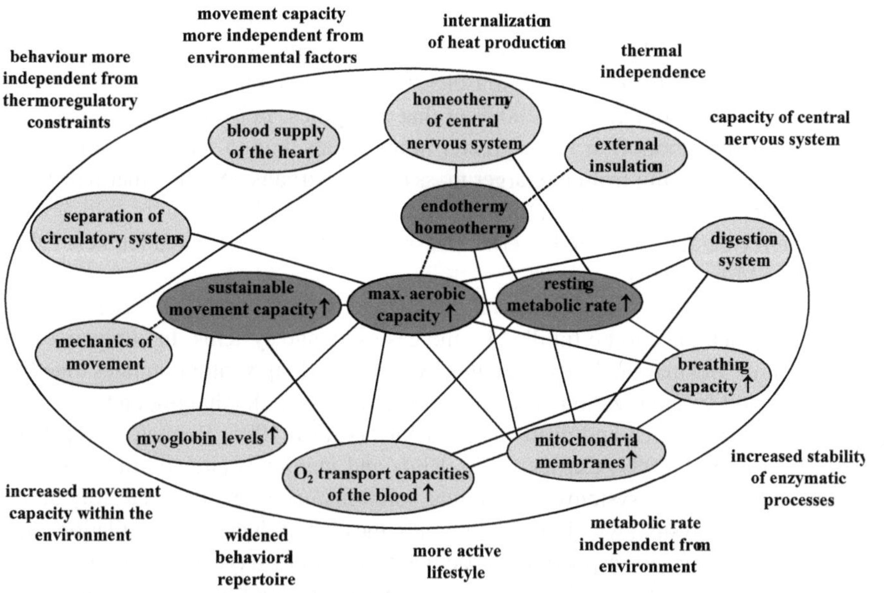


**Fig. 9.1** Graphic representation of the functional complex view. The characters at the outside of the *oval line* each indicate a feature of autonomy

depend on the respective viewpoint regarding which factor is given priority. However, this difficulty does not necessarily render the hypotheses useless but rather shows the insufficiency of single-cause explanations. Each approach may reveal elements of the whole complex. However, a more synthetic view of the various factors involved in this transition may deliver a picture that is more compelling.

Figure 9.1 summarizes the main evolutionary changes that occurred during the transition from reptilian physiology to the physiology of mammals and birds. The lines indicate interrelationships as found and usually described in comparative physiology of extant animals. In today's animals, they are elements of a complex system of interrelated and interdependent features. However, this may only be a rudimentary image of the real interdependencies. Nevertheless, the meshwork gives an illustration of the whole jigsaw puzzle as a synthetic functional complex.

Reptiles have a limited capacity for aerobically supported movements. Whereas slow movements are supported by aerobic metabolism, at higher levels of activity 50–98 % of the energy comes from anaerobic metabolism, leading to fast exhaustion, long recovery periods, and limited movement abilities. High rates of exercise can be maintained only for about 2–5 min. Recovery can take up to several hours. One evolutionary innovation of endothermic mammals and birds is the greatly increased capacity for aerobically supported movements. This resulted in substantially increased routine activity levels, stamina, and endurance (see center of Fig. 9.1). The physiological relation between maximal aerobic capacity and movement capacities is strong and cannot be separated because the former is the physiological basis for the latter. Likewise, the relation to the total amount of mitochondrial membrane surface area is equally strong, as this is the structural basis of aerobic capacity. In mammalian skeletal and cardiac muscles, the respective mitochondrial membrane surface areas are about 220 and 290 % of that of reptiles. In liver, kidney, heart, and brain tissue, it is about four times greater in mammals than in reptiles, with half of the difference deriving from the larger mass of these organs and the other half from greater mitochondrial density in these tissues. In addition, there is more cytochrome oxidase enzyme activity in endotherms (Nespolo et al. 2011; Else and Hulbert 1985; Hulbert and Else 1989).

Every increased in mitochondrial membrane surface area increases oxygen demands, which have to be fulfilled by the gas exchange systems. In comparison to reptiles, birds and mammals achieved this by increased lung ventilation rates together with expanded vascularization of the lung, integration of a diaphragm, and increased pulmonary surface area and diffusion capacities (Ruben 1995; Bennett 1991; Perry 1985; Ruben et al. 1987; Carrier 1987). Birds and mammals attained the increased effectiveness of these systems by different solutions, using either a one-way aerial flow system or a system with blind ending sacs, the alveoli. Increased ventilation rates are useless without corresponding capacities to transport oxygen in the blood and delivery systems within the tissues. Thus, increased oxygen transport abilities of the blood, as well as increased vascularization of oxygen-consuming tissues, myoglobin levels, and oxygen extraction rates can be expected to be closely related to increased aerobic capacities (Schweitzer and Marshall 2001; Pough 1980; Ruben et al. 2003).

The complete structural separation of the cardiac ventricles occurred during this transition, and maximal cardiac output has increased greatly in mammals and birds. This is the basis for the establishment of the high arterial pressure system, again contributing to the high demands of aerobically working tissues. As this requires an efficient and powerful heart muscle with compactly arranged myocytes, the blood supply of the heart is now completely provided by coronary circulation.

Less clear is the relation between maximal aerobic capacities and resting metabolic rates, as discussed previously. An independent process of generation during evolution cannot be excluded. However, in comparative physiological studies of extant animals, some constant relation has regularly been found, and as the high resting metabolic rates again need all the other factors, there are distinct relations within the network. A mildly elevated basic metabolic rate together with endothermy would require increased ATP production with all the consequences. Thus, endothermy, with its strong relation to the basic metabolic rate, is also integrated into the network relations, as represented in Fig. 9.1. Another strong relation can be expected between these features and increased capacities of the alimentary tract and other visceral organs (i.e., the liver, kidneys, and heart), as significantly more energy is needed by mitochondria for ATP production (Pough 1980; Koteja 2000; Karasov and Diamond 1985).

Fur and feathers, and in aquatic mammals subcutaneous fat deposits, considerably reduce rates of conductive heat loss, thus showing a clear physiological relation to endothermic physiology in extant animals. However, this correlation in evolutionary origin is less distinct. Ruben and Jones (2000) argue on an uncorrelated origin of elevated metabolic rates and external insulation layers. Their reconstruction reveals that a complex insulative fur coat probably appeared first in the earliest mammals long after some therapsids had attained near-mammalian metabolic rates.

General rearrangements of skeletal elements led to two different solutions for movement characteristics in the avian and the mammalian lineage. However, both exhibit the mechanical basis for greatly enhanced possibilities for vigorous and wide-ranging movements. Movement by erect limb posture within the mammalian line, as well as the ability to fly within the ancestry of birds, may not have been correlated with increased aerobic capacities, but the use of these mechanical innovations for long-ranging movements in time and space again depends on appropriate energy supplies. Track evidence of erect posture, running, and even climbing sand dunes also indicates high activity levels among synapsids in Permian times (Kemp 2005; Lockley and Hunt 1995).

## 9.3 A Functional Complex for Autonomy

If the described physiological and morphological elements evolved as a functional complex with various correlations and interconnections, it is logical to ask whether a common general trait integrates or underlies this fundamental functional shift. What is the general difference between low- and high-energy lifestyles? The answer might be found in reconsideration of the various effects of the functional complex on the organism/environment relationship and the lifestyle of animals when comparing ectotherms and endotherms (characters at the outside of the oval line in Fig. 9.1).

First and basically, endothermy increases the thermal and metabolic independence from temporal fluctuations within the environment. Whereas metabolic processes and rates in ectotherms are predominantly affected by changes in ambient temperatures, endotherms establish homeostasis that is internally stabilized and controlled and thus is more autonomous than in ectotherms. This independence results in the potential that many temperature-sensitive physiological processes proceed at relatively stable rates over a wide range of ambient temperatures and allow fine-tuning of metabolism (Nespolo et al. 2011; Clarke and Pörtner 2010).

Many reptiles can increase their body temperatures above ambient levels to carry on some of their biological activities, including high performance during movement. However, they obtain large amounts of the necessary energy from outside their bodies, thus depending on external sources. In contrast, birds and mammals internally generate most of the energy needed for thermoregulation. The source of the energy used to regulate body temperature is a significantly more trenchant difference between the groups than are the body temperatures maintained (Pough 1980). The internalization of the energy source provides emancipation from external sources and temperature factors and makes warmth available more constantly. Movement capacity becomes independent from local temperatures as an ecological factor. Thus, the potential for movement within the environment is in principle always available. Endothermy allows the animal to be "instantly" ready for additional high-rate activity.

The substantial increase of sustainable movement capacities serves a greater potential for activities. A reptilian-type physiology is well able to support fast movements for a short duration at speeds comparable to fast endotherms. However, physiological restrictions limit the availability of these movements in time as well as in range. In mammals and birds, the availability of movements is largely expanded with significantly longer periods of intense activity, including prolonged nocturnal activity. Large-range and long-duration movements, including migrations, become possible for the individual in many species. In summary, this leads to a more active lifestyle (Ruben et al. 1987) with expanded degrees of freedom.

The behavior patterns of many reptiles and amphibians seem to reflect their physiological characteristics. Thus, a great deal of behavior possible at higher body temperatures is not available at lower ones (Hertz et al. 1988; Bennett 1989). On the contrary, the more or less continuously available activity capacity generated by the functional complex of endothermy allows the generation of behavioral and social repertoires that are more complex.

Some ectothermic organisms, such as many lizards, employ behavioral thermoregulation through basking for solar radiation utilization and microhabitat selection. Thus, their behavior is constrained by the necessary movements within thermally suitable zones of their habitat, which are in turn constrained by many factors, such as daily temperature patterns. The behavior of endotherms is to some extent freed from these regulatory restrictions.

By far the majority of lizards and snakes are restricted to microhabitats with burrows, bushes, or rocks, where they can hide. After phases of high anaerobic energy production, they can be vulnerable for a long time, until the oxygen debt is repaid. Thus, the hiding place is essential for survival. Aerobic movement physiology combined with independence from local thermal constraints overcome these restrictions to various degrees and lead to enhanced ecological possibilities, including access to new environments and ecological niches or to the maintenance of large home ranges. However, because many small mammals also prefer certain microhabitats, body size plays a crucial role and is an additional factor in environmental emancipation. This is demonstrated by large ectotherms, which may expand their environmental emancipation, as has been found in species of *Varanus* (McNab and Auffenberg 1976; Bennett 1972).

These different and profound changes are all elements of the establishment of a more autonomous lifestyle for the individual endothermic organism. This is the general difference between low-energy and high-energy lifestyles and includes an extension in energy intensiveness in the sense of Vermeij (1999) and Milewski and Mills (2010).

The ectotherm low-energy lifestyle has various advantages. It can work in a wide range of body sizes and usually has less need for water, oxygen, and food supply. This works well in a range of environments and microniches. Ectotherms can use a larger proportion of their energy budget to produce new biomass and to reproduce, with less investment in day-to-day maintenance (Hayes 2010; Willmer et al. 2000; Pough 1980). In nature, lizards have annual energy budgets that are only about a 40th of those of birds and mammals of equivalent size, and reptiles can often survive long periods without energy input. However, the endothermic lifestyle provides new flexibility for the individual within the possible ecological systems and significantly increased physiological and behavioral plasticity. As this can be described in many details, as summarized previously, this gain in autonomy is a factor in the evolution of endothermy and needs to be considered among other factors, such as adaptation or energy budgets and the like.

The proposed autonomy concept does not contradict the other scenarios cited. The synthesis in the functional complex view (Fig. 9.1) brings together various different viewpoints. The possibility of such a synthetic view has occasionally been indicated previously. Bennett (1991, p. 16) concedes that a combination of factors might have been involved simultaneously or sequentially and that "unifactorial explanations for the evolution of any complex character are almost certainly incomplete." Schweitzer and Marshall (2001, p. 318) suppose that endothermy is an "'emergent' trait, deriving from higher-level organismal integration," and that discrete evolutionary steps must have been connected to and dependent on one another, generating "modular units." Taigen (1983, p. 106) saw the elements involved as "parts of an integrated complex" and states that "evolutionary models that consider only one character in isolation from the others may be of limited utility." Also, Clarke and Pörtner (2010) favor an integrated package. The recent phylogenetic analysis by Lovegrove (2012) also assumes an integrative view. He calls it macrophysiological modeling and examines different possible stages in the evolution of endothermy during the Cenozoic.

Kemp (2006) proposes with his "correlated progression model" the most consequent attempt to regard the interdependencies within the system. He states that the characteristics of an organism are highly integrated with one another, so that no single one can evolve by more than a small increment at any time. Therefore, over the course of the evolution of new features, all the structures and associated functions must evolve by respective sequences of small steps in loose correlation with each other to maintain continuous functional integration. According to Kemp (p. 480): "Under this model, explaining the course of evolution of any particular case of megaevolution, such as the origin of endotherms, becomes a matter of understanding the nature of the integration between all the structures and functions involved in the evolutionary transition, rather than identifying one particular structure or function as paramount over the rest as far as selection is concerned."

However, understanding the nature of the integration also requires perceiving the essential characteristic that emerges from all these profound changes. Thus, my proposal is basically similar to the cited integrative views but focuses more on the general result as the unifying principle than on the question of how tight or loose the single functions were connected during evolution. I think that the relation of these functions to each other concerns a subsequential question after first comprehending what the organisms gained with endothermy.

During evolution, three strategies developed to gain stabilization of body temperatures together with widened movement capacities. One way was temperature regulation by behavior, as it is widely found in modern reptiles. Another was by thermal inertia. In large reptiles, reduced surface-to-volume ratios and larger body masses reduce heat loss and allow some stabilization of body temperatures (Gillooly et al. 2006; Ruben et al. 2003; O'Connor and Dodson 1999; Spotila et al. 1991).

This is in principle found within the largest extant reptiles, such as *Varanus komodoensis* (McNab and Auffenberg 1976), crocodiles (Seebacher et al. 1999), and leatherback turtles (Paladino et al. 1990).

Ruben et al. (2003) present arguments for a possible "inertial homeothermy" in dinosaurs while maintaining reptile-like, ectothermic metabolic rates. Large dinosaurs are expected to have had high body temperatures and might have been relatively unaffected by diurnal temperature fluctuations. Smaller ones might have relied on a combination with behavioral regulations. This would have enabled many dinosaurs to have rather active lifestyles and to maintain large home ranges. Both principles contain certain degrees of thermal autonomy and independence of movement capacities (remember that dinosaurs also developed a way to have their limbs more vertically under their bodies; see Chap. 5). These two types of physiology were different evolutionary attempts to gain physiological emancipation from environmental restrictions. However, the third way included further stabilization of temperature regulation together with the internalization of heat production and thus increased the degree of autonomization. As different types of physiology attempted to reach similar results, it can be assumed that increasing autonomy was a central factor. If this assumption is right, one may speculate that the generation of thermal inertia by extremely high body masses was a blind alley because at the same time it restricted movement capacities by creating sluggishness that was counterproductive or simply by putting some inherent physical limit on size and mass. Indeed, such size/mass limitations are often discussed in connection with dinosaurs – and because these animals included the largest land vertebrates known, one can argue that they are the "empirical" expression of a natural limit.

Although homeothermy has often been cited as an example for homeostatic autonomy and independence (e.g., Bernard 1859, 1878; Bertalanffy 1949; Beurlen 1949; Huxley 1974; Kipp 1948; McNab 1978; McNab and Auffenberg 1976; Rensch 1959; Schad 1992, 1997; Starck 1978; Waddington 1961; Wake 1986; Clarke and Pörtner 2010), the full relevance of the trait for evolution on the whole has not yet been recognized, and the relationship to the different factors of the functional complex has not been taken into account. However, because the feature is clearly describable, it must have consequences for organism/environment relations.

In summary, the general difference between low-energy and high-energy physiologies, as it is established in endotherm animals, corresponds to the criteria of Definition 2:

- Endothermy establishes a relative thermal and metabolic independence from temporal fluctuations within the environment.
- Endothermy establishes homeostasis that is internally stabilized and controlled, so that many temperature-sensitive physiological processes proceed at relatively stable rates.

(continued)

(continued)

- The internalization of energy sources and heat production emancipates from external sources and temperature factors and makes warmth available more constantly.
- The movement capacity becomes largely independent from local temperatures.
- Increased sustainable movement capacities substantially expand the potential for and duration of activities and thus enable a significantly more active lifestyle.
- Behavioral and social repertoires can potentially be more complex and are less constrained by behavioral thermoregulation.
- Aerobic movement physiology combined with independence from local thermal constraints overcome the restrictions of living in suitable micro-habitats and lead to enhanced ecological possibilities.

These different and profound changes (see Fig. 9.1) are all elements that establish a more autonomous lifestyle for the individual endothermic organism. They provide a new flexibility within the possible ecological systems and vastly increased physiological and behavioral plasticity. As this can be described in many details, the gain in autonomy is a factor in the evolution of endothermy and needs to be considered along with other factors.

# Chapter 10
# The Evolution of Brains and Behavior: Is There a Trend?

*If we were to attempt to define a mammal briefly, it could perhaps be done in two words – activity and intelligence.*

(Romer 1967, p. 1635)

Conflicting views on the relation of brain organization in vertebrates and differences in their behavior have been burdened by three theoretical assumptions. The classic view saw an ascending lineage of brain organization from fishes to amphibians, reptiles, and mammals. Combined with the conceptual background of natural selection, it was assumed that the selection pressure generated these increasingly larger and more efficient brains. As behavior is essential during adaptation of animals to their environment, the components of the brain might have developed according to behavioral adaptations. Organisms with more numerous and more complex neuronal networks would thus have increased potential for survival.

A problem with this interpretation is that animals with relatively simple brains also survive and even might have certain advantages, such as, the lower metabolic costs for the brain. However, if larger brains had no general advantages for survival, the different forms of central nervous systems just seem to be adaptations to different niches. A general trend and a selection pressure toward larger brains and more complex behavior patterns do not exist; therefore, a comparison between groups or evolutionary levels concerning brain size and brain capacity seems to be superfluous. This is a second theoretical assumption, and from its perspective, the first view has been criticized as a "scala naturae view," and in neurophysiology as well as in ethology, there are strong tendencies to see the capacities of differently organized brains exclusively under the aspect of the respective adaptations and the resulting divergence (Butler 2008; Pearce 1997; Alcock 1993; McFarland 2007; Shettleworth 1998; Butler and Hodos 1996; Deacon 1990). This dominates a large part of the literature on this subject. The differences seem to be unimportant as long as survival

B. Rosslenbroich, *On the Origin of Autonomy: A New Look at the Major Transitions in Evolution*, History, Philosophy and Theory of the Life Sciences 5, DOI 10.1007/978-3-319-04141-4_10, © Springer International Publishing Switzerland 2014

and reproductive success are ensured, or they are epiphenomena. Applied to ethology, Pearce (1997, p. 4), for example, states:

> … it is no more possible to conclude that one species is more intelligent than another than it is to say that one is more evolved than the other. All that can be said is that the species have developed different intellectual abilities that enable them to survive in their particular environments.

According to this argument, the intelligence of animals is often defined as the capacity to behave adaptively. Then, a comparison of intelligence between different species would apparently be senseless because each species has to deal with a different set of environmental problems. How would it be possible to state whether a rat or a dog, a newt or a chimp is better adapted according to behavior?

However, with this argument one loses any access to an explanation for the differences that are present and have been described in detail by morphology, physiology, and ethology: Because everything is adaptive and a comparison of capacities between different groups is declared to be irrelevant, the central question of evolutionary change is excluded from scientific reasoning, which is an epistemological blind alley.

According to the third theoretical assumption, which is especially supported by MacPhail (1982, 2001; MacPhail and Bolhuis 2001), there are no differences in the intelligence of vertebrates; only man has a special position. Within animals, there is nothing comparable, especially with respect to language.

Finally, the classical ideas about the evolution of brain size in vertebrates have also been questioned. Accordingly, the view supports that there is nothing like a progressive increase of brain size and that there is a wide variety of increases of different parts of the brain and with this also of the overall size of brains. Thus, within the bony fishes, there are, for example, some with an enlarged cerebellum, and within cartilaginous fishes, there are some with an enlarged forebrain without the generation of a cerebral cortex (McFarland 2007).

However, in opposition to these assumptions, there exists extensive literature that compares the allometric characteristics of central nervous systems within the groups of vertebrates and relates them to the respective capacities of behavior (for summaries, see, e.g., Roth and Wullimann 2001; Jerison 1973; Jerison and Jerison 1988; Striedter 2005) or examines the origin of the intelligence of humans from the evolution of vertebrates (e.g., Byrne 1995; Gärdenfors 2003; Eibl-Eibesfeldt 1999; Roth and Dicke 2005; Barton 2006; Gibson 2002).

In this chapter, I support the view that there is an overall tendency of increases in relative brain size over the course of vertebrate evolution, although this increase is not linear but complex and diverse. In addition, I state that the differences of size, features of component parts, and the neuronal fine wiring of different brains led to different possibilities of behavior that need to be interpreted in their specific quality. This meets with the repeated requests of Bullock (1986, 1993, 1995) to compare the qualitative differences of neurological and cognitive capacities of different taxa in more detail.

All the facts presented here are well known in neurophysiology and in ethology. However, they are often interpreted under certain theoretical presumptions, thus losing sight of specific qualities and hiding essential patterns of evolution. I propose that by means of the idea of biological autonomy a new starting point is gained to understand the evolutionary context.

## 10.1 Allometric Studies of the Brain of Vertebrates

During the nineteenth century, it became clear that the brain size of animals should preferably be compared to body size. However, comparison of brain sizes is still problematic and a subject of ongoing discussion (Striedter 2005). Jerison (1973) developed a form of representation showing the degrees of encephalization within different groups in relation to the respective body size. His assumption was that the amount of information the brain is able to process should be proportional to the number of connections of the neurons and that it should be possible to estimate the number of neurons and synapses according to the size of the brain and compare different species (Jerison 2001).

Besides measuring the brain size of extant vertebrates, in some cases it is also possible to measure the size of endocranial casts of fossils to obtain a rough estimate of the size of their brains. Figure 10.1 shows the classical graph of Jerison. Each polygon shows, for every class of vertebrates, a characteristic, surrounded area within which brain size varies. The polygons are oriented upward, indicating the necessary relationship between brain size and body size: Larger-bodied animals

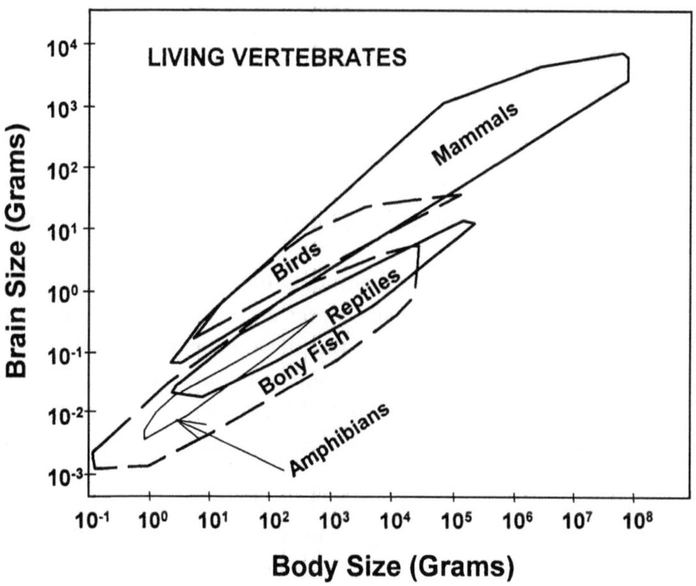


**Fig. 10.1** Brain-body relation in living vertebrate species enclosed in minimum convex polygons (Modified from Jerison 2001)

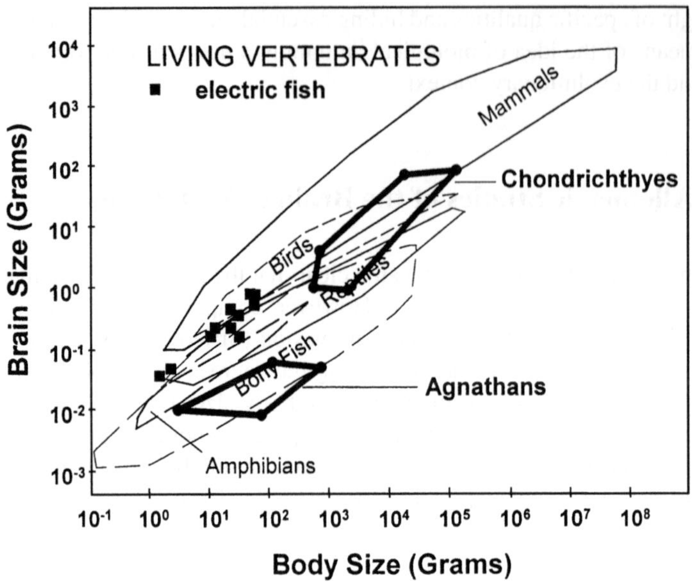


**Fig. 10.2** Polygons on Chondrichthyes (sharks, rays, and skates) and class Agnatha (lamprey and hagfish) and data on 11 electric fish, including mormyrids (Modified from Jerison 2001)

have to have larger brains, but the polygons of birds and mammals are in a higher cluster than those of bony fishes, amphibians, and reptiles. The vertical shift is the effect of encephalization. Higher vertebrates generally have several times as much brain as lower vertebrates per unit body weight (Jerison 2001).

It has been assumed that agnathans have some kind of minimal cerebral equipment for vertebrates. In some older graphs, a polygon for agnathans was far beneath the one of bony fishes. However, the graph in Fig. 10.2 includes some more extensive data for bony fishes that enlarge their area so that the data of agnathans overlap with the lower area of bony fishes. Thus, agnathans might be in the lower area of bony fishes (Jerison 2001).

Concerning brain size, the cartilaginous fishes are especially interesting (Striedter 2005). Their relative brain size varies enormously. Some have brains that are larger than those of bony fishes and thus have encephalization comparable to some mammals (Fig. 10.2). Some carcharhiniform sharks (e.g., hammerhead and requiem sharks) have brains that are about 5–10 times larger than the brains of other sharks with the same body size. Similarly, large brains are found in mylobatiform rays (e.g., stingrays and devil or manta rays). This correlates with complex behavior of some sharks, whose possibilities might have been underestimated up to now. In contrast, some large filter-feeding sharks have surprisingly small brains. Possibly phylogenetic decreases in relative brain size have taken place. Striedter (2005) assumes that relative brain size increased and decreased independently several times within cartilaginous fishes.

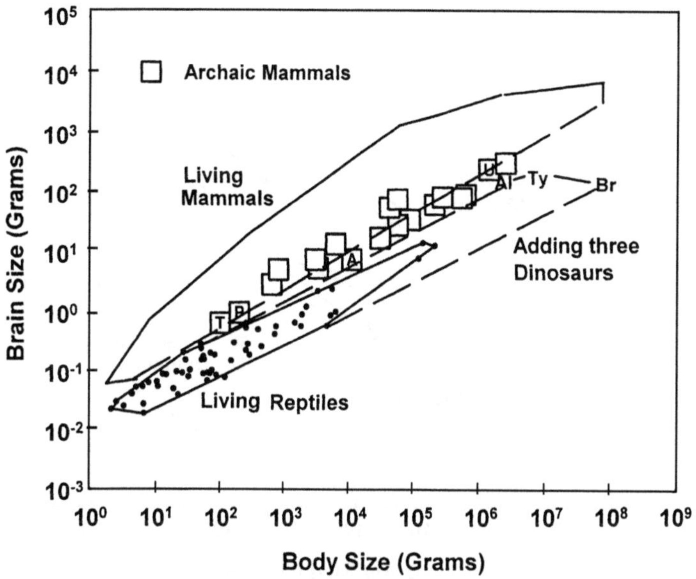


**Fig. 10.3** Brain-body relations in archaic fossil and living mammals and reptiles, each class enclosed in a minimum convex polygon (Modified from Jerison 2001)

Figure 10.3 shows data of the brain-body relation of some early mammals. Mesozoic mammals were on average about fourfold more encephalized than their relatives within reptiles (Jerison 2001). During the Paleocene era, radiation of mammals occurred, during which relative brain size initially stayed at the lower margin of the polygon for mammals. Only subsequent evolution brought forth forms with enlarged brains, as we know them today.

With the exception of primates, whose encephalization possibly began during the Paleocene era, the further encephalization of many mammals above the lower margin of mammals was a phenomenon of the past 55 million years. It began during the transition from Paleocene to Eocene and followed a time of relative stasis concerning brain size within the early mammals (Jerison 2001).

Additional data on dinosaurs show that dinosaurs did not have especially small brains as sometimes has been assumed. Their data fit well within the reptilian polygon.

Figure 10.4 again displays the graphs of the earlier figures with some additional single data. The figure shows that there are extant mammals that do not have larger brains than the earliest mammals. The opossum (*Didelphis marsupialis*) is shown as an example. Within mammals, encephalization is prominent among primates and dolphins (Byrne 1995; Falk and Gibbson 2001; Martin 1990; Marino 2002). However, the brain of dolphins is in some respects organized differently from the brain of primates. Dolphins have a thinner neocortex, the layers of the brain are less well divided from each other, and there are only about two thirds the number of neurons per volume compared to the brain of land mammals (Jerison 2001).

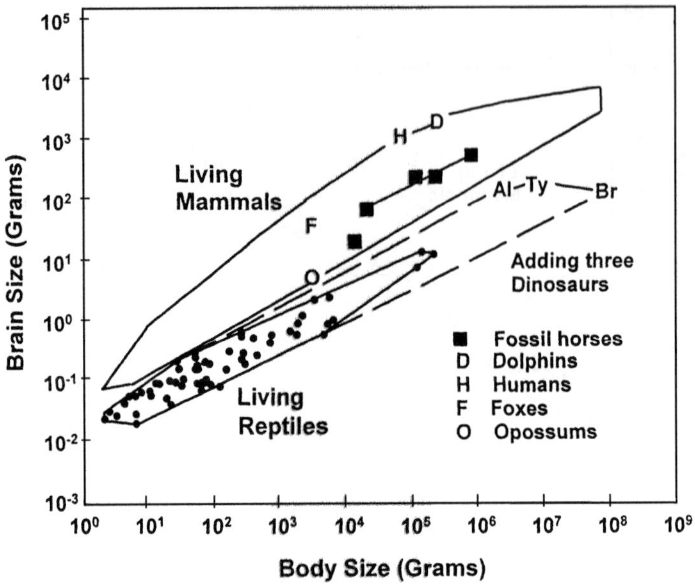


**Fig. 10.4** Brain-body polygons with data on fossil equids and some living mammals added (Modified from Jerison 2001)

The encephalization quotient (EQ) determines the relation between the actual relative brain size of a species and the relative brain size expected in the mean according to body size. The EQ for the brain of human beings is about seven times larger than the mean value within mammals and approximately three times larger than that of chimpanzees if they had the same body size as human beings (Jerison 1973, 1991). One of the most distinct features of recent human evolution is the trend toward increasingly large brains over the Plio-Pleistocene era. Early hominin Australopithecines had a cranial capacity slightly larger than that of extant apes. Over the subsequent three million years, average brain size trebled (Shultz et al. 2012; Robson and Wood 2008; Sherwood et al. 2008). Thus, humans have been characterized as brain or cortex organisms.

Although human beings are unique within primates in this respect, similar evolutionary processes must have occurred within whales (Marino 2002). Toothed whales, and especially dolphins, have EQs that are beyond those of primates, except that of human beings (Fig. 10.5). The other group with enormously enlarged brains is elephants. They also have a thicker cortex and the typical structure with six layers. However, generally not much is known about the brains of elephants (Roth 2001).

There are different opinions about which criterion for allometry of the brain is adequate, and some other criteria have been proposed, such as the comparison of defined parts of the brain to its overall size or the weight of the forebrain relative to body weight. Besides this, brain size is only one parameter among several possibilities. Thus, the neuronal fine wiring may be more important than mere size (Rehkämper et al. 2001).

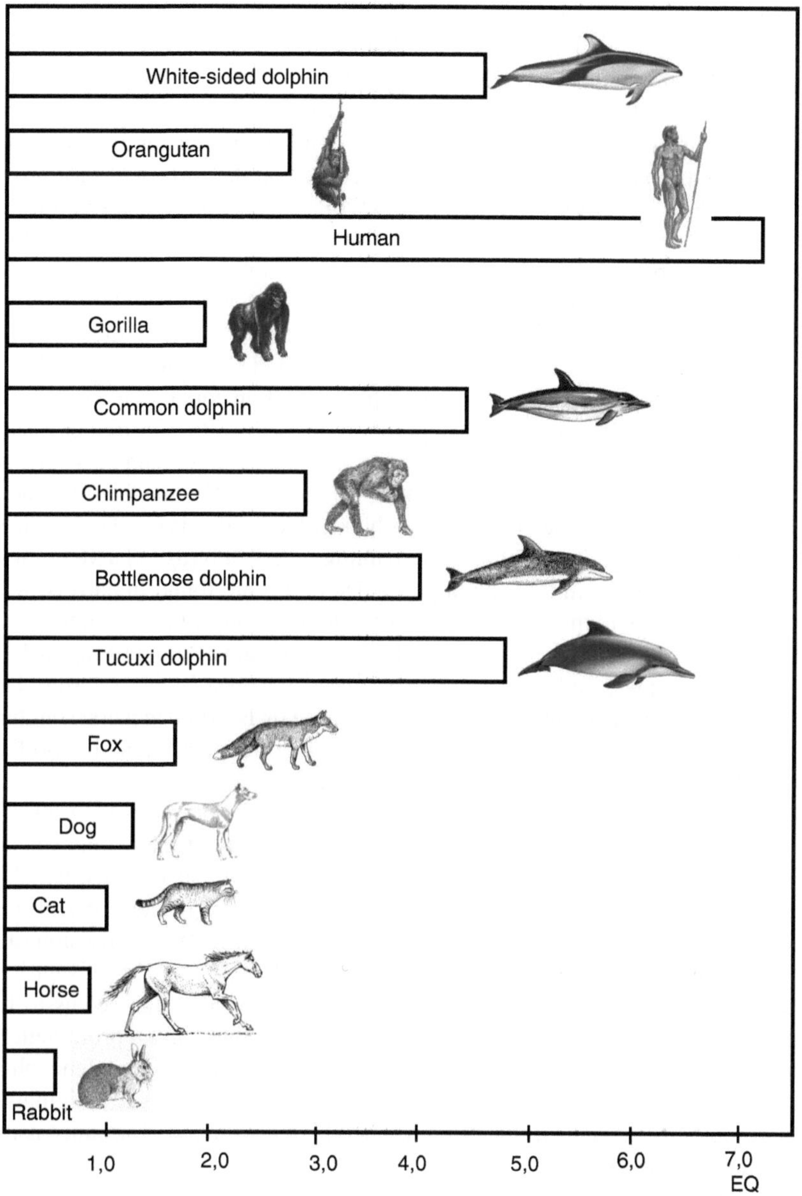


**Fig. 10.5** Encephalization quotients compared (Modified from Marino 2002 and Jerison 1973)

In any case, there is an expansion of the forebrain in relation to the whole brain during evolution of mammals and birds. However, the forebrain portion varies in mammals between 49 % in the insectivore *Geogale*, which has allometrically the smallest brain of all mammals, to 85 % in human beings (Rehkämper et al. 2001).

All primate brains have unusually large neocortices. Both absolute brain size and the proportional size of the neocortex increased enormously within primates, particularly within our own phylogenetic branch (Striedter 2005).

With the size of the neocortex in mammals, the surface of the brain also increases by building furrows. This widens the circuitry for an increasing number of neurons. Especially in primates, this process is amplified, so that the surface of the cortex in human beings is about four times larger than that of chimpanzees. Hominids and human beings have been called cortex organisms (Heldmaier and Neuweiler 2003). In addition, the brain is "ruinously expensive": Although it only amounts to about 2 % of body weight, it consumes about 20 % of the metabolic energy.

Birds, the "flying dinosaurs," have brains that are significantly larger than those of reptiles with the same body size (Striedter 2005). The brain of an ostrich, for example, is roughly four times as large as the brain of an equally heavy crocodile. Because ostriches belong to the oldest group of birds, the ratites, Striedter assumes that the earliest birds had relative brain sizes considerably larger than those of extant reptiles. This is supported by the available fossil record and by the fact that another primitive group of birds, the Galliformes (chicken-like birds), is similar to ratites in terms of relative brain size. Within the remaining orders of birds, relative brain size increased again several times. The largest avian brains are found in parrots, which are well known for high performance on intelligence tests. Owls, woodpeckers, and songbirds (passerines) also have relatively large brains. Among songbirds, the corvids (crows, jays, and their relatives) have the largest brains (Karten 1991; Delius et al. 2001; Rehkämper et al. 2001). Because owls, woodpeckers, songbirds, and parrots are only distantly related to one another, relative brain size probably increased independently in each of these four lineages.

In birds, the portion of the brain built by the telencephalon varies between 50 % in gallinaceous birds and 80 % in certain species of songbirds. Here, an expansion of the forebrain occurred by way of the evolution of neuronal systems, which are constructed differently from the cortex of mammals, but reached performance comparable to that of mammals in many cases. It is a different subdivision of the forebrain in birds that is mainly responsible for this, the so-called dorsal ventricular ridge (DVR). The DVR resembles the mammalian neocortex and seems to have comparable functions in several respects. Another area that plays a prominent role in some birds is the so-called Wulst, positioned like a crest atop the telencephalon (Striedter 2005). However, the organization of avian brains is still enigmatic in many respects, but it is clear that birds reached high neuronal performance using a different evolutionary path.

Striedter (2005, p. 109) summarizes:

> "Considering all these increases and decreases in relative brain size, we may ask whether any general pattern of change can be discerned. Clearly, relative brain size did not increase linearly "from fish to man", as Huxley and others had supposed. Rather, it increased independently in several different lineages and it decreased in others. The pattern is complex. However, if we step back and look for an overall trend, we can see that, on average, relative brain size has increased over the course of vertebrate evolution. A closer look suggests that this trend exists not because the increases are, on average, larger than the decreases but because the increases outnumber the decreases."

## 10.2 The Relation Between Brain Organization and Flexible Behavior

Now, the crucial question is the following: What significance do large brains with high capacities have? Often, large brains are related to "higher intelligence" or to more "complex behavior." However, this has also been challenged (Butler and Hodos 1996; MacPhail and Bolhuis 2001). According to Butler and Hodos (1996), the relation between brain size and intelligence is not clear. However, judgments on this relation depend on the definition of intelligence. Terms such as "intelligence," "cognition," or "associative functions" are used in relation to animals but are difficult to define and are not always used with the same meaning (Delius et al. 2001; Byrne 1995; Menzel 1986).

When ethologists observe an animal as it behaves within its natural environment, it is usually assumed that its behavior evolved through adaptation during evolution. This does not necessarily mean that behavior is considered exclusively genetically programmed. Ontogenesis of the behavior of many animals includes a subtle interaction between inherited functions, developmental processes, and environment. All living animals are well adapted to their environments; this is proven by their survival. However, if "intelligent" means the same as "well adapted to the environment" (Pearce 1997; Sibley and McFarland 1976; Slobodkin and Rapoport 1974), then all animals are intelligent in their own special way. Given such an assumption, it is not surprising that some authors doubt whether intelligence is a relevant measure at all. Environmental adaptations are of great biological relevance, but to call them intelligent leads nowhere. Behavioral programs of relatively simple organisms can be complex and adaptive, for example, with respect to their reproduction, but this cannot be claimed to be a product of intelligence.

Thus, following Byrne (1995), it is proposed here to regard *flexibility* as an essential component of the intelligence of animals. Flexibility of behavior allows the individual to find its own solution for problems and tasks or even to act independently from external necessities, as is the case during play. The essential point is that new actions can be generated and practiced. This position also takes into account that the variety of possible answers to the same environment is larger than necessary at any particular moment. On the contrary, inherent or early imprinted forms of behavior are mainly fixed and relatively inflexible, especially if they are well adjusted to environmental circumstances.

Intelligence, however, may include more than flexibility. Abilities such as memory, intentionality, representation, and many more can be viewed as essential to intelligence. These abilities are so complex and diverse that it is still difficult to describe them adequately. Beyond that, it will not be possible to reduce them to a single point of view. Within the argument developed here, it is only emphasized that flexibility is but one central feature among others.

Increased flexibility of behavior was generated during evolution by way of the principle of uncoupling. Beginning with the generation of simple nervous systems, sensory stimuli were uncoupled from reactions, which subsequently could be increasingly modulated. Mere reflexes are still closely coupled to sensory input.

However, by the evolution of brains, uncoupling and modulation increased tremendously; thus, greater autonomy was attained, as I argued in Chap. 8. This can be illustrated by an example from Gärdenfors (2003).

Many mammals and some birds are able to generate internal representations, which can be dealt with more or less independently from the external world, that is, they are uncoupled from it. A cat, which follows a mouse and sees it disappearing behind the curtain, is able to foresee that the mouse will reappear on the other side of the curtain. Thus, the cat is able to draw conclusions about the mouse even if it gets no direct signals from the mouse. The cat must have some sort of an inner picture of the mouse, even when the cat cannot see it. The internal representation generates expectations regarding the mouse. Such representations seem to be lacking in reptiles, at least nothing comparable – as in mammals or in birds – could be demonstrated by previous studies. Thus, the way in which a snake and a cat hunt a mouse differs fundamentally (Gärdenfors 2003; Sjölander 1993). Both the snake and the cat use a combination of senses during hunting, but in a somewhat different manner. If a poisonous snake attacks a mouse, it uses its visual sense or sometimes heat detectors on its head. After being bitten, the mouse usually runs some distance before the poison becomes effective. To find the dead mouse, the snake uses only its sense of smell. Even if the mouse happens to lie directly in front of the snake, it detects the mouse only by the scent. Before the snake can swallow the mouse, it must first find the head because headfirst is the only direction in which the mouse will go down the snake's throat. The snake could have accomplished this by sight or smell, but it only uses its sense of touch to solve the problem. The snake thus uses three different senses to catch and eat a mouse. However, there are few connections between these senses employed in the various stages of the process. The snake reacts only to these sensations, but it seems to have no perception of the mouse as an object and hence no representation of it.

In contrast to the snake, the cat applies several sense organs simultaneously: eyes, ears, nose, whiskers, and paws. The information from the different senses is integrated into a coherent inner picture, which is present even if the mouse disappears.

> "The cat has object permanence and can therefore predict that a mouse that runs under one side of an armchair will come out the other side. A snake could never manage that. The cat can "think" of the mouse even when it is receiving no signals from its senses: it can, for example, wait outside a mouse hole. A snake never does that. The decisive difference is that the cat has a representation – a comprehensive image of the mouse – that combines the input from its various senses. The representation is in the cat's head even when it is not receiving any direct sensations from the mouse. It is this representation that makes the cat's way of hunting much more flexible and efficient than the snake's is" (Gärdenfors 2003, p. 39).

In the following, some abilities of animals and human beings are examined with respect to flexibility and uncoupling: learning, play, imitation, use of tools, insight, empathy, self-awareness, and language. The idea is that flexibility, as it is exhibited by these forms of behavior and abilities, is an essential element of increasing autonomy, as defined in Definition 2 (Chap. 3).

## 10.3 Learning

Much of the behavior of animals on all phylogenetic levels is mainly performed according to some largely fixed patterns. Thus, they are called *fixed action patterns* (FAPs). Usually, they are considered innate. There are, however, discussions about the usefulness of the innate/acquired distinction (Griffiths 2009). The belief that a trait is innate is today commonly expressed by saying, "It is in the genes." However, although genes do play some role in the production of every trait, many other factors, such as epigenetic or environmental factors, are involved as well. Developmental systems theory (Oyama 2000a; Oyama et al. 2001; Sterelny and Griffiths 1999) attempts to treat this problem from a more dynamic perspective (see Chap. 3).

The essential point for the argument here is that there is a spectrum of behavior from being more fixed in standard actions to less fixed with a broader range of possibilities. The sharp distinction between instinctive and acquired behavior, which has been described by the school of Konrad Lorenz, may need some extension, as there is a more continuous spectrum between fixed and less-fixed behavior, wherever this fixation is located. This is what empirical ethology considers, producing an overwhelming amount of examples. Thus, Marler (2004, cited in Griffiths 2009) states that placing behavior on a lability continuum definitely has some value, with certain behavior more changeable and variable and other actions more stereotypical and resistant to change. In the same sense, Mayr (1974) distinguishes between "open behavior programs" and "closed behavior programs." He sees closed programs mainly realized in lower animals, whereas in many vertebrates and especially in mammals and birds, some more flexible actions are involved in many different kinds of behavior (see also Conway Morris 2003; Eibl-Eibesfeldt 1999).

FAPs typically are initiated by a specific key stimulus. Such a key stimulus may also initiate a sequence of actions if the stimulus is no longer present. All members of a species (of comparable age and sex) will respond to the stimulus with nearly identical behavior, even if they had no previous experience with it. A classical example is the fixed sequence of actions in the larva of the silk moth when it begins to spin a cocoon after it has finished growing. FAPs are the basis for the behavior of all animals. However, the FAPs can be supplemented and widened to different degrees by learning, a modification of behavior by experience (Eckert 2000; Eibl-Eibesfeldt 1999). This introduces a certain degree of plasticity in action and reaction in general, as different solutions for environmental problems become possible.

During imprinting, learning is restricted to a special sensitive time during development. However, here also there is widened flexibility because the knowledge of the characters of the species is not completely fixed, which is often the case in mammals and birds. This belongs to the spectrum between fixed and less-fixed behaviors. It is also an example of a behavior that at first is not fixed (the knowledge of the conspecific) and becomes fixed after the sensitive time is over.

Learning takes place in lower animals. Even in protists, some primitive learning abilities in regard to meeting obstacles have been described. In many invertebrates, there are specialized learning dispositions. Thus, bees learn optical and olfactoral traits of a food source well, and they also learn the way that leads to it by memorizing objects and directions in the environment. However, other actions of bees are basically fixed and hardly indicate possibilities of flexible combinations. Scientists have used the specific learning capacity of the marine snail, *Aplysia californica*, for its defensive gill and siphon reflex to gain insights into neurophysiological principles of learning (Eckert 2000).

Many authors (e.g., Conway Morris 2003; Delius et al. 2001; Eckert 2000; Heldmaier and Neuweiler 2003; Mayr 1974) see a correlation between an extensive potential for learning and complex nervous systems, including larger brains. *Octopus vulgaris* is an impressive example of this. The octopus has a highly concentrated, relatively large brain, which was generated by the fusion of ganglia. At the same time, it has highly flexible behavior with an extensive capacity for learning and an impressive memory (Young 1991; Fiorito and Scotto 1992; Boal et al. 2000; Hanlon and Messenger 1996). Fiorito and Scotto show how one octopus was able to learn by observing another one. In the experiment, two octopuses were kept in neighboring tanks. One of them had been trained to choose between two objects with different colors. While he performed this several times with the appropriate solution, the other one could observe him. When the untrained octopus was subsequently prompted to find the appropriate object, he more often selected the same color as his model. Another experiment was more difficult. This time the "pupil" watched through the glass how his neighbor skillfully opened a box containing food. When a similar box appeared in his own tank, he opened it much faster than another specimen without such an experience. Generally, the pupil used the same technique as his model.

Centralized nervous systems are characteristic for vertebrates. With increasing centralization and complexity of brains, a greater amount of learned behavior is to be expected (Conway Morris 2003; Eibl-Eibesfeldt 1999; Franck 1997). Thus, most research that focused on learning has been conducted with birds and mammals.

Beyond this, comprehensive learning dispositions in animals are mainly related to certain tasks that are pertinent for them. For example, whereas sounds produced by pigeons, woodpeckers, and cuckoos are fixed, nearly all songbirds must learn from older conspecific birds to acquire their species-typical song (Streffer 2009). Thus, European chaffinches (*Fringilla coelebs*) are predisposed to learn a certain song pattern in which the first part is fixed, whereas they have to learn the second part with a quaver from an elder specimen. Young birds memorize the songs they hear during their first year, and the next spring, when they are grown up, they try step by step to adapt their own song to the one in their memory. Males raised in acoustic isolation are able to acquire alien sounds but have a higher learning disposition for conspecific songs. The internal sound picture first needs to be stimulated. Other species, such as the common starling (*Sturnus vulgaris*), the Eurasian blackbird (*Turdus merula*), the nightingale (*Luscinia megarhynchos*), the marsh warbler (*Acrocephalus palustris*), or the song thrush (*Turdus philomelos*), not only learn conspecific and alien bird songs but also often imitate sounds that do not come from other birds. Some parrots easily learn to imitate foreign sounds. In an unusual extent, this is true for the gray parrot (*Psittacus erithacus*) from central Africa, which Irene Pepperberg demonstrated with her famous Alex (Pepperberg 1990,

1994a, b) and was previously described by the nineteenth century German naturalist Alfred Brehm. Alex not only learned to pronounce the names of 50 different objects and sorts of food but also identified them correctly and was able to categorize them according to several properties.

In mammals, there are many behavior patterns that have to be formed by learning. The range varies not only from species to species but also in relation to functions within the same species. Thus, in European squirrels (*Sciurus vulgaris*), the ability to hide food is fixed. Squirrels raised without the possibility to gain experience hiding food were able to show the behavior as soon as they were allowed. Experiments showed that the sequence of actions consists of a rigid follow-up of behavioral elements. However, opening a nut requires a relatively higher component of learning. The movements of gnawing and breaking are fixed, but the animals have to learn how to apply these movements to be successful. This generates individual differences in opening nuts that are a result of the plasticity in learning. In rats, all action necessary to build a nest is fixed, but they learn the most practical sequence of movements. An inexperienced raven masters certain movements for building nests but has to learn which materials to use. Polecats (*Putorius putorius*) have to learn how to kill animals that resist the attack.

The differentiation between obligatory and facultative learning hints at another degree of plasticity in learning. Obligatory learning is necessary for species-typical behavior such as avoiding enemies, acquiring food, or sexual behavior. Female rhesus monkeys, for example, need experiences with a caring mother during their own youth; otherwise, they are not able to care for their own babies later. On the other hand, facultative learning is possible but not necessary and does not usually occur under natural conditions. Such behavior has larger degrees of freedom and is less fixed by predispositions. Well-known examples of animals with distinct facultative learning capacities are dolphins that play with a ball, parrots that close the door of their cage, ravens that train to catch prey during flight, or monkeys that are able to combine symbols with tasks or objects (Fig. 10.6).

An especially expanded disposition for learning is also expressed in behavior arising from curiosity. When a new object is placed in a room inhabited by rats, the rats begin to explore it. First, one after the other cautiously creeps up to the object, then it is sniffed, gnawed on, climbed over, and marked. Then, they lose interest in it. Similarly, all higher mammals also examine a new object within their environment. How an animal approaches such an object is typical, which Eibl-Eibesfeldt (1999, p. 401) describes as follows:

> The animal is attracted by the object, but does not engage in a certain rigid behavior but rather has the possibility to leave the object temporarily. It is this ability to distance oneself from the object that is the prerequisite for every examination in the sense of a dialogue with the object. It is typical for investigation out of curiosity and even more so for play. During ontogenesis of human children, one can pursue how this ability develops step by step. When the child begins to reach for objects, it first follows a rigid sequence of actions. It grasps the object, directs it to the mouth and begins to suck on it. … Soon the child is able to remove the object from its mouth, look at it, suck on it again and possibly grasp it with the other hand. Now the rigid sequence of action is broken and the child is able to investigate. … within this capacity to gain distance anew lies one of the roots of human freedom.

**Fig. 10.6** Learning in dolphins (Redrawn from Gould and Gould 1994)

Most mammals, at least juvenile ones, demonstrate a distinct behavior out of curiosity and search actively for new situations. Of course, groups differ in quality as well as in quantity. Primates, dolphins, and carnivores, for example, are more curious than rodents. Apes look at the objects and test them with their hands. However, even some fishes and birds are curious.

Ethologists, who follow an adaptationist view, tend to deny a difference in the learning capacity between animals of different phylogenetic levels (McFarland 2007; Shettleworth 1998; Pearce 1997). However, even McFarland concedes that "most contemporary psychologists recognize a spectrum of learning abilities, which reaches from simple forms of learning in more primitive, ancestral animals up to the cognitive capabilities of human beings" (p. 305).

This does not deny that some species (e.g., within fishes) also have expanded learning capacity. Experiments on learning with sharks, which have a complex central nervous system, still have to be conducted. It is possible that the learning capacity of some of them remains underestimated. However, this flexibility is more expanded within mammals and birds. Researchers may try to train a frog or a lizard abilities that are no problem for dogs, chimpanzees, or dolphins. Yet, there is no linear evolutionary sequence of these abilities. The patterns of learned and fixed behavior show different combinations in each species, although general tendencies are present.

Moore (2004) published an attempt to reconstruct the phylogeny of learning. The hierarchical sequence of different forms of learning within his cladogram reaches from simple forms such as habituation and conditioning to imitation, learning of simple symbols, full speech, and reading and writing. He seeks a rational basis to compare "intelligence" between different species. What he describes, without mentioning it, is a hierarchy of learning dispositions and flexibility of behavior that correlates with the different phylogenetic levels.

## 10.4 Play

It clearly has been demonstrated that most mammals and some birds play (Bekoff and Byers 1998; Fagen 1981, 1986; Ortega and Bekoff 1987; Burghardt 2005). Although reptiles generally do not play, there are some remarkable exceptions. Thus, there are observations of play-like behaviors from the monitor lizards. In particular, the Komodo dragon (*Varanus komodoensis*) has been observed in captivity playing with objects. Also, there are some observations of aquatic turtles playing with balls, sticks, hoops of hose, and other objects, and locomotor play seems to occur. In addition, there is a series of indications for play-like behavior in some groups of fish. Burghardt (2005) assumes that play might be more disseminated than previously thought, although generally it is most common within endotherms.

It is still difficult to explain the function of play behavior. That young animals practice behavior they need later might be involved; it does not, however, explain its evolutionary occurrence. An adaptive function has not clearly been demonstrated so far (Bekoff and Allen 1998; Burghardt 1998, 2005). The usual cost-benefit calculations fail as play is costly in energy and time and potentially risky. If the main function of play would be practice of behavioral sequences, it would still be unclear why such behavior is not fixed so that it could be performed more reliably as soon as it is needed. This is the case in most nonplaying groups of vertebrates. Because of the difficulty of explaining play by adaptation, there were only a few attempts to study its evolutionary origin, so that Burghardt (1998, p. 3) writes: "The origins of vertebrate play are obscure." Or, it is even described as a "biological paradox" (Fagen 1986, p. VII): "Animal play behavior addresses a major biological paradox. Why do young and old animals of many species spend time and energy, and even risk physical injury, performing the apparently unproductive behaviors colloquially called play? What makes this 'useless' activity so important that animals literally risk their lives for it? And, even more curiously, why are humans both enchanted and enraged by play?" In some modern adaptationistically oriented books on ethology, the topic is simply omitted (McFarland 2007; Shettleworth 1998).

Different definitions of animal play exist and are discussed in some detail by Burghardt (2005), before he provides five key criteria that proved to be helpful in identifying play behavior. Although Burghardt dismisses aspects of flexibility, I think that his criteria well contain them. The first criterion for recognizing play is that the performance of the behavior is not fully functional for coping with an actual problem of life such as foraging, reproduction, or defense. However, this means that play is uncoupled from direct needs and environmental challenges. The action becomes independent from the necessities of life, and at least there is no direct relation to them. The second criterion is that behavior is spontaneous, voluntary, intentional, pleasurable, rewarding, reinforcing, or autotelic ("done for its own sake"), as Burghardt formulates. This means that it is detached from instinctive coercions and contains elements of a constitutive autonomy.

The third criterion is that it differs from the "serious" performance of ethotypic behavior structurally and temporally. It can be incomplete, exaggerated, awkward, or precocious, or it involves behavior patterns with modified form, sequencing, or targeting. For the fourth criterion, the behavior is performed repeatedly in a similar, but not rigidly stereotyped, form. Both of these criteria mean that elements from the behavioral repertoire can be produced in forms detached from each other. They can occur in largely varied sequences, and the sequences can have new combinations and time structures. In this sense, they contain clear elements of flexibility compared to many FAPs. That they can occur in repetition does not contradict this element of versatility.

The fifth criterion is that the behavior is initiated when an animal is in a "relaxed field": It is adequately fed, healthy, and free from stress. Thus, it is liberated for some time from the necessities of life maintenance. In this sense, play behavior contains clear degrees of freedom, so that it is part of the autonomy of endotherms, not a conundrum but rather to be expected.

Animal play can be categorized roughly into locomotor play, predatory play, object play, and social play. During play, new movements are invented or behavior patterns or fragments of behavior appear that would be serious in another context, such as escaping, attacking, or hunting prey. However, during play they can be combined flexibly rather than integrated into fixed sequences. If play has a function of practicing at all, this would be performed on the basis of a relatively flexible and plastic process, so that such behavior can have individual components and differences. Many mammals perform complex and acrobatic movements during play. The prerequisite for this is, again, the general flexibility of movements as described in Chap. 8 concerning its neuromotoric basis and in Chap. 9 regarding the functional complex of endothermy.

In many species, play is mainly restricted to juveniles. In others, the adults also play, which is especially true for carnivores, rodents, primates, and whales. Especially, adult dolphins tend to play intensively during their whole lifetime (Fig. 10.7). Within birds, reports of play have concentrated particularly on ravens and parrots; other accounts exist, for example, regarding species of falcons. However, when birds of prey soar to enormous heights using thermals, this is independent from direct biological necessities and thus is not fully functional, is spontaneous and voluntary, and seems to take place in a relaxed field. In this sense, this behavior is also a form of play.

Keas (*Nestor notabilis*), large parrots from New Zealand, "may be the most playful avian species in the world, even among a family with such cognitive stars as gray parrots" (Burghardt 2005, p. 255). Playing around, they are notorious for their attraction to various human objects, such as backpacks, tents, windshield wipers on cars, or the equipment of film teams trying to film them, often destroying what they manage to obtain. They steal television antennas from houses, let the air out of automobile tires, or remove covers from garbage cans, exhibiting some sort of "demolishing play," as Burghardt formulates. In locomotor play, they stand on their heads, turn somersaults, land upside down, use branches as swings, and make and push around snowballs.

Play is part of the singing behavior of many songbirds as well. Streffer (2009) convincingly argues that the interpretation of birdsongs as territorial marking with

**Fig. 10.7** Playing dolphins (Redrawn from Conway Morris 2003)

adaptive functions is extremely one sided and thus conceals a deeper insight into the phenomenon. His starting point is to consider birdsongs in their own musical quality. From this point of view, singing behavior seems to be subject to biological necessities only in certain situations. A large part of birdsongs is emancipated from the constraints of biological necessities and elaborates into a world of an amazingly high flexibility of dealing with vocal possibilities in the respective species. The most flexible vocal capabilities are presented by the best singers, for example, the nightingale (*Luscinia megarhynchos*) or the Asian shama (*Copsychus malabaricus*). Whereas mammals elaborate their playing behavior via body movements, many birds celebrate their flexibility via their voice. This is yet another example of how different features of autonomy are developed during evolution. A few authors earlier recognized some examples of birds' playful use of their voice (Hassenstein 1969; Lorenz 1935), but a new study by Streffer is the first systematic account of this marvelous phenomenon. Also, Burghardt (2005) discusses "vocal play," and it would be interesting for future research to apply his five criteria for play regarding Streffer's view on singing behaviors in birds.

During the time period young birds and mammals are sheltered by their parents, a situation is created in which the young animals can play. It leaves them independent from tasks that are important for survival, giving them time for learning and developing flexible behavior. Young reptiles, on the other hand, behave from the beginning nearly like miniature adults.

Especially in primates, rather individual play behavior has been studied, during which only vaguely determined behavioral sequences are practiced. Thus, individualized solutions and habits can be trained because some sort of a creative component may be involved. Occasionally, this has also been observed with birds. Franck (1997) describes how during the play of young garden warblers (*Sylvia borin*), which had been raised by hand, a pebble fell by chance, tinkling into a bowl made of glass. After this, the birds carried the pebble repeatedly into the air and let it fall, watching it and listening to its clear ring. They did not move before they heard it ring. After 2 weeks, they began to vary the game by experimenting with other objects.

Play includes a certain amount of mental flexibility, as it includes pretending behavior (e.g., to pretend fighting). Because the cat does not eat the toy mouse, it "knows" the mouse is not real and thus is taking an "as-if" stand toward reality (Burghardt 2005). Once again, there is the detachment in representations between the real object and its function during play. In social play, the animals must be capable of recognizing pretense in a partner.

Play by human children is significantly characterized by pretend actions. This ability develops at the age of 2 years and is practiced for many years during childhood. A predisposition for this is the uncoupling of two representations that are involved: For objects, there is the real representation and at the same time an imagined version of it (the "cake" in the sandbox) or, in relation to actions, the real experience of oneself and the imagined action, which is clearly experienced as "only play" (Leslie 1987). As animals at play can also perform pretend actions, it must be assumed that they also are able to uncouple these two representations.

Beyond this, in animals it has been regularly observed that larger or stronger playing partners restrain themselves as if they want to keep the play fair. Ethologists call this "self-handicapping" (Fagen 1986; Bekoff 2004). All this needs a certain degree of cognitive ability and thus is in relation to the capacities of the central nervous system (Bekoff and Allen 1998). In this connection, elements of communication, intention, role play, and cooperation, all of which are also elements of cognition, are often involved. Although there are still many uncertainties concerning play behavior, some authors assume that from its study more knowledge about the cognitive capacities of animals will be available than from the study of any other behavior (Bekoff and Byers 1998).

In mammals, Burghardt (2005) sees a correlation between play behavior and large brains, intelligence, high metabolic rates, parental care, altriciality, sociality, and some other factors, but not in a strict and regular sense. They can be combined in different associations and can be found in different subtypes. He finds these criteria also within the most playful birds: "The playfulness of these avian play champions seems to be associated with large brains, prolonged postnatal development, a complex behavioral repertoire and fine motor control, high metabolic rate and endothermy, generalized foraging abilities, and even, in some species, complex sociality. Parental care is highly developed and protects the young from a hostile environment. Overall, play is less ubiquitous and complex than in placental mammals, but many more of the latter have slow development and larger body sizes" (p. 279).

Thus, several features of increased autonomy are involved when some groups generate playful behavior: behavioral and mental flexibility, independence from

**Fig. 10.8** Young elephants playing football (Scene redrawn from a WDR-Documentation [Germany]: "Die Natur im Waisenhaus" [Sept. 2012])

external necessities, degrees of freedom from instinctive coercions, and a variety of movement possibilities, combined with physiological features of autonomy such as endothermy and generally high-energy intensiveness (Fig. 10.8).

This becomes even more interesting regarding those animals within ectotherms that exhibit some play or play-like behaviors (Burghardt 2005). Thus, the monitor lizards (*Varanus*) have metabolic rates and aerobic scopes that are unusually high within the reptiles and tend to some endothermy. Furthermore, monitors seem to have overcome a restriction of lizards, compared with mammals, in obtaining enough oxygen for continuous activity: the absence of a diaphragm. Monitors have a large throat pouch that can expand and force air into their lungs and provide more oxygen for aerobic activity. A comparable feature seems to occur within fish: The great white shark (*Carcharodon carcharias*), for example, has been observed playing with objects offered to them from boats. However, these animals seem to be more intelligent than has been assumed previously, have a relatively large brain, tend to some functional endothermy, and are viviparous. Some mormyrids (weakly electric fish) have also been described as conspicuously playful, intelligent, and personable when held in an aquarium. Because of an especially large cerebellum, their overall relative brain size is beyond that of usual fishes (Fig. 10.2).

## 10.5 Imitation

In modern ethology, imitation is now assessed differently from earlier times (Miklósi 2004). It does not occur as frequently as assumed previously. To identify real imitation, Byrne (1991, 1993, 1995) differentiates two forms of behavior, which can be induced by an example. The first form is induced by the example, but the resulting action is already part of the available repertoire. The example merely increases the probability of activating a certain behavior, which otherwise is seldom used. What is observable may impressively look like imitation. The other form, real imitation, evokes completely new actions. These actions are not part of a fixed or previously learned behavior but are newly generated directly with the imitation. Imitation is applied to perform new behavior or new sequences of old behavior only by way of observation.

In addition, Byrne distinguishes three levels of real imitation. On the first level, a familiar movement is imitated within a new context (contextual imitation). For a second level, simple new actions or their sequences are imitated (production imitation). The third level indicates imitations that reproduce the structure of a complex action in its sequence (program-level imitation). It is the observational learning of the organizational gist of a task (Bates and Byrne 2010). Here, the fine details often are unimportant as long as the correct result is obtained, and it may lead to an individual solution concerning the exact performance.

This likewise demonstrates different levels of flexibility. First, any imitation needs some detachment from FAPs as there is the orientation toward an external example for the new behavior; at the same time, some flexibility in neuronal control as well as motoric abilities are necessary. On the second level, the flexibility is widened in the form of new actions; therefore, new behavioral combinations become possible. Especially, program-level imitation requires the ability to understand and copy the logical organization of programs of actions and to copy detailed action patterns from others. The complex sequence has to be assembled without reliable guidance from experience or fixed behavior and thus requires a high degree of neuronal freedom.

The observation of real imitation in animals is difficult. Often, it cannot be truly recognized, whether or not already available actions are involved in some form or another. The regularly cited example of Japanese macaques, which wash sand off potatoes before they eat them, could possibly be the result of a fixed action that is only activated by way of example from another individual (Visalberghi and Fragaszy 1990).

A clear example of program-level imitation comes from the African gray parrot (*Psittacus erithacus*). The parrot was regularly shown a videotape depicting a person who performed a few stereotyped actions accompanied by words or phrases as labels (Byrne 1995; Moore 1993). One such routine, labeled "Ciao!" consisted of opening the door with the left hand while waving with the right, then closing it with the right while waving with the left. The parrot copied all of this, waving first with its right foot and then with the left, meanwhile vocalizing "Ciao!" and finally closing the door with a click. The parrot also imitated other routines. Such sequences are not behavior that in some form could be fixed and are relatively foreign to a parrot.

Within the great apes, there are also examples of program-level imitation. There are many reports about chimpanzees that lived in captivity near human beings. In chimpanzees, the flexibility can even go so far that they understand the concept of imitation itself. That is, they understand that they are expected to imitate something

**Fig. 10.9** This orangutan has, without human help, tied a hammock between two trees. He failed to fix the second end, so that he holds it in his hand (Redrawn from Byrne 1995)

("do as I do" task). Thus, chimpanzees have been trained successfully to imitate movements of their trainer after a single observation (Miklósi 2004).

There are also distinct observations of program-level imitation with orangutans and gorillas. Thus, orangutans, during rehabilitation programs into the wild, copied rather elaborate human actions (the following examples were reported by Russon and Galdikas 1993, 1995 and discussed by Byrne 1995). A camp assistant, for example, used to chop weeds from the edges of forest paths and sweep them up into a row of piles. One day, an adult female orangutan was seen reproducing the entire process, using a stick the assistant had missed to chop weeds and making extra piles, similarly in a line in the center of the path. Other orangutans in the camp were seen imitating a range of other human activities: anointing the body with pilfered insect repellent or attempting to siphon liquid with a hose. Also, there were attempts to hang a hammock between two trees. The orangutan wound the rope around each tree several times without human encouragement. However, no orangutan was able to tie a knot properly (Fig. 10.9). An adult female tried to start a fire and copied exactly the procedures of the camp cooks, but finally she was not successful. Another female was observed washing clothes, including dipping into the water, soaping, rubbing with a brush, and finally wringing out the clothes. Sometimes, the animals repeated their imitations, specifically when they were unsuccessful the first time. Thus, the same female that tried to make fire, once imitated hammering nails into wood after watching construction work. She copied the actions closely but used insufficient power so that her nails fell out. She then tried repeatedly over several days after she had gathered together the necessary nails, hammer, and wood. This form of imitation needs a mental symbolic representation of the goal strived for, especially as preparations need to be made before the task is carried out.

Impressive examples of imitation by dolphins have been reported. Bottlenose dolphins (*Tursiops* sp.), for example, mimicked the behavior of humans or of a sea lion, which was living in their aquarium pool. An adult dolphin imitated the swimming mode, grooming actions, and sleeping posture of the sea lion. When a human at a pool's viewing glass blew a cloud of cigarette smoke just as an infant dolphin was watching him, the young dolphin swam off to its mother, returned, and released a mouthful of milk, having much the same effect as the cigarette smoke had. Another dolphin mimicked the scraping of the pool's observation window by a diver, even copying the sound of the pillar valve on the diving apparatus while releasing a stream of bubbles from its blowhole. All these actions are entirely foreign to a dolphin (Taylor and Saayman 1973; Byrne 1995).

In specific tests, dolphins also were able to deal with the do as I do task. Thus, they learned to imitate instantly movements such as turns from a trainer or another dolphin. They were also successful even if other parts of their body had to be used than those in the human example. If the trainer, for example, threw a ball into a basket with his hands, the dolphins imitated it with the help of their snout. Simple gestures of the trainer were also copied. If the trainer leaned forward and lowered his head toward the ground, the dolphin moved along the water surface and bowed his head as far as his anatomical possibilities allowed. Miklósi (2004, p. 304) summarizes:

> These results not only suggest that dolphins have a concept of imitation, but also that they understand something about the functional significance of the action; that is, they do not aim for an accurate reproduction of the demonstration (which is impossible because of the anatomical differences between dolphins and humans), but they try to imitate the significant invariant aspect of the action.

Imitation is also common among many songbirds (Streffer 2009). Here, three different types are distinguished, which again represent different levels of flexibility: A first type is copying calls and songs, which are adequate to the songbird's voice and thus correspond to the species' norms (e.g., great tit, lesser whitethroat). The second type is the copying of songs or parts of songs of foreign species of birds. Some species are able to copy calls and songs relatively exactly (e.g., song thrush, common starling). Sometimes, the imitation consists of several examples that can be mixed, so that the species' own singing pattern is hardly discernible (e.g., icterine warbler, marsh warbler), or foreign patterns are not imitated exactly but extensively changed (e.g., garden warbler, Eurasian blackbird, northern mockingbird). Beyond this, the Eurasian blackbird and the northern mockingbird can invent new patterns by themselves. Thus, the flexibility that is basically needed for imitation is extended so that individual variations of the model are performed (theme and variation), exhibiting further degrees of flexibility. In some cases, even technical sounds and music are imitated.

To a third group, other species can be classified that have extensively increased capacities of imitation. Especially impressive examples are the Australian lyrebird (*Menura superba*), which is able to imitate any sound. Besides other birdsongs, which it combines with its own beautiful song, it imitates a variety of different sounds, such as sounds of barking hounds, meowing cats, a buzz saw, the horn of a car, or a flying swarm of birds.

Examples of "program-level imitation" are direct imitations without time delay. The Asian white-rumped shama (*Copsychus malabaricus*) is a striking example of this. Within its own songs, which are extremely rich in variations, patterns of a neighbor not only can be imitated directly but also can be enriched by additional musical elements. In this way, "counter songs" can be created, during which the neighbors try to surpass each other in variations. Shamas in captivity were even able to sing parts of classical music they had heard from a tape recorder. This is a summit in flexibility with regard to musical capacities.

Bates and Byrne (2010) identify another level of imitation, namely, "rational imitation of actions." They describe the principle in 18-month-old infants: When in an experiment they are shown how to turn on a light by an adult who leans forward to press the switch with the forehead, they usually copy it, although they could use a hand. But, when the task was modified so that the adult's hands were occupied by something else, the infants were much less likely to copy the use of the forehead than they were when the adult used the forehead despite having nothing in his hands. It seems that the child is able to understand that the forehead is only needed if the hands are occupied. The infants obviously evaluate the causal necessity of the model's actions and only copy those actions they consider to be necessary and thus rational in attaining the goal. It is remarkable that something similar to rational imitation has been seen in chimpanzees. In experiments, they were presented with a demonstration of how to obtain food from a puzzle box that was either transparent or opaque: use a tool first to tap the outside of the box and then insert it into the box to retrieve a reward. When the box was opaque, the chimpanzees copied both the tapping and inserting actions used by the experimenter, but when it was transparent, revealing that the first action made no contact with the food, they ignored it and copied only stick insertion.

In another experiment, chimpanzees observed a human model operating switches with the feet or head when their hands were either free or occupied with carrying a bucket. Chimpanzees that saw the model's hand was occupied used their own hand to operate the switch, but if they observed an unexpected body part used by a model with hands unconstrained, they often copied the use of this body part. Bates and Byrne assume the following: "It seems that, like preverbal infants, chimpanzees imitate rationally: in some way, they can understand the logic of how actions achieve their ends, and choose to copy only those components of action that seem to be necessary." They mention that even in dogs there are some observations of such an ability.

## 10.6 Tool Use

Use of tools by animals has been studied extensively in hope of obtaining some insights into their understanding of mechanical procedures as a part of intelligence (Griffin 2001). However, many examples of tool use show relatively fixed behaviors. Thus, many ants produce sponges from leaves and use them to transport liquids from fruits. Some sand wasps use little stones to fix sand and pieces of wood, which they use to close the entrance of their nests. Sea otters (*Enhydra lutris*) use stones to open mussel shells, and Egyptian vultures (*Neophron percnopterus*) throw stones on ostrich eggs to open them.

Galapagos woodpecker finches (*Camarhynchus pallidus*) use cactus spines as probes to find insects under bark. The tool is employed for no other purpose, and no other Galapagos finch uses such tools. They use the spines in a manner that is comparable to the use of the tongue in European woodpeckers. This activity does not need some special intelligence or an insight into cause-and-effect relationships. This is also true for tool use in most other animals. Typically, one type of tool is used for one certain action, and all individuals of a species use the tool in the same manner, provided the opportunity is the same.

However, for some species among primates and among corvids, it has been observed that they can handle a range of tools for a scope of purposes, choosing between methods (Santos 2004; Byrne 1995). They also can use objects for tasks that have been unknown to them before. This requires increased flexibility and more extensive uncoupling from FAPs. Evidence of increased flexibility is also the modification or production of the tool. In dolphins , there are also some hints of flexible use of tools (Smolker 2004). Hart and Hart (2004) describe forms of flexible use of tools with the help of the dexterous trunk of elephants and see a relation to their large brain. The impressive versatility of the trunk, which can be developed by elephants in captivity, is well known. However, in elephants living wild, these actions are only scarcely shown.

In all great apes, flexible tool use has been observed (Seed and Byrne 2010). Chimpanzees, for example, used posts as a ladder and short sticks as hooks for climbing to escape from captivity. To reach food, chimpanzees use objects from their surroundings in extremely varied ways, so that it could be studied how far their insight into cause-and-effect relationships may reach.

Wild chimpanzees also use tools for different purposes (Fig. 10.10). Thus, they use suitable sticks to probe for ants or termites or to search for honey. These sticks can be prepared carefully. They chew bundles of leaves so that they can be used as sponges. Rocks are placed so that, with another stone, nuts can be cracked on them. Leaves are used to wipe blood from wounds or feces from the bottom, and banana leaves are used as umbrellas. Chimpanzees have also been observed as they broke up pieces of bark to use them under their feet to climb thorny trees to reach fruits. Summaries of this "chimpanzee technology" show how flexible and manifold tools can be used (Tomasello and Call 1997; McGrew 2010). This flexibility can be determined not only by the variety of tool applications but also by the invention of innovations.

Gorillas demonstrate less tendency to use tools than chimpanzees. However, Dian Fossey repeatedly observed gorillas throwing sticks. She also noticed how an old male broke off a long-stemmed flower to tickle a youngster gorilla with it. Other researchers saw them using sticks to collect fruits, which were otherwise unreachable. Observations of wild gorillas documented their use of sticks to wade through water and to stabilize themselves on marshy grounds (Breuer et al. 2005). Generally, primates stand out among mammals as the most frequent tool users. Like chimpanzees, capuchin monkeys use stones for nut cracking, digging, and many more tasks (Seed and Byrne 2010).

A high degree of flexibility in using tools has also been demonstrated by Weir et al. (2002) in Caledonian crows (*Corvus moneduloides*) in captivity.

**Fig. 10.10** Tool use: Nut cracking is difficult for chimpanzees to learn. This 3-year-old female uses her mother's wooden hammer but has failed to position the nut on an anvil, so she will not be successful (Redrawn from Byrne 1995)

The scientists put a container with food into a transparent plastic tube. To retrieve the container from the tube, the crows had some pieces of wire available without having experience with the wire. One female crow was able to solve the problem: She bent a piece of wire so that it could be used as a hook to fish out the container. These results suggest that crows, similar to chimpanzees, are able to modify objects relatively flexibly and use them spontaneously to solve new tasks. This exemplifies that they are also able to produce completely new tools for new tasks. The experiments have been inspired by the observation of flexible applications of tools by these animals in the wild (Santos 2004; Kacelnik et al. 2004; Kenward et al. 2005). It has been found in general that, among birds, "true" tool users have larger brains relative to their bodies than proto-tool users, which use tools less flexibly (Lefebvre et al. 2002).

Using a tool as an extension of the body seems to have particular consequences for psychological processes such as perception, attention, and cognition because the periphery of the body is thereby changed in mechanical and sensory capabilities. Research suggests that, in humans and monkeys, this extended motor capability is followed by changes in specific neural networks that hold an updated map of body shape and posture (Maravita and Iriki 2004). Seed and Byrne (2010) hold that tool use of chimpanzees, capuchin monkeys, and New Caledonian crows, at least, display the hallmarks of goal directedness, which requires some form of an internal representation of the goal in mind and knowledge that its actions will cause it to follow.

## 10.7 Insight

Insight requires at least two components: First, there must be a certain understanding of a relationship (i.e., some understanding of cause-and-effect relations). Second, some foresight is needed. It must be possible to simulate an action mentally at least in a minimal manner without simultaneously performing it. Often, longer chains of actions must be foreseen. Thus, here also flexibility is involved, although flexibility is only one part of the abilities needed (Bekoff et al. 2002; Seed and Byrne 2010; Heinrich and Bugnyar 2007).

Flexibility might be related to the elements of action being uncoupled from each other: the independence of the mental representation from immediate actions and the novelty of actions without prior attempts or imitation. The mental simulation of intended actions is performed independently from their realization in the environment. The purported action is internalized into the psychical realm of the individual by mental representation; therefore, it is highly emancipated from the environment. In some few animals, there are traces of such capabilities.

The classical example of behavior that might be directed by insight originated with Wolfgang Köhler (Eibl-Eibesfeldt 1999), who observed a chimpanzee combining two short sticks to a long one to reach food. In another setting, chimpanzees placed boxes on top of each other to reach the banana under the ceiling. A prerequisite for this activity is that the tools as well as the desired objects are within the same field of vision. Consequently, the perception must propose the solution. It is possible that there is not yet free internal imagination. Beyond this, it cannot be ruled out that the animals may have only used behavior that they had trained before during play or in other situations (Pearce 1997). Then, the behavior would not really be driven by insight, but merely be part of the already-available repertoire. However, the act needs some rudimentary insight into physical relations, and the animal must understand that a certain action is applicable to solve the problem.

Nonetheless, greater flexibility would be involved if a truly new action is performed that definitely was not part of an already-present repertoire. There are many observations that suggest the presence of such abilities in the great apes (Byrne 1995). Within the described frame of tool use, wild chimpanzees carefully select the objects that can be used as tools for a special purpose. This shows that the needed function of the tool must be present in advance in some form in the imagination. If, for example, chimpanzees search for stones to use as hammer and anvil to open nuts, they are considerably selective because they know that they will only be successful with special forms of stones. Even if this had been acquired previously by means of play or trial and error, now the correct relationship is discerned, and the action is performed in a goal-directed manner.

Other tools are prepared, for example, by way of removing parts from them. Thus, a stick to probe for ants is selected and broken off, and side branches and leaves are removed. The tool is prepared prior to probing with it, sometimes far away from the place of use. This demonstrates that the chimpanzees must have some representation of the properties of the tool needed.

**Fig. 10.11** One of the ravens of Heinrich during problem solving (Redrawn from Gould and Gould 1994)

In the previous example of an orangutan trying to make fire or to hang up a hammock, some form of mental simulation must have been involved. When an orangutan collects a hammer and nails to implement its plan, the orangutan must have a representation of the action in advance. The animal simulates psychically the physical process.

In complicated learning experiments with a female chimpanzee, Rensch (1973) demonstrated that the animal was able mentally to perform different possibilities of an action during a "planning phase." One task was to pull an iron ring with the help of a magnet through a labyrinth that was covered by Plexiglas. At the starting point, the chimpanzee had to choose one of two possible paths. The correct one led toward an exit where the ring was exchanged for a food reward; the wrong one was blocked at varying points. The task was mastered after a long period of training. Now, the chimpanzee first scrutinized the arrangement for a while before it performed the task in one continuous action.

More recent experiments demonstrate hints of insightful behavior in ravens (*Corvus corax*). Within an aviary, Heinrich (1999) used a cord on a branch to hang pieces of meat (Fig. 10.11). The cord was so long that the birds could not reach the meat while sitting on the branch. Then, they tried to remove the meat during flight, but the pieces were fastened too tightly. After some futile attempts, they gave up and seemingly disregarded the food. However, some hours later, one of the ravens suddenly came up with the solution and at once showed the whole sequence of action to solve the problem: It grasped the cord with its beak as far down as it could reach, pulled the piece up to the branch and fixed it with its foot. Then, the raven repeated the action until the meat was lifted to the branch so that the raven was able to remove it from the cord. Some days later, a second raven found a different solution: It pulled a piece of the cord up to the branch and moved sideways along the branch until the whole cord was extended along the branch and the meat came up to it. Two other ravens laid large loops of cord and held them with their feet. One raven never reached the goal and was the only one that did not learn that it was not possible to fly away with the tied meat. Heinrich concluded that the birds understood the problem, especially as they solved it without previous attempts or intermediate steps. Similar problems posed to other animals either remained unsolved or required extensive trial and error.

Interestingly, birds need the direct visual presence of the objects: When deprived of visual feedback through the use of a horizontal occluder with a hole just large enough for the reward to pass through, naïve New Caledonian crows tugged at the string but did not pull up the reward. Even the performance of the experienced subjects was disrupted (Taylor et al. 2010). It is possible this shows that insight in this case is still narrowly tied to the direct perception of the problem rather than further detached from the visual input. But, some understanding of the problem is still required. Heinrich summarizes the experience from his work with different species of ravens as follows: "Corvids are widely, and probably correctly, perceived as the brains of the bird world, and a number of species are of great interest because of their sociality, phenomenal memory, food hoarding, elaborate play, ingenuity, and tool use" (Heinrich 2004, p. 445).

A good overview of cognitive abilities is provided by Emery (2006). Emery summarizes that crows and parrots have consistently demonstrated intellectual skills that are qualitatively and quantitatively more sophisticated than have been demonstrated by other birds and in many domains comparable to monkeys and apes.

## 10.8 Empathy

Humans routinely are able to understand mental states of other people. We assume that other individuals, just like ourselves, are conscious beings with beliefs, feelings, intentions, and so on. On the basis of what we believe people are thinking or feeling, we can perceive – to however a limited extent – what they are thinking and feeling or even what they are going to do. This ability has been called empathy. It is the perception of others as beings with emotions and intentions that are similar to our own, an understanding of what others know, think, or feel. A question of comparative ethology is whether some animals have this ability to at least a certain extent (Byrne 1995; McFarland 2007; de Waal 2009, 2012; Decety and Svetlova 2012). The concept of empathy in animals, however, has had a difficult history, marked by disagreement and discrepancy. Although it has been realized for a long time, a scientific investigation has been hampered by its seemingly subjective nature.

Preston and de Waal (2002) propose a "perception-action model." It states that the perception of an object's state activates the subject's corresponding representations, which in turn activate somatic responses. The object is referred to as the primary individual who experienced the emotion or state; the subject is the individual who secondarily experienced or understood the emotion/state of the object through empathy. This supports basic behaviors such as alarm, social facilitation, vicariousness of emotions, mother-infant responsiveness, and the modeling of competitors and predators. The authors view the term *empathy* as “any process where the attended perception of the object’s state generates a state in the subject that is more applicable to the object’s state or situation than to the subject’s own prior state or situation” (p. 4). This is a broad definition that focuses on the process and makes empathy a superordinate category that includes several subclasses of phenomena sharing the same process. It includes emotional contagion, sympathy, cognitive empathy, helping behavior, and many others.

This again involves some degree of uncoupling or detachment: In the subject, there is not only the immediate experience of an emotion or state, just being caught within the subject’s internal processes, but also there is rather a secondary internal process that enables the subject to follow the represented internal states of the object. The subject experiences this process as related to the object besides its own internal state. We do not know how conscious animals can experience this difference; humans, however, are able to perceive the difference consciously.

From analogies with the generated representations of motoric actions in observed objects (the neurophysiological basis of which seems to be the mirror neurons), Preston and de Waal (2002) assume that also in empathy a comparable process might be involved. For motoric actions, it is likely that the representational responses of the subject are always generated, but the prefrontal cortex inhibits the motoric response. Early imitations in babies may be less controlled and thus performed at once, while in the growing child the control through the prefrontal cortex increases. This also leads to the ability to distinguish self from other. The activity in a certain area of the prefrontal cortex is able to distinguish activity from observing an act in another from self-generated activity. Applied to empathy, with an understanding of the way representations change with experience, this process might explain differences between and within individuals in the level of empathy expressed because development of self-other differentiation is correlated with the development of empathy, and both are correlated with development of the prefrontal cortex. However, the basic principle is detachment: The representational response is increasingly detached from a direct external reaction, and different degrees of this may be found within differently evolved mammals.

Empathy has especially been described from mammals, whereby the ability of empathy via direct perception-action processes corresponds to the similarity between subject and object as well as the amount of experiences with the object. In experiments, laboratory rats as well as rhesus monkeys showed physiological changes when they had to watch distressed conspecifics. Many mammals react to the perception of distress, illness, or wounding of others.

This applies mainly for members of the same species but may also go far beyond that. Thus, there are reports of related behaviors in apes helping an injured bird or showing consideration for humans in their surrounding or of dolphins helping people. Preston and de Waal (2002) hold that empathy facilitates group living and is especially crucial in mammalian mother-offspring bonds.

As an elaborated case, Preston and de Waal (2002) define *cognitive empathy* as a more controlled, actively cognitive, process. The subject is thought to use perspective-taking processes to imagine or project into the place of the object. Cognitive empathy appears to emerge developmentally and phylogenetically with other markers of mind, such as mirror self-recognition, deception, and tool use. The behavioral complexity and flexibility of these behaviors are greatly increased in humans and apes relative to other primates and most mammals.

> According to the [perception-action model], these processes were augmented by prefrontal capacities to increase the flexibility and control. 'Markers of mind' and cognitive empathy are associated with a larger proportional prefrontal region. The protracted development of Hominoid species increases the extent of learning before adulthood and is speculated to result in the disproportionate increase in the prefrontal cortex. … Prefrontal functions facilitate cognitive empathy through increased inhibition, increased working memory, and an increased ability to assess short- and long-term goals before responding. (Preston and de Waal 2002, p. 19)

The authors propose a developmental and phylogenetic sequence in which in early forms automatic processes cause the state of the object to elicit a similar or relevant state in the subject. This limits empathic processes to ones such as social facilitation, alarm, and emotional contagion because the subject cannot distinguish personal distress from the object's distress and has less control over emotional reactivity. Longer life spans increase the base of knowledge of individuals and situations, allowing individuals to better predict and understand the situations that cause distress in particular objects. Prolonged development increases the period in which individuals may learn (neuronally and subjectively) how to distinguish distress directly caused by personal insult from distress caused indirectly by insult to the object.

Extended prenatal and perinatal development is correlated with expansions of the prefrontal cortex, increasing working memory, planning, and inhibition. With this, individuals can hold information in mind and manipulate the information to predict, compare possible outcomes, and decide on an appropriate course of action. Working memory also increases imaginative processes that allow individuals to evoke empathic processes in the absence of the object. With increased inhibition, as a form of detachment, the subject can avoid becoming contagiously distressed from the object. The subject can inhibit the processes that normally augment personal distress, such as attention to the distress, expression of the distress, and elaboration of the distress, and can react in a more self-determined manner.

In a broad sense, these processes may also be involved when an individual understands some of the mental states of others. Humans assume that other individuals, just like ourselves, are animate agents with minds, and we treat them as having mental states. Furthermore, this "works" in the sense that, on the basis of what we think other people are thinking, we can predict to some extent what they are going to do, and we may be able to influence them. This has been called a "theory of mind" (Byrne 1995).

Some animals can be observed that act as if other individuals have mental states, so that it can be expected that they have a theory of mind, that is, some representation of the mental state of others. This has been tested with clear positive results in chimpanzees (Call and Tomasello 2008; Kaminski et al. 2008).

Examples of an especially flexible use of mental representations come from the behavior of some apes, which make use of so-called tactical deceptions. In such cases, other individuals are deceived (e.g., by faking wrong circumstances to manipulate their behavior). This requires that the manipulating ape has an idea of what the other ape feels or perceives or how the ape will react. There are reliable reports of such behavior especially from the two chimpanzee species and from gorillas and orangutans (Byrne 1995).

## 10.9 Self-Awareness

A theory of mind can also be applied to oneself, which may be a form of self-awareness or even a form of self-consciousness, however rudimentary it may be (Gallup 1997, 1998; Gould and Gould 1994; Bekoff et al. 2002). In accordance with some other researchers, Gallup showed that some animals are able to recognize themselves in a mirror. Whereas other animals react toward their mirror image as if it were a conspecific, chimpanzees and orangutans are able to relate the image to themselves. During experiments in which the animals were discretely marked with a spot on a part of their body that they could only see in a mirror, they interpreted the mirror image correctly and tried to remove the spot from themselves (Fig. 10.12). However, not every individual is able to pass this mirror test and needs a couple of prerequisites, such as being acquainted with mirrors.

Similar observations were also made for dolphins (Marino 2002; Reiss and Marino 2001). Hess (1989) watched how young, wild mountain gorillas spontaneously observed their mirror image on the surface of water. Recently, mirror recognition has also been demonstrated in experiments with magpies (*Pica pica*) (Prior et al. 2008) and with an Asian elephant (*Elephas maximus*) (Plotnik et al. 2006).

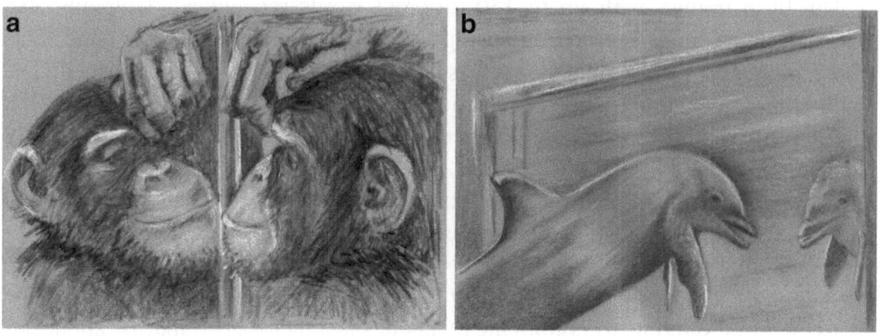


**Fig. 10.12** Mirror selfrecognition in dolphins and chimpanzees (From Byrne 1995; Reiss and Marino 2001)

The indications of empathy and self-awareness in animals are discussed controversially. Without debating here the interesting questions of cognition and consciousness in animals (Griffin 1982, 1984; Griffin and Speck 2004; Ristau 1991; Roth 2001; Stamp Dawkins 1993), it can be stated that the observed abilities, however limited they might be, are an expression of increased mental flexibility dealing with perceptions. The animals are in their attentiveness not bound to the direct perception and an immediate reaction to it. In empathy, they can rather process the signals they receive from other individuals in a way that they can make a conclusion regarding the emotions of others and use the signals for flexible behavior. They must be able to uncouple representations from the direct perception. Self-awareness even needs a twofold process of uncoupling: first the uncoupling from the immediate sensory input and then the uncoupling from the perception of any animal by seeing itself. This is some kind of referential ability by means of far-reaching uncouplings.

## 10.10 Language

For a long time, there have been attempts to elicit some form of language from animals. In this context, animals were trained, for example, with simple systems of signs or gestures. One method was to have a keyboard with signs for actions or objects. Chimpanzees were able not only to understand and use these signs but also to combine them to short sequences for a message. In this manner, also games were practiced with chimpanzees. For example, a player with a tray of delicacies was sitting on one side and a chimpanzee with a keyboard on the other side, with the signs on the keyboard standing for a certain delicacy. When the appropriate sign was chosen, both were allowed to eat the respective delicacy. The roles could be exchanged. First, the chimps were trained by humans, but later they began to play the game with each other, exhibiting a remarkable amount of self-control in regard to the tray with delicacies (Savage-Rumbaugh 1986; Savage-Rumbaugh and McDonald 1988).

In some settings, also sign language used by deaf persons was learned or words spoken by a trainer, with up to 150 different words recognized. Dolphins also learned a simple signaling structure using computer-generated sounds or hand signs from a trainer for certain objects, such as ball or ring. These signs were even understood within new combinations and sequences (Herman et al. 1984, 1999, 2001). The impressive abilities of Irene Pepperberg's gray parrot, Alex, have previously been mentioned.

These abilities need high mental flexibility. Between the external situation and personal reactions, a signal can be processed flexibly. It is also necessary to represent the signal internally. However, this is not a real language. To establish the difference, the definition of Gärdenfors (2003) is helpful. He understands language as a system of symbols to communicate about our inner world, that is, about our imaginations, memories, plans, and dreams. We can communicate about things that may not be perceivable or even do not exist. Thus, symbolic language needs a rich and complex mental world. Signals as well as symbols are means of communication. However, the main difference is that *symbols* relate to representations, which are potentially completely detached from the environment or from actual feelings and desires, whereas *signals* are about perceptions from the environment or immediate internal states such as fear, hunger, or annoyance.

All attempts to train some communication with primates or dolphins use systems of signals. However, these signals can be dealt with by way of amazing flexibility. Yet, the signals are bound to external objects or immediate desires and cannot be detached from them.

> It is neither the complexity of grammar nor the problems of learning a large system that makes language difficult to reach for other animal species, but the fact that it is symbolic. The inner worlds of other animals are not sufficiently rich to manage the complexity of detached representations that language refers to. (Gärdenfors 2003, p. 143)

Sjölander (1993) explains the difference (taken from Gärdenfors 2003, p. 143):

> The predominant function of language is to communicate about that which is not here and not now. A dog can 'say': I am angry, I want water, I want to go out, I like you, etc. But it has no communicative means enabling it to 'say': I was angry yesterday, nor can it 'say': I will be angry if you lock me up tonight again, and I will chew up the carpet. Likewise, the dog can 'say': There is a rat here! but it cannot 'say': There is a rat in the next room. … Clearly, if you live in the present, communicating mainly about how you feel and what you want to do in the moment, the biological signals inherent in each species are sufficient. A language is needed only to communicate your internal representation of what could be, what has been, and of those things and happenings that are not present in the vicinity.

There are many descriptions of communicative flexibility among animals, also within their natural environments and behavior (Kimbrough Oller and Griebel 2008), but again they are all within the range of signals. The bees' dance also consists of signals, even if they seem to contain some sort of grammar. In an admirably differentiated and precise way, the bees describe to each other the best places to find nectar. Nevertheless, they exclusively use their dance in relation to the external circumstances and use it only immediately after having returned from a flight. Goodall (1986) notes that the cries of chimpanzees were always closely related to their actual emotions. Generating sounds independent from their immediate state of being seems to be impossible for them.

Beyond this, the language of humans includes the dimension of time. We are able to talk about objects and occurrences of the past and about plans for the future. On the basis of the mental representation, we are able to detach ourselves not only from directly present objects but also from the present itself and communicate about the contents of these representations. Animals, on the contrary, live exclusively in the present. "Homo sapiens seems to be the only species that uses language in a totally detached manner" (Gärdenfors 2003, p. 158).

The mental capacities of some apes and dolphins enable flexible use of signals, far beyond what fixed structures for communication offer. However, none of the studies reveals clues to an ability to deal with symbols, which are not within the actual context. The impressive abilities of the famous bonobo Kanzi (Savage-Rumbaugh and Lewin 1994; Savage-Rumbaugh et al. 1998) also stayed within an extensively sophisticated use of signals. In addition, animals lack any form of syntax or grammar. Generally, a grammar of two or three words was the maximum reached. These abilities have been compared with the two-word stage of small human children. The mental capacity of chimpanzees does not exceed this stage, whereas children leave it soon and develop their abilities for language far beyond. Attempts to teach chimpanzees to speak with a voice failed. They lack the anatomic prerequisites and the neuronal equipment for the necessary flexibility and control of tongue, mouth, and breathing.

## 10.11 Brain Size and Flexibility of Behavior

Behavior can be called complex in the sense that it is possible to generate new combinations of actions. In this sense, the discussion of complex behavior by Conway Morris (2003) is identical with what is presented here.

Any form of complex behavior emancipates organisms from their environment as they not only react to immediate environmental influences but also actively act within their surroundings in a more self-determined manner. Thus, complex behaviors can be part of an interactive autonomy. In addition, behavior can be generated that does not directly cope with an environmental problem, thus showing constitutive autonomy as defined in Chap. 3.

Autonomy in the sense of behavioral flexibility can be largely enhanced by several abilities, as discussed in the previous paragraphs: Learning introduces different degrees of plasticity of behavior as different individual solutions for environmental problems become possible, including changes in the interaction with the environment and the social surroundings. Play is an expression of pronounced flexibility of behavior, uncoupled from direct needs and environmental challenges. The actions reveal highly varied sequences of movement and behavior, and the sequences can have new combinations and time structures. Imitation reveals flexibility in behavioral possibilities to follow an external example for some new behavior. It requires far-reaching detachments from FAPs and variability in neuronal control. Flexible tool use and insight also require detachments from FAPs and some neuronal possibilities to understand the function of a tool or the context of a task or a problem. Extensive degrees of neuronal flexibility are needed for abilities such as empathy and self-awareness. Detachment allows the subject to experience an observed process as related to an object, besides its own internal state, or even to experience itself in some objective form. Signals can be used in a flexible manner by several animals with complex neuronal systems. However, the symbolic language of humans relates to representations, which are potentially completely detached from the environment or from actual feelings and desires.

Flexible behavior emancipates the organism in different degrees from the constraints of FAPs. Hassenstein (1969) calls this a gradual liberation of behavior and summarizes it as "aspects of 'freedom' in animal behavior." Regarding ethological reports of flexible behavior, it comes without exception from animals with large and complex brains having extensive neuronal connection structures. Within mammals, large brains exist especially in primates, elephants, and dolphins (Byrne 1995; Marino 1998, 2002; Reader and Laland 2002; Roth and Dicke 2005; Barton 2006) and within birds in ravens and parrots (Delius et al. 2001; Rehkämper et al. 2001; Lefebvre et al. 2002; Nektaria and Lefebvre 2000; Sol et al. 2002). This is in accordance with many studies about questions of intelligence in vertebrate animals, in which extensive sensory functions, focused attention, expanded memories, planning with foresight, skillfulness in problem solving, and an expanded social memory are attributed to large brains (Stamp Dawkins 1993; Parker et al. 1994; Byrne 1993, 1995; Roth 2001; Delius et al. 2001; Heldmaier and Neuweiler 2003; Jerison 2001; Reader and Laland 2002; Lefebvre et al. 2002).

Because the relation of brain volume and intelligence is questioned by some authors, Reader and Laland (2002) conducted a meta-analysis on reports from the literature concerning this relation. They defined three parameters as a measure for flexibility of behavior: the generation of behavioral innovations (the tendency to discover novel solutions to environmental or social problems), social learning (to learn skills and acquire information from others), and use of tools. The incidence of ethological reports about these types of behavior correlated respectively with the relative size of the forebrain (using forebrain volume in relation to brain stem volume) of the animals studied. The method followed the work of Lefebvre, who showed a comparable relation within birds (Lefebvre et al. 1997, 1998).

That there is a relation of large brains and flexible behavior is impressively shown by the example of the octopus. However, at the same time, it shows that there is no simple linear evolution toward these capabilities. Rather, there are convergences (Conway Morris 2003). The evolution of dolphins and primates also went different ways, at least since the end of the Mesozoic. Early fossil whales still had relatively small brains (Marino 1998), so that the increase in brain size to that of present-day dolphins must have taken place independently from that of the primates. During this evolution, different parts of the brain were enlarged, and there are considerable differences in the cytoarchitecture. Another example for convergence is the brain of birds and of mammals, which have different brain structures responsible for behavioral flexibility.

Within the groups, there is, however, considerable variation as well, so that the respective potentials seem to specialize adaptively (Barton 2006). By means of convergences and adaptations, different realizations and combinations of flexible behavior were generated. A large part of the critique on the standpoint of larger brains correlating to increasing intelligence could have its origin in this mosaic of different specializations based on the general organization. Possibly a more detailed analysis of the respective combinations of features of autonomy on the one hand and adaptation to the environment on the other hand would contribute to a solution. By means of these different combinations, the considerable variety of behavioral possibilities also may have been generated. As each animal is equipped with different prerequisites, it is difficult to compare different species in uniform tests of intelligence, as attempted in extensive studies. Consequently, such studies have been criticized (Shettleworth 1998; McFarland 2007).

**Fig. 10.13** Multitasking: motoric dexterity and flexible problem solving in a chimpanzee in the zoo in Gelsenkirchen, Germany

Increasing intelligence correlates at least in part with an increase of motoric flexibility (Chap. 8). Thus, especially the primates have manual dexterity thanks to the increased generation of the pyramidal tract (Fig. 10.13). In a fairly different way, parrots acquired some dexterity of their legs, using them to manipulate objects. Dolphins, on the other hand, are adapted to an aquatic lifestyle, so manual dexterity receded into the background. This again is a hint to the different combinations of flexibility and adaptation. However, motoric intelligence is constantly underestimated (Heldmaier and Neuweiler 2003).

## 10.12 Humans

Essential for representations, such as that of a mouse by the cat, is the cerebral cortex. The cerebellum is responsible for the fine-tuning of learned movements. Both the cortex and the cerebellum are responsible for most of the enlargement of the skull during the evolution of hominids.

One function of the cortex is to generate representations of objects and events from the environment. Internal representations in the brain can be used instead of the object itself, for example, to prepare an action. Gärdenfors (2003) calls this "detached representations" and compares them to the plan of an engineer designing a bridge. The plan is detached from the real world, which delivers the necessary data. Then, the plan is worked out, all elements are combined on a trial basis, and only if the combinations promise success will the plan be transferred to reality. Similar to this planning on paper, the brain can simulate actions in its representations prior to realization.

Calvin (1994) assumes that the evolution of this ability in man could have been in relation to his ability to throw objects well directed, which humans do better than all other primates (Roach et al. 2013). Human beings can simulate the throw in advance. During this simulation, we can estimate the expected movement of the object without having feedback from a real throw. There is a kind of simulator detached from the real feedback (Gärdenfors 2003). The simulator quickly estimates what the anticipated result of the signals to the muscles will be. The signals, which leave the motor part of the cortex, are sent by the cerebellum to both the arm and the simulator. A calculation is made in the brain of what is about to happen in the arm, and the result is sent back to the cerebellum, which adjusts the arm's continuing movement. The calculation loop in the simulator is faster than the loop that moves via the body's muscles. In this way, the simulator manages to adjust the arm's movements during the throw more quickly than what can be accomplished with muscular feedback. The faster we want to throw an object, the less time we have to steer our arm. Without such a simulator, we would not be able to solve the control problem involved in aiming. However, it is not known in detail how the simulator is implemented in the brain.

Gärdenfors (2003) makes a distinction between "cued representations" and "detached representations." Cued representations stand for a perception from the present environmental situation. A detached representation, on the other hand, is not linked to such a present perception and can be used independently from external signals. He further subdivides detached representations into "dependent representations," which are dependent on an external referent, even if the object is not immediately present in the surroundings, and "independent representations," which have no correlation to any external referent. Fantasies, imaginations, and abstract notions are representations, which can be handled independently. Detached representations are necessary for planning and for all other higher cognitive functions. His hypothesis is that this also represents the sequence of the evolution of these abilities (cued representations → dependent representations → independent representations), and that the development of thinking can be described as the detachment of increasing representations. When different representations can be used, then different ways to reach a goal can also be taken into consideration. This is one aspect of our potential to act using reasoning.

In the same sense, sequences of actions can be planned regarding the future. Thus, we can plan for future needs and foresee that we will be hungry tomorrow and put aside some of our food. We can also foreknow that it will be cold in the winter, so we build a shelter in advance. This anticipatory planning is yet another example of how human thought can be detached from present situations (Gärdenfors 2003). Generation of a concept of time might be closely connected to our ability to plan. Therefore, the human capacity to think has at least one origin in the principle of detachment that was generated throughout evolution.

Especially, the increase of secondary areas (association areas) of the brain is crucial within mammals. These are areas that are not imposed by certain motoric or sensory fields but have flexible integrative functions. The prefrontal cortex particularly is understood as the area in which associations are generated and that enable

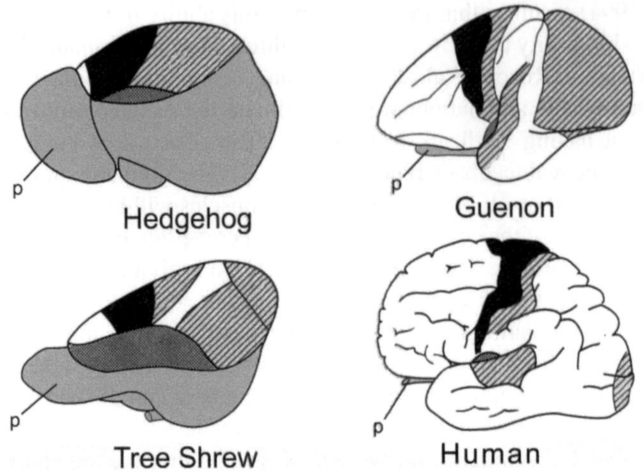


**Fig. 10.14** Association areas (*white*) in the neocortex of some mammals and humans. *Black*: motor areas. *Striped*: sensory areas. *Dark gray*: older nonneocortex areas. *p* palaeocortex (Modified from Heldmaier and Neuweiler 2003)

thinking and planning. These areas are not bound to monomodal functions but rather are the basis for multiple connections, contemplating alternative scenarios of how to interact with the external world and to find novel solutions for behavioral tasks. This provides the physiological basis for the plasticity and for the flexibility of behavior. In primates, especially in man, the prefrontal cortex and the association areas are much larger than in other mammals (Sherwood et al. 2008) (Fig. 10.14). Striedter assumes that especially "the enlargement of the lateral prefrontal cortex probably helped to increase the ability of humans to suppress reflexive responses to stimuli. This behavioral 'freedom' probably helped *Homo sapiens* evolve symbolic language" (Striedter 2005, p. 13).

The prefrontal cortex integrates different neuronal processes and combines them to sequences of action. Areas of the prefrontal cortex are, for example, responsible for learning new sequences of action. Apes with lesions in the prefrontal area are no longer able to learn new procedures, whereas procedures learned before the lesion was set are still performed. From observations such as these, it is known today that in humans the prefrontal cortex is essential for the plasticity of behavior and action. Here, elements of memory are retrieved and combined to new sequences in time. This flexible connectivity is used not only for actions within the environment but also for virtual (notional) acting out during planning, for thinking, and for speech (Heldmaier and Neuweiler 2003; Kandel et al. 1995; Sherwood et al. 2008).

Behaviorists have denied these abilities. They tried to reduce animals – and to an essential degree, even man – to mere stimulus-response machines. However, their explanation was inappropriate because, by means of representations, situations can be simulated before a first attempt at a solution is even made. Thus, behaviorism was not able to grasp the decisive point within evolutionary changes. Meanwhile,

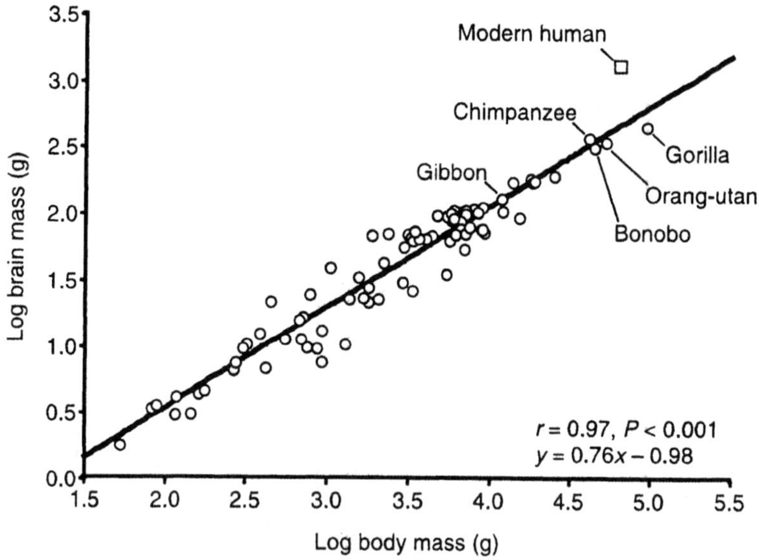


**Fig. 10.15** Allometric scaling plot of brain mass versus body mass in 86 species of primates, showing the least-squares regression line fit to the nonhuman data. Modern humans have brains that are approximately three times larger than would be predicted for a primate of the same body mass (From Sherwood et al. 2008, with permission)

this is well accepted today. Nonetheless, a new fallacy arises from sociobiology and evolutionary psychology, which try to explain animal and human behavior as fixed by neuronal modules. Each of these modules is a way of behavior and thinking that was shaped by natural selection to solve a particular type of problem faced by our Stone Age ancestors, for example, communicating, prey stalking, disease avoidance, mate choice, and coalition formation. The potential of man to act – at least in part – independently from his biological constraints is largely ignored by this approach. However, there are philosophers as well as neurologists who criticize evolutionary psychology and develop alternative views (Heyes 2012a; Richardson 2007).

There is more about the evolution of nervous systems in animals and humans than just the successful occupation of niches, the adaptation to environmental conditions, or the species' strategy for survival. The description of these differences, however, in spite of the amazingly clear evidence from the data of physiology and comparative neurology, is often neglected. Other publications point at new dimensions of flexibility that were generated by complex nervous systems (Neuweiler 2008; Barrett 2012; Gibson 2002; Heyes 2012a, b; Sherwood et al. 2008) (Fig. 10.15).

In humans, detachments by means of the high grade of encephalization lead to far-reaching flexibility of behavior. *Learning* is a central component during child development. There is only a small amount of fixed behavior remaining on which we can rely. On the other hand, there exists a learning ability that lasts a lifetime. This is also widened to cultural learning, which is thought to enable cumulative cultural evolution, the nongenetic inheritance of information in a way that allows individuals and groups to achieve a progressively better fit with the demands of the social and physical environment and serves as a basis for further creativity. Many researchers interested in the evolution of human cognition believe that this type of cultural inheritance is what makes the lives of contemporary humans – with our built environment, science, technology, art, political and economic systems – so different from the lives of animals, including those of our closest living relatives (Heyes 2012b; Richerson and Boyd 2005; Shea 2012; Lewis and Laland 2012).

*Imitation* (or imitation learning) is a type of social learning, not only in children, that is also thought to play an especially important role in cultural inheritance (Heyes 2012b; Richerson and Boyd 2005). It is social learning in which the observer acquires new abilities and skills. Noting that skills such as flint knapping and basket weaving require new ways of moving the hands and fingers, relative to one another and to materials, many researchers regard imitation as crucial for the cultural inheritance of instrumental-technological skills. Imitation also seems to be indispensable in the development of communicative-gestural skills, in learning the postures, gestures, and ritualistic movement patterns, such as those used in dance and many other social interactions. The range and precision of our imitation of body movements far outstrip anything found elsewhere in the animal kingdom.

Using *language and speech*, we are not only able to communicate with others in a highly flexible and complex manner but also can communicate about past and present and about subjects completely detached from the external world. No animal is able to pass on experiences to other individuals or to following generations to such a degree and to generate a collective memory independent from the genetic prerequisites as human beings can. Language is tethered to sociality and is an important characteristic in human cognitive evolution. It increases an individual's understanding of the world because the individual is no longer limited to its own experiences. Apart from the ability to share intentional thoughts, language also allows the exchange of information, ultimately leading to distributed cognition, and provides an important substrate for symbolic thought (Shultz et al. 2012; Sterelny 2012).

*Tool use* is elaborated to computers, cars, and airplanes (Nowell and Davidson 2010; Lewis and Laland 2012). Our social communication and cooperation, including the high good of humanity, are based on the ability of *empathy* (de Waal 2009, 2012).

It is characteristic that *play* is such an important part of our culture (Buchsbaum et al. 2012). In children, play is related to creativity, behavioral plasticity, physical fitness, and increased abilities to innovate (Nowell 2014). But, adults also engage in play: From chess to boules, we cultivate our emancipation. During sports such as baseball, skiing, or tennis, up to the Olympic "games" and in the admired acrobatic skills during circus performances, play combines with the cultivation of the autonomy of movements. Play is also an integral part of the arts, such as in theatre, dance, and many others, but here further cultural dimensions are involved.

This flexibility and independence furnish the essential basis for the human potential of freedom (Eibl-Eibesfeldt 1999; Hassenstein 1969; Heldmaier and Neuweiler 2003; Neuweiler 2008; Heilinger 2007). Biological restrictions still accompany us and impose limitations on our degrees of freedom, not only in our physiological needs, but also in our behavior and in some of our social structures. However, at the

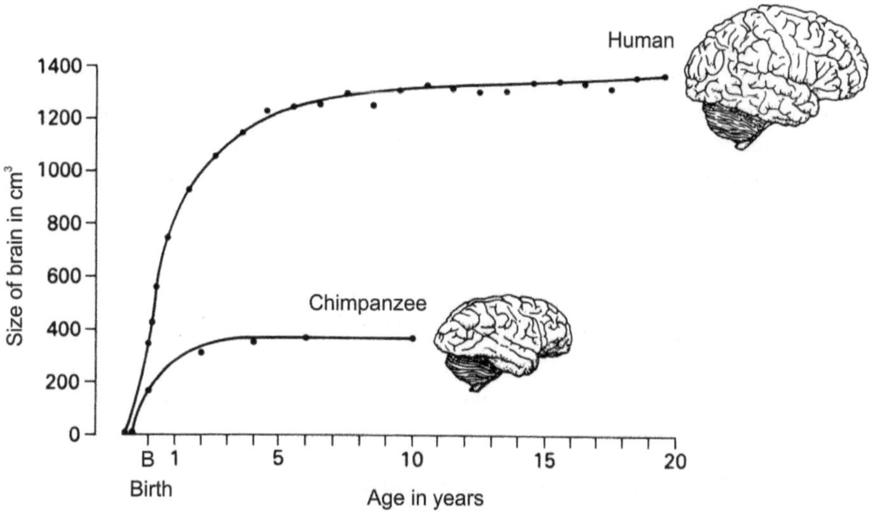


**Fig. 10.16** Growth patterns in chimpanzee and human brains (From Passingham 1975, with permission)

same time our biological organization enables the generation of a world of play, speech, and culture in which we have emancipated ourselves from determinants of pure biological necessities to a large degree.

The largely prolonged development, the enhanced time for growth and maturation, as well as an extended period of plasticity of cortical synapses play an essential role in human beings (Buchsbaum et al. 2012; Nowell 2014; Roth 2001). In prosimians, the maturation of the brain is completed by the age of about 2 years, in the great apes by the age of 6–7 years. However, the brain of man continues to mature up to the age of 20 years (Fig. 10.16), which is much longer than in any other primate species. More slowly maturing neuronal pathways provided an opportunity for learning in a social environment to strongly shape plastic changes in the developing brain (Kipp 1980; Nowell 2014; Sherwood et al. 2008). A crucial phase during the development of the brain in human children is around the age of 2.5 years. At this time, the growth of larger anatomical structures within the associative cortex comes to an end, and a phase of fine wiring of neuronal connections begins (Mrzljak et al. 1990). As demonstrated previously, this is also about the same time when the developing cognitive abilities of humans leave those of other primates far behind. The elongated time of maturation as well as the possibility of maturation by way of practice, play, and experience widen the plasticity of cortical functions drastically (Roth 2001). In addition, recent research proposes longer malleability of the adult human brain, essentially as supple during the whole lifetime (Robson 2013).

Buchsbaum et al. (2012) see a close connection between the longer protected periods of learning during childhood, play, and adult cognition. They assume that changes in these abilities originated in immature proto-humans enjoying longer protected periods of learning and, in particular, engaging more extensively in the free exploration found in play. They propose that this developmental change created the context for the application of more powerful learning processes. In particular, these learning processes included a newly sophisticated ability and motivation to learn about causation and to construct causal models. Those models, in turn, support sophisticated inference and planning by allowing consideration of a wide range of alternative possible future outcomes. The result was a set of new abilities, including more sophisticated tool use for foraging and more sophisticated social intelligence for cooperative child rearing. Those abilities in turn allowed for still greater caregiving investment and a still longer childhood and so on. However, they describe that children also engage in a particularly distinctive kind of pretend or symbolic play. In this type of play, children go beyond simply practicing actions they will require later or manipulating objects to discover their causal features. Instead, they work out elaborate unreal scenarios, often with the aid of language, props, and gestures.

## 10.13 A Bridge Between Motoric and Mental Play?

Let me follow the argument of a strong relation between motoric and mental actions a bit further. Burghardt (2005) presents some fine considerations on the relation between behavioral play and mental mobility in humans. Play in humans involves strong mental elements, including pretense, feinting, phantasy, predicting, and anticipating the actions of objects or other players. Visual-motor integration may be heavily involved, as other sensory systems may be operating as well. Burghardt poses the question of whether these aspects of play could be linked to planning, mental rehearsal, imagination, and creativity in humans. For a possible answer, he cites recent neurophysiological work, which is increasingly uncovering relations among motoric acts, mental imagery, and the brain (for references, see Burghardt 2005, p. 393).

Circuits in the basal ganglia are involved in both motor learning and cognitive behavior, suggesting that the links between motor performance and cognitive processes either are ancient or are modifications from ancestral vertebrate systems. Striatal projection neurons depend on afferent and loop circuits in both the neocortex (especially premotor, parietal, and prefrontal cortex) and the thalamus. Now, it is also known that new neurons can be added to adult brains; thus, continuity from childhood play may have a physiological basis.

A more direct cognitive link is being made by the description of the so-called mirror neurons. A research group in Italy showed that neurons in the premotor cortex fired when the studied macacas performed a goal-directed action, such as reaching for and grasping a piece of food. Some of these same neurons also fired when the monkeys simply observed the experimenter doing the same thing. These neurons then were called mirror neurons as they mirrored what the monkey observed (Gallese et al. 1996; Agnew et al. 2007).

Several types of these neurons were identified. Some were specific to single actions by the monkey or the observed experimenter, such as grasping, holding, manipulating, and placing. Other mirror neurons were specific to two or three actions in combination.

The area of the premotor cortex involved has been considered homologous with Broca's area. Thus, it has been argued that such mirror neuronal systems in humans could be involved in the recognition of both actions and sounds. According to Burghardt, these studies suggest that mental imagery, physical movements, and perception can be linked in certain areas of the brain even down to involvement of the same neuron, and that imagery might be a bridge between perception and motor control.

With accumulating evidence on the synaptic changes that can rapidly occur in the brain, practice and repetition of similar, but not identical, behavior during play might serve a role in the shifting mental states involved in anticipating, predicting, and controlling one's own behavior in relation to external stimuli. Further work on monkeys identified areas in the premotor cortex and prefrontal cortex that integrate information on serial sensory information, motor sequences, goals, and rewards. Thus, as the cortex develops, processes evolve to channel and "use" the sensory and motor activities generated by play via instinctive behavior generated in more ancient regions of the brain.

> The studies cited here model how we could move from play as a motoric, active response to stimuli, to active play involving pretense and make believe and eventually to behavior largely divorced from physical actions such as imagination, fantasy, new ideas, and complex social assessments.... These phenomena may all be more closely linked than our theories suggest, perhaps because of the enduring mind-body dualism that disparages mere animate movement.... (Burghardt 2005, p. 394)

Other motoric actions can also be closely related to mental processes. Thus, we tend to draw, sound out, or act out ideas to "test them." The process of writing (possibly even more so with a pencil on paper rather than with the computer) also seems to facilitate the process to settle thoughts or concepts, and gesturing seems useful in facilitating verbal expression of ideas.

In the opposite direction, motor imagery is the process of imagining behavioral actions. When in experiments people were asked to imagine movements, the neurophysiological responses in various regions of the brain resembled those made when such movements were actually executed. When people were asked to imagine writing a letter and then actually wrote the letter, the same areas in the cerebellum, the prefrontal cortex and the supplementary motor cortex, were activated. There have even been reports of experiments in which the merely imagined exercise had beneficial effects on subsequent physical ability and strength of movements. Thus, it is possible that there is a functional link between mental imagery and physical performance.

If merely imagining activity is functional, the next step is to mentally rehearse different actions to choose among alternatives before making a behavioral commitment. Linking these mental rehearsals with possible outcomes is essential for creativity, innovation, and social actions.

Burghardt (2005, p. 397) comes to the following conclusion:

> "Behavioral (physical, motoric, sensory) play may be an important developmental precursor to mental play involving rehearsal, prediction, planning, imagination, problem solving and creativity in realms such as social adeptness, language and communication, cognitive (decision-making) abilities, and emotional (empathic and "mind-reading") processes.

> Thus physical play with self, partners, and objects leads to learning about how to effectively respond to unpredictable, unexpected, or complex features of the world. … It is an essential evolved and developmental process in mental rehearsal, where one can try out scenarios internally before choosing which, if any, would be most productive if physically carried out."

In my paragraph on play, I stated that the interpretation of play as a mere practice of actions that might be needed later as an adult is too simple. Now, we can state more precisely: Play is an expression of an increased degree of freedom in general as well as an exercise and means of creativity, flexibility, and individualized learning that may produce individual and possibly new solutions of physical, cognitive, and social tasks.

In recent years, a new field in neurobiology and philosophy called embodied cognitive science has emerged. Whereas traditional neurology rests on a fixed inside-outside distinction, in which subjectivity is seen as a mere by-product of the brain's activity as a symbol-manipulating machine or an information processor, the embodied cognition perspective views mind and brain as a biological system rooted in body experience and interaction with other individuals. Embodiment refers to both the embedding of cognitive processes in brain circuitry and to the origin of these processes in an organism's sensorimotor experience. Thus, action and perception are not described in terms of the classic mind-body dualism but rather as closely interlinked. In this context, the brain is interpreted as an organ of modulation and transformation that mediates the cycles of organism-environment interaction (Fuchs 2009a, b, 2012; Thompson 2007; Thompson and Varela 2001).

The assumptions made by embodied cognitive science are far reaching concerning the understanding of the relation of biological life and the phenomenon of consciousness: Conventional neuroscience regards the mind as somehow localized in, caused by, or identifiable with the brain. Consciousness becomes an epiphenomenon of the neuronal machinery that, operating behind our backs, creates the illusion of a continuous self and an autonomous will. The body with its sensors and effectors becomes a mere input-output device in the brain's service. This concept is the core of the so-called mind-body problem, the long-standing Cartesian divide. Hence, there is no way to close the gap between mind and life.

The embodied approach, however, proposes a way out of this divide, referring to both the embedding of mental processes in the living organism and the origin of these processes in an organism's sensorimotor experience (Fuchs 2009b). The brain is primarily seen as an organ of the living being and only by this becomes an organ of the mind. Both life and mind are seen as essentially related, dependent on the continuous exchange with their environment. Just as respiration cannot be regarded as restricted to the lungs but only functions in a systemic unity with the whole body and the environment, so the individual mind cannot be restricted to the brain. Consciousness is not seen as an object or state that can be localized but rather as a process of relating to something: a perceiving of, remembering of, wishing for, aiming at, and so on. This dynamic and intentional character of consciousness is not covered by the concept of single "mental events" that could be translated into corresponding brain states. Therefore, the neurocognitive system cannot be grasped separately either; it exists only enmeshed in the world in which we move and live with others through our bodily existence.

According to Fuchs (2009b, p. 224):

> "Apart from inner regulation, the main task of the nervous system is to establish the sensorimotor cycles that connect organism and environment. Here embodiment implies the inherent connection of perception and bodily action, as already developed in the concepts of von Uexküll's (1973) Funktionskreis and von Weizsäcker's (1986) Gestaltkreis. What the organism senses is a function of how it moves, and how it moves is a function of what it senses. … Perceptual space is not a pre-given external container, but rather a medium or working-space, moulded by our sensing and moving bodies from undifferentiated visual stimuli. This means that living systems do not operate on the basis of internal representations of an external world. Rather, they enact an environment inseparable from their own structure and actions, an Umwelt in von Uexküll's sense."

Considering this concept of embodiment, my previous discussion on behavioral flexibility and intelligence in animals may need a more integrated approach to the different components involved. It should be possible to study the intelligence of animals within the close correlations between neuronal, motoric, sensory, and other physiological systems to understand the truly specific intelligence of these animals within their specific worlds. As yet, the concept of embodied cognition has predominantly been discussed in human neurobiology, referring only to some animal experiments as scientific models. Comparative work among animals, however, using this framework would add the evolutionary dimension to learn more about the origins of these systems, research that still has to be performed.

## 10.14 Emotions

There is no reason to assume that animals may not have mental processes that are to some extent comparable to those of humans, embodied within their special organizations. This also includes the emotional lives of animals, which may also show phylogenetic differences according to the degree of their overall autonomy.

Aspects of the mental lives of animals and their significance for their behavior are studied by the modern field of cognitive ethology, which performs comparative, evolutionary, and ecological studies of animal minds, including what animals feel and their emotions, consciousness, and self-awareness (Griffin and Speck 2004; Bekoff 2004; Bekoff et al. 2002). Bekoff (2007), for example, comes to the conclusion that animals have a rich inner world with feelings that in many aspects resemble our own.

For those of us who lived and worked with animals or observed them intensively, it is natural that a dog, a cat, or a parrot have emotions we can perceive, describe, and influence in many cases. Anyone who has been working with young horses knows that to reach them emotionally and gain their confidence is a central task of the training, and that no two horses are the same in the structure of their individuality. But, for a long time science denied that it would be possible to study emotions in animals, or even that these emotions exist, a result of the often extremely reductionistic orientation of science. Thus, cognitive ethology has had a hard time placing this aspect in science, and it was often subjected to objections.

We can reliably perceive animal emotions through empathy, and cognitive ethology showed convincingly that this can be included in scientific studies. It is an extremely important and fruitful area of research, as the inner world of animals is something similar to an open book regarding not only their true characters but also the patterns of evolution that shaped these features. It belongs to the most interesting miracles of how and why evolution generated affection, friendship, joy, and empathy. Last but not least, knowledge of the richness of animal emotions is essential for how we treat animals in domestication as well as in the wild (Bekoff 2007; de Waal 2009, 2012).

Bekoff prefers to attribute emotions to all animals, not only to those with a large relative brain size. One of his arguments is that especially some older parts of the brain (i.e., the limbic system, including the almond-shaped structure called the amygdala) are important in human emotions, and they are shared by many animals. More complex emotions involve higher brain centers in the cerebral cortex, some of which also are found in animals.

Basically, Bekoff might be right to attribute emotions to all animals, but if we look further into the evolutionary past, there are animals that at least do not show such elaborate and nuanced emotions as dogs and chimpanzees. There might be some emotional capacities also in lower vertebrates. Fishes, for example, may experience pain and fear. But, how about a clam or a sea urchin? They are clearly sensitive and have some degree of self-determined nervous circuits, which move them far above what an apparatus is. Of course, one difficulty is that we cannot really know as our empathy does not reach so far to experience more about perhaps some basic and simple, but for the animal no less important, emotions within its world. Nevertheless, there seem to be degrees and differences, and it could be important one day to study them, when the field will be methodologically more developed. However, the essential principles of sensibility, detachment, and relative flexibility, and thus the prerequisites, are already involved in early steps of evolution.

If the theory is correct that more elaborated nervous systems with increasingly self-determined reactions and self-referential modulations are part of an increasing autonomy of individuals, then the evolution of richer and more differentiated emotions must be in some relation to this autonomy. The detachment from environmental stimuli and elaboration of self-referential neuronal processes, which increased in complexity and vivacity, are examples for constitutive autonomy as defined in Chap. 3. An increasingly autonomous and richer internal world is generated. Thus, internalization gains a new dimension with emotional and cognitive qualities. Finally, these self-generated inner worlds start to communicate and to interact with each other. They begin to experience that their buddies also have such an inner world, which can be experienced sympathetically or antipathetically.

However, in humans a new quality of autonomy emerges concerning emotions: Animals have unfiltered, direct, and spontaneous emotions. But, in humans, thinking about the emotion allows flexibility of response in changing situations after evaluating which of a variety of actions would be the most appropriate to perform. We learn to think before acting. Thinking allows us to control and direct feelings and to control the resulting actions, so that, depending on the social situation, we might do the right thing. We are mostly able to detach thinking from our emotions. However, sometimes our emotions can become out of control, but usually we try to be master in our house. This is an essential part of human autonomy.

In summary, it can be stated that any form of complex behavior enables organisms to answer in a self-determined and flexible manner to signals and conditions of their environment. This self-determination can be enhanced in some organisms, so that the answers become more flexible in the sense that the type of reaction is less fixed, and it is possible to generate new combinations of actions and reactions. The variety of possible answers to the environment is larger than necessary at any particular moment (interactive autonomy). This correlates with the evolution of more complex and sophisticated central nervous systems in different varieties of design, which increases the scope of self-referential, intrinsic functions within the system (constitutive autonomy) as more sophisticated internal processing becomes possible (see Chap. 8). This fulfills criteria of Definition 2 and thus contributes to the overall autonomy of more evolved animals and humans as well.

Examples of flexible behavior are as follows:

- Learning introduces different degrees of plasticity of behavior as different solutions for environmental problems become possible.
- Play is an expression of pronounced flexibility of behavior, uncoupled from direct needs and environmental challenges. The actions reveal highly varied sequences of movement and behavior, and the sequences can even have new combinations and time structures.
- Imitation reveals a flexibility in behavioral possibilities in order to follow an external example for some new behavior. It requires far-reaching detachments from FAPs and variability in neuronal control.
- Flexible tool use and insight also require detachments from FAPs and some neuronal possibilities to understand the function of a tool or the context of a task or a problem.
- Extensive degrees of neuronal flexibility are needed for abilities such as empathy and self-awareness. Detachment allows the subject to experience an observed process as related to an object beyond its own internal state or even to experience itself in some objective form.
- Signals can be used in a flexible manner by several animals with complex neuronal systems. However, the symbolic language of humans relates to representations, which are potentially completely detached from the environment or from actual feelings and desires.

# Chapter 11
# The Evolution of Man

Chapter 10 described that humans have far-reaching degrees of behavioral and motoric flexibility. However, man is not the most autonomous organism in general. We do not have the physiological autonomy to exist at –30 °C in Greenland like the musk ox. We do not have the running capacity of the cheetah. We did not emancipate our earthbound movement capacity to enable us to fly through the air as most birds do, and we admire the arctic tern, which is able to fly around the world once every year, a form of emancipation that is not achievable for us. Beyond this, we are not able to move through treetops in such a skillful manner as many other primates do. Instead, humans have a special combination of features of autonomy.

This combination generates some essential degrees of flexibility and possibilities. First, there are those features we share with all mammals that have been discussed in the previous chapters: skin, which simultaneously closes relatively tight against the environment and is highly flexible and light; endothermy combined with a high aerobic capacity, enabling movements with endurance and to a large extent emancipating mammals from variations in environmental temperatures; effectively stabilized fluid management, including refined processes for homeostasis and highly efficient renal functions; a medium body size, which supports homeostatic functions but does not generate a larger burden for movements; and an extremely refined immune system.

In addition, there are features we share with other primates: The extremities of nonhuman as well as human primates have an exceptionally wide scope of movements in all directions. The shoulder and the pelvic girdle allow wide-ranging excursions. For the shoulder, the preservation of the long clavicula is important. It has been forfeited by many mammals in favor of specializations in quadruped terrestrial runners, which have legs that are narrow and under the body, more or less restricting the scope of their excursions. The extended possibilities of movement of the extremities of primates is also present in the extensive possibilities to rotate the distal extremities (pronation and supination), the range of movements in the hand and foot, and the independence of single fingers and toes from each other. There are special adaptations that can reduce this flexibility among nonhuman primates, but in

B. Rosslenbroich, *On the Origin of Autonomy: A New Look at the Major Transitions in Evolution*, History, Philosophy and Theory of the Life Sciences 5,
DOI 10.1007/978-3-319-04141-4_11, 

general, there is a trend throughout the evolution of primates to retain and to increase these varied possibilities. However, they lead not only to a broad spectrum of locomotory possibilities but also to new functions independent from locomotion.

The thumb and the big toe especially gain increasing independence, first being able to stretch out further and then to rotate and become opposable to the other fingers and toes. This feature supports exact grasping abilities not only during locomotion but also during manipulation of objects, which generally becomes possible.

There is also a special sensibilization of primate hands and feet. On their surfaces, the padding of the balls of the feet decreases during primate evolution, and uniform palms and soles are developed. They are covered by a sensory system that is extremely sensitive to touch and is formed in an especially differentiated way on the pads of the fingers and toes, thus becoming exceptionally sensitive for manipulation and exploring.

The hands increasingly take over further functions supported by an upright posture during sitting or squatting, which is typical for primates and emancipates the hands for different tasks. Thus, functions are extended exceptionally for feeding, extensive exploration and manipulation of objects, carrying and holding, and grooming of skin and fur, whether their own or of their fellows as part of social interaction. Finally, hands and arms also gain functions for communication and become organs of expression even on a subhuman level of primate evolution.

Within the different sensory functions, the optical sense has an eminent significance in primates. The optical sense allows the best spatial orientation within complex structured environments. In nonhuman primates, this is important for survival in the tops of trees, where fast movement and exact orientation are crucially important. Visual acuity and the ability to see stereoscopically especially are both improved. The eyes move into a frontal position, resulting in parallelization of the optical axes as a prerequisite for three-dimensional sight. In this connection, within primate evolution concerning the crossing of the optic nerves in the optic chiasm, there is a trend toward a balanced relation of crossing and noncrossing fibers. In humans, both halves of the brain receive information through the slightly different angle of vision from both eyes from the same points perceived. From this the brain is able to reconstruct the impression of a three-dimensional situation of space.

The retina of mammals contains two major types of light-sensitive photoreceptor cells used for vision: rods and cones. Rods have a special sensibility for light-dark distinctions and moving objects but only weak discrimination. Cones are specialized in visual acuity and discrimination of colors. During primate evolution, there is a trend to concentrate cones within the center of the retina, whereas the periphery contains predominantly rods. Thus, the fovea develops in the center of the retina, as it is known in humans. This improvement of the optical apparatus correlates with an enlargement of the visual cortex. It enables an exact, three-dimensional orientation and might also have some relation to the highly sensitive hands, which are involved in inspecting fine objects.

The profound internalization of embryonic development (see Chap. 7) can also be found within nonhuman primates. The most intensive form of metabolic exchange between mother and fetus is reached in the hemochorial placenta, in which blood vessels of the mother's placenta are opened so that the mother's blood directly washes around the fetal chorionic villi. Thus, birth leads to a bleeding wound within the wall of the uterus (decidual placenta). Whereas most prosimians still have several tissue layers between the blood of the mother and the fetus, all simians and humans have a hemochorial placenta so that embryonic development is maximally internalized.

## 11.1 Special Features of Autonomy in Humans

All of these features, which were generated during the evolution of primates, are elaborated to a special degree in humans. Man's hands are completely freed from locomotor functions because of our upright posture and have nearly unlimited possibilities of flexibility and dexterity. They are able to assume a variety of positions, to resist high levels of mechanical force, and to perform a great variety of tasks. This all arises from a combination of flexion of the fingers at the joints between the finger bones and the wrist bones and finger rotation, mainly of the thumb, but also of the other fingers. The thumb is sufficiently long in relation to the other fingers to allow the fingertips to touch, whether palm to palm or with the thumb on the sides of the other fingers. These grip positions are enhanced by our ability to rotate the palmar side of the forefinger toward the thumb. In addition, the joint at the base of the middle finger, between the carpal and metacarpal bones, is obliquely oriented to resist forces generated by the thumb pressing objects against the other fingers. In apes, it is transverse because here it primarily resists the forces produced during locomotion. Human fingertips are large to resist a force applied to them. We are also able to position our whole hands precisely through flexion and extension at the wrist, by rotation of the forearm, flexion and extension of the elbow, and extensive movement at the open ball-and-socket joint of the shoulder. These patterns of mobility of the human shoulder and arm are of course shared with apes. However, the degree of mobility of the human hand is far greater than in all other primates. Furthermore, this condition is supported by fine neural control and hand-to-eye coordination. Man has greatly expanded the motor and sensory portions of his brain for perception and for control of his hands; this is reflected in highly dense nerve endings in the muscles, joints, and skin of the hand (Trinkaus 1992; Striedter 2005).

Remains of human evolution, as far as we recognize them today, show that the upright posture is an exceedingly old innovation of man. *Ororin tugenensis* (6 million years ago) and *Ardipithecus ramidus* (4.4 million years ago) already had an upright posture. Thus, the first steps toward the emancipation of the forehand evolved before most enlargement of the brain took place. However, the flexibility we now possess was not present to the full extent in early hominids. The hands of *Australopithecus afarensis* show an anatomical pattern with only slight modifications for manipulation. The first major changes in human manipulative prowess appeared with the emergence of the genus *Homo*. Fossilized hand bones indicate a modern human pattern of joint mobility at the thumb in *Homo habilis*. The powerful finger flexion musculature was maintained but was associated with fingertips even larger than those of modern humans and with straight finger bones. This pattern emerged parallel to the first Oldowan tools. However, the size of the spinal cord remained limited compared to *Homo sapiens*, implying relatively crude motor control. Therefore, despite changes in the patterns of hand movement, with a greater range of grip positions in *Homo habilis* and *Homo erectus*, the fine control we associate with human manipulation must have come later. Neanderthals had hands that were more comparable to the modern human hand. However, their hands were predominantly shaped for a powerful grip. Manipulative skills similar to our own did not appear before the emergence and spread of anatomically modern humans approximately 100,000 years ago. The remains of hands and arms of these early modern humans are indistinguishable from those of contemporary athletic persons. These anatomical changes are associated subsequently with several technological changes, such as those of the Upper Paleolithic industries (Trinkaus 1992; Klein 2009; Nowell and Davidson 2010).

A precise grip was one prerequisite for developing the dexterity needed to knap stones. This development must have been accompanied by further elaboration of the pyramidal tract, which in no other primate is as prominent as in humans and is the neurological basis for precise movements of human hands.

Upright posture and walking were early innovations that enabled hominids to reach wider areas within the savanna and finally to leave their African homeland. Humans are able to walk with stamina and are easily able to manage distances of more than 30 km per day. The combination of upright walking together with hands, which can carry things and babies and which become increasingly able to perform techniques to provide oneself with food and objects needed for life, enabled man to undertake considerable migrations. As far as we know, *Homo erectus* was the first to reach Asia, Europe, and Indonesia. Much speculation is possible about the reasons why the population began to undertake such arduous migrations, partly into the inhospitable climates of the ice ages, but we know virtually nothing for certain. However, this might also be an expression of increasing capacities of flexibility and autonomy within the environment.

One aspect of man's mobile flexibility is also the fact that man is a skilled runner with endurance. Bramble and Lieberman (2004) show how well humans perform at sustained long-distance running and give a review of the physiological and anatomical prerequisites.

Man became able to buffer seasonal changes either by techniques or by avoidance using migration. Today, each of us carries a tropical microenvironment in the form of clothes and dwellings, emancipating us from hostile and changing temperatures.

One of the successful techniques involved in contending with the environment and for survival was the ability to ignite and control a fire. First traces of this ability are about 0.5 million years old and were left by *Homo erectus*. This dexterity correlates with other new abilities of these persons related to the growing brain. In Chap. 10, I discussed how anticipatory planning is an example of how human thought can be detached from present situations and thus widens the capacities of flexible behavior. Gärdenfors (2003, p. 76) presents the art of keeping a fire alive as an example for a sequence of actions, which – although relatively easy for us today –needs complex anticipatory planning. This ability presupposes the following thought components:

1. The insight that the fire consumes the fuel.
2. The conclusion that new fuel must be added if the fire is to go on burning.
3. The awareness that, when the fire goes out, I will be cold (or predators will come close).
4. The insight that this creates a future need, which requires me to act now.
5. The conclusion that I must start gathering firewood right now. Once the fire has gone out, it is too late.

The crucial step is imagining oneself freezing when the fire dies or, in more general terms, the ability to conceive one's needs in a future situation. This is one of the essential techniques to oppose the rigors of the environment; thus, it emancipates man from environmental influences to some degree. The use of fire also enables him to cook food, increasing its recoverable nutritive value, which is crucial as increased autonomy is often coupled with the need for more energy. The largely growing brain especially needs a great deal of energy (20 % of the total energy consumed by a human today).

Insights into some mental abilities of early hominids are delivered by examining the shape of stone tools, which are found from *Homo rudolfensis* onward (Nowell and Davidson 2010). The oldest are about 2.5 million years old. To learn how to produce stone tools and to found a tradition, one must be able to imitate someone else, to remember a series of actions in the right order, and to perform purposeful training. Whereas early tools were predominantly formed according to the material and shape of the stones used (early Oldowan), later products of *Homo erectus*, as with the Levallois technique, required that an idea of the final form must be conceived in advance and a plan developed on how to put it into praxis. Tools such as these indicate a fairly advanced capacity for thought (Fig. 11.1).

The history of stone tool activities gives evidence about ancient hominids' abilities to plan ahead. Reconstructions of knapped flint from over 2 million years ago show a detailed sequence of detachments that can only result from mentally "looking ahead" in the manufacturing process (Roche et al. 2009).

It is known that a few animals use simple forms of tools and demonstrate a certain degree of flexibility doing so. However, humans elaborated these capabilities to a variety and skillfulness unparalleled among animals. The Oldowan stone tools reflect an ability to flake stone, which living chimpanzees cannot achieve. Later stone tools show refined shaping and retouching – at least beginning with the Acheulean culture.

In Africa, the Acheulean hand ax tradition has two stages: an early one, when the hand axes were relatively thick, weakly trimmed, and relatively unsymmetrical; and a later phase, when they were often much thinner, far more extensively trimmed, and much more symmetrical in both plan and side views. Late Acheulean collections of artifacts also tend to contain a wider range of well-made flake tools, which anticipate those of the succeeding Middle Stone Age.

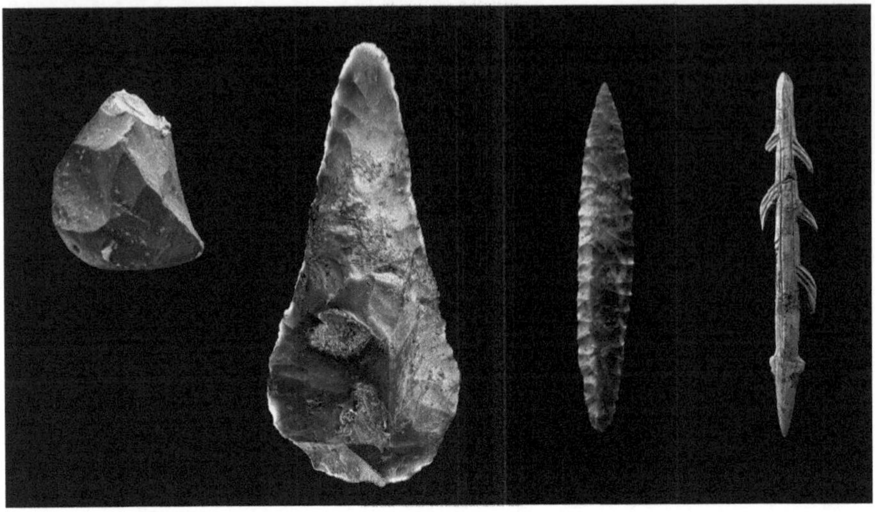

**Fig. 11.1** Artifacts of different epochs. *Left to right*: Oldowan culture, Acheulean culture, Solutréen culture. On the *right side*, a harpoon with barbed hooks from the Magdalénien of France is shown (From Johanson and Edgar 2000)

Late Acheulean people, who lived around 700,000 years ago, often produced hand axes, which appeal to the modern eye for their remarkable symmetry in both plan form and side view. The manufacturers formed these hand axes extensively and meticulously. Moreover, it is credible that they were guided by an evolving aesthetic sensibility (Klein 2009; Le Tensorer 2001). Thus, there do exist older objects exhibiting simple artistic elements. However, objects whose artistic meaning is unequivocal became commonplace only 50,000 years before today. These are associated with the origins and dispersion of fully modern humans from Africa. At that point, the freedom to create art, detached from the requirements of daily care, was developed.

At present, it is still unclear at what time during the evolution of man comprehensive speech aptitude emerged (Sterelny 2012). Some scientists think that early members of the genus *Homo* were already able to speak; others consider *Homo sapiens* as the first species to begin to use speech in such an extremely flexible manner as we do today. In Chap. 10, I characterized a fully elaborated language as a system of symbols. It builds on representations, which can be totally detached from actual circumstances. It is possible that the question regarding the origin of speech will not be solved easily. We do know examples of a fascinatingly flexible use of signals among some higher animals. Thus, it is probable that early hominids had at least some form of highly flexible use of signals that played an essential role in social interaction. At what point communication grew beyond the use of signals to become a symbolic language, however, will be hard to determine. It is unlikely, though, that this occurred at only one point or at a special time. The capability of speech must have developed gradually and in a close interrelationship with other skills, which were acquired step by step, such as manufacturing hand axes and creating artwork or using tools and fire. Such abilities must have had mutual influences on each other. Also, it will be impossible to differentiate retrospectively whether the neuronal basis for all of this activity, including the extension and elaboration of the brain, was generated first or whether the training of such abilities was the driving force behind neuronal extension.

Social learning was another reason for man's increasing flexibility (Buchsbaum et al. 2012; Barrett et al. 2012; Whiten and Erdal 2012). Tomasello (2009) describes this process as the "ratchet effect": Some individual or group of individuals first invented a primitive version of the artifact or practice, and then some later user or users made a modification, an improvement, which others then adopted – perhaps without change for many generations. At some point in time, another individual or group of individuals made a further modification, which was then learned and used by others – and so on over historical time. This process of cumulative cultural evolution requires not only creative invention but also faithful social transmission, which can work as a ratchet to prevent backward slippage. Thus, the newly invented artifact or practice may retain its new and improved form relatively faithfully until a further modification or improvement is devised. The outcome is that human beings are able to pool their cognitive resources in ways that animals cannot match. Tomasello describes in detail how contemporary children learn these cultural traditions during their development and education. Recent neurobiological findings demonstrate how the fine whirring of the brain emerges in an extremely flexible manner during this learning process whereby a symbolic language promotes the process.

Taking into account how many cultural traditions exist throughout the world emphasizes the flexibility of this process. However, even though this capability is truly beneficial to man, it also contains potential for conflicts and wars as long as people do not recognize that different cultures can be complementary to one another.

## 11.2 Neuronal Specialties

Humans are unique among primates in terms of brain size, as described in Chap. 10. This indicates that a significant amount of brain mass increase occurred in the hominine lineage since its origination from a supposed common ancestor. Fossil evidence indicates that in the hominine lineage there have been periods of gradual increases in cranial capacity that were occasionally accompanied by increases in body mass. However, beginning with *Homo erectus*, brain expansion in hominines occurred at a more rapid pace (Sherwood et al. 2008).

Singer (2006) poses the question of whether our present-day brains differ from those of our ancestors some 30,000 years ago. He assumes that there cannot be major differences among the general features with their molecular composition, anatomical structures, and basic connectivity patterns. However, he also thinks that the essential changes, which led to modern humans, might have been dependent on experience, which in turn brought about epigenetic changes.

At least during individual development, there is extremely high plasticity. Whereas the gross structure of the nervous system is most likely set up by the demarcation of pathways of nerve growth, which must have some genetic background, the fine structuring of the brain is not determined. Rather, there is a process Gerhart and Kirschner (1997) call an "exploratory mechanism": During maturation, the brain initially generates an excessively large number of neurons and excess neuronal connections. Then, which ones will survive depends on the connections actually made and their use. Cell bodies of neurons die if their axons fail to reach the tissue they will innervate. Also, only those particular axonal projections that are used will be preserved. This optimizes the connectivity of the nervous system and renders it ideally close to the function the system shall perform.

Thus, as the neuronal-level processes gain experience and learning, they modify the architecture of neuronal connectivity by path finding and stabilizing pathways, mainly up to puberty. After puberty, the brain is also still highly flexible, but changes are more restricted to alterations in the efficacy of the repertoire of connections. Thus, this central organ of humans is characterized by extremely high flexibility, not only in its daily functioning but also in the course of its maturation during individual development and possibly during its recent evolution.

According to Singer, our cultural competence results from the evolutionary development of certain cognitive functions. One of these is our ability to generate abstract, symbolic metarepresentations of cognitive contents by subjecting the results of first-order cognitive operations to further cognitive processing of a higher order (Fig. 11.2). This competence requires the ability to bind the distributed cognitive processes together and to represent the results of these binding operations anew at higher processing levels. Because these higher-order descriptions keep track of the brain's own cognitive operations, they can be considered the substratum for our ability to be aware of our own sensations and intentions as well as those of others. In turn, this awareness is probably the origin of our unique ability to use a symbolic communication system and to generate a theory of mind. In my terms of Chap. 8, this is another level of the principle of detachment, which I traced throughout the evolution of the organization of nerves: the ability to generate cognitive contents on a meta-level and neuronal connections, which are not just in line with first-order representations.

Humans seem to be the only species capable of imagining the thoughts and emotions of another person, Singer continues his argument. We are capable of entering into dialogues of the form, "I know that you know that I know," or "I know that you know how I feel." Such dialogues not only permit a deeper understanding of the respective other person but also allow one to experience one's own cognitive functions in the perceptions of others. Thus, the ability to generate a theory of mind has probably also been instrumental in shaping concepts of self and personal autonomy endowed with intentionality and free will and in the creation of social realities such as value systems. These cultural constructs are as real as the precultural realities. They are part of the environment in which human beings evolve and hence are likely to have as important a role in the epigenetic shaping of the brain's functional architecture as other environmental factors.

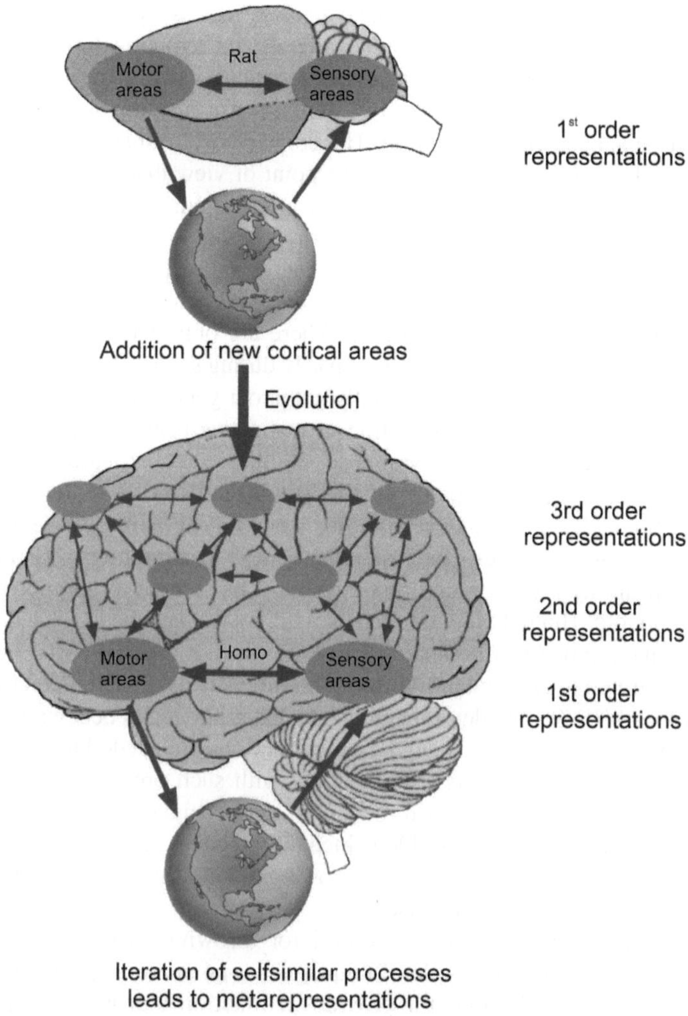


**Fig. 11.2** Singer's (2006) representation of the evolution of the mammalian brain, which he sees characterized by an increase in the surface of the cerebral cortex and the addition of new cortical areas (From Singer 2006, Figure 9.1, page 186, © 2005 Massachusetts Institute of Technology, by permission of the MIT Press)

Singer's description has profound consequences, as cultural embedding is able to influence the phenotype. And, because it does so as a function of experience gathered by preceding generations (Tomasello 2006), cultural evolution might follow its own principles and might not be reducible to purely biological factors, as sociobiology and evolutionary psychology claim.

According to Singer, the new functions of the brain must have been realized simply by the addition of further areas of the neocortex and by the rearrangement of connections among neocortical areas. The additional areas differ from the more ancient ones in the way in which they are connected to sensory systems and effector organs. The new areas in the occipital, parietal, and temporal lobes seem to be involved primarily in the refinement of sensory functions, whereas the new areas in the frontal lobes subserve more executive functions, such as planning action, short-term storage, and management of attention. The connectivity of these phylogenetically recent cortical areas is compatible with the point of view that they reevaluate and bind the distributed results of primary cognitive operations and thereby provide the substratum for the generation of higher-order representations.

With all this, a feature emerged that is clearly unparalleled within the animal world: the ability of self-control and willful, conscious behavior, with all the related qualities that make a responsible person. There are only a few situations left in which we are guided by instincts, such as partly during sexual behavior, mainly for the sexual intercourse itself, but essentially not in our general behavior between the different sexes in general. The reduction of instinctive behaviors is paralleled by large degrees of freedom in behavioral possibilities, such as learning capacities, imitation, tool use, insight, empathy, self-reflection, and speech.

Refined brain capacities also serve a precise ability to aim at something when throwing a spear or stones. In Schöningen near Braunschweig (Brunswick), Germany, spears have been found that archeologists assume to be approximately 400,000 years old. Hunted animals such as wild horses have been related to them. The spears have properties comparable to those used today in sports, which has been demonstrated in experiments with javelin throwers using reconstructed copies of the old spears. If we consider for a moment what is altogether necessary to handle a spear, it becomes clear how many flexible functions are needed to manage such a weapon. No other primate is able to throw with such precision. Chimpanzees occasionally fling sticks, but their throwing is vague. Humans possess the obvious dexterity of hands, arms, and shoulders, including the legs and the rest of the body as well. The whole body is involved in throwing a spear, and all parts of the body must be able to cooperate in a precise way.

The extremely refined neuronal control for throwing, which I described in Chap. 10, is a fascinating example of human neuronal and motoric flexibility and precision. The neuronal simulator provides a function that helps us guide the direction and force of our limbs' movements with great accuracy. It is important not only to aim when we throw, but also to knap a piece of flint into an axe, for example. We need precise guidance of our hand and of the correct knapping action as well as much practice to obtain a well-formed axe. The well-made wooden spears from Schöningen illustrate the point because they are associated with the production of flake-and-chopper stone tools, yet they reflect at least as much forethought and control in manufacturing as any hand ax. The voluntary availability of these complex processes is an essential expansion of our independence and flexibility.

Increased flexibility might have been involved when certain hominids were able to evolve to *Homo sapiens*, whereas others disappeared from the scene. Thus, it is assumed that the robust australopithecines were more specialized concerning food and other features of their lifestyle. Schrenk (1998) assumes that *Homo rudolfensis* showed more adaptability regarding food and possibly in all of his behavior, thus emancipating him further from the environmental conditions than the australopithecines. Also, *Homo sapiens* might have been more flexible compared to Neanderthals concerning the capacity of the prefrontal cortex and dexterity of his hands.

Enhanced cultural abilities evolved rapidly once the first steps were taken in this direction. Archeology describes many of them. However, let us take a quick look at the principle of detachment, which again was involved when man produced those wonderful paintings in caves of France and Spain, beginning around 40,000 years ago. The illustrations have been produced at some distance from the natural specimens of living animals and people, within the darkness of the cave, presumably using some sort of artificial light. This means that all of these often relatively exact pictures, showing certain species of animals with their respective characteristics and including many details, movements, typical situations of their life, and much more, were produced by memory and from the imagination of the painter, detached from the direct example outside the cave. Is it possible that this was a means to train cognitive detachment?

Probably, it is also no coincidence that around the same time new types of artifacts made of stones, bones, ivory, and antlers emerged, as well as objects that obviously had symbolic meaning of social relevance, such as pieces of jewelry or hunting spears with ornamentations. This marks the beginning of the late Paleolithic Aurignacien epoch of culture. Now, figurines depicting natural impressions, and more importantly representing abstract aesthetic images, became possible. These artifacts demonstrate a developing ability to think in pictures and to communicate about them. Forms and structures from nature could now be dealt with in an abstract and metaphoric manner.

This newly acquired ability to isolate the properties of objects in the mind and to place them into a new context had profound consequences. Properties such as "pointed" or "hooked" could now be combined and used technically. Once one is able to think in terms of "prong-like" or "hook-like" detached from the natural object, then it is possible to imagine the effect of tips of spears with barbed hooks (Fig. 11.1). The concepts do not develop their effects visibly. Also, composite tools, which combined separate pieces of material, were manufactured, again implying the ability of much higher abstraction than was necessary for hand axes, whose impact could be experienced directly.

Whereas Neanderthals remained basically on the same cultural level for more than 100,000 years, the Aurignacien people were able to explore new possibilities in many areas of life within a relatively short period of time. This took place with increasing speed and continued until our own times. To a large extent, this development is associated with increasing abilities of perception, handling, and communication of mental representations.

Late Paleolithic people might also have acquired some understanding of time, representing an imaginative detachment from the present. Thus, the representation of forthcoming needs becomes possible, perhaps much more elaborately than the first users of fire were capable of, as discussed previously. However, some of the upcoming signs of culture might be interpreted especially in this sense.

The well-known "Venus figurines" with exaggerated buttocks and breasts seem to be fertility symbols, indicating some reasoning about the origin and the beginnings of life. The remains of graves might also indicate commencing consciousness about the end of life on Earth and ideas about the possibility of life in the hereafter. The first graves known are from earlier Mousterien sites, but unequivocal indications for some kind of burial ritual, including valued items or funeral offerings, come from the late Paleolithic. Some graves are elaborate, and individuals were often buried with special bone, shell, or stone artifacts. Clusters of perforated seashells, animal teeth, and dense concentrations of ocher characterize numerous graves. Two neonates buried at a site in Austria were laid on a bed of red ocher, and their grave, 40 cm long, was covered by a mammoth scapula supported by a tusk. More than 30 ivory beds accompanied one of the bodies. These and even more elaborate graves suggest a burial ritual and perhaps notions of an afterlife similar to those recorded among historic hunter-gatherers (Klein 2009).

Later, during the Neolithicum, humans achieved further emancipation by way of agriculture and the domestication of animals, thus becoming more independent from direct supplies in the natural environment. With settlement, humans organized the basis for the fulfillment of their needs themselves. Huts and houses became increasingly planned means to retreat better from the influences of climate and weather. Further emancipation was accomplished by way of technical equipment, reaching well into our modern society and thus carrying on the long trend of biological autonomy. However, this emancipation in many cases goes so far that man even loses contact with his natural environment and starts to destroy it. Increasingly, humans have even become more dependent on their social structures and technology than on environmental influences.

Man has a greatly prolonged time of development and youth, and this also makes sense regarding autonomy: Prolonged time for learning and maturing results in increased capacities of flexibility and variety. The slowing of the maturation process in humans adds additional years for behavioral experience and enhances developmental plasticity to learn, practice, transmit, modify, and innovate regarding aspects of their culture. Children's play during this time might have been a key element of the evolution of unique aspects of human cognition and behavior (Buchsbaum et al. 2012; Nowell 2014). Thus, for young people, it is essential to have enough time for individual development. Any attempts at premature education or acceleration of training will reduce creative capacities. The history of mankind must be told in terms of autonomy, increasing flexibility, and degrees of freedom.

## 11.3 Animals and Man

A long tradition of attempts to define the decisive difference of man from animals by some single criterion exists; these attempts either often met objections or at least seemed to solve only part of the problem. For example, the focus was frequently on only one intellectual capacity, such as language, tool making, or theory of mind, as if that capacity existed in isolation from others.

Instead, several abilities might be involved simultaneously: liberal mindedness, curiosity, universality, speech, insight, reflection and self-reflection, culture, self-control, tools, moral responsibility, and artistic abilities. They can all be characterized by the relative freedom from direct instincts based on natural conditions. The survey in this book showed that much of what makes us human in this sense has been prepared during evolution.

Man, however, is definitely on a special evolutionary track. He was able to widen his natural abilities so far that they became the basis for culture and humanities. They are the result of synergistic interactions among several evolutionary developments. Thus, not a single, but a whole complex of features has set us free from determinants of the purely biological organization – if even only to a restricted degree. We are still bound to many requirements of our physical body. At least when we are hungry, we experience this. Besides such biological constraints, there are behavioral ones that carry relicts from our evolutionary past, but they do not constitute the whole human.

Nearly all of our biological inheritance is superimposed by cultural, moral, and individual standards and motives. We also integrate bodily needs into culturally shaped events. For example, the necessity for nutrition is integrated into a business lunch or a candlelit dinner, thus raising the act far above animal-like satisfaction of hunger. Mating and marriage cannot be simply reduced to reproduction. In humans, this behavior is heavily driven by emotional affection, and we marry the partner with whom we have common interests and ideals. Of course, there are residual instincts, but we expect that a mature person is able to control them and to integrate them into cultural traditions and social responsibility. In this sense, the theory of autonomy can be an important component of the answer to the outstanding question of how man and his cultural capacities can be linked to the history of biological evolution.

Discussions of the evolution of human intelligence often ignore motor and perceptual abilities, preferring instead to focus on higher cognitive phenomena. However, as I showed in Chaps. 8 and 10, our independence also concerns our flexibility and versatility of movements, which become part of our culture in play, sports, handicraft, and many more instances. In addition, the majority of us can communicate complex information and ideas, using complex motor skills such as speech, gestural languages, handwriting, and typing. We can make practical use of creative ideas only because we have the manual dexterity needed to create tools, architecture, and art. Our motor skills thus serve as intellectual and creative channels (Gibson 2002). Basically, all the features of autonomy, as they are described in the present text, contribute to the range of possibilities we have from our biological organization.

The theory of "embodied cognitive neurobiology," briefly discussed in Chap. 10, deals with the consequences of the close correlations not only between neuronal functions and motor abilities but also with brain functions and the overall body physiology in general. It views mind and brain as a biological system that is rooted in body experience and interaction with the environment. Embodiment refers to both the embedding of cognitive processes in brain circuitry and the origin of these processes in an organism's sensorimotor experience and sees the principle of autonomy as one essential prerequisite (Fuchs 2009a, b, 2012; Thompson 2007; Thompson and Varela 2001). Against the background of this concept, motor skills such as throwing a spear and manufacturing stone tools must be seen in a new light concerning the evolution of humans, a field, however, that will need more research in the future.

Our autonomy and freedom are endangered if we have handicaps, if we are sick, if there are perilous influences, or if there are circumstances such as hunger or poverty, which prevent the unfolding of these capacities. Thus, I propose that health should be defined in medicine and sociology as the capacity for individual autonomy and freedom.

However, human potential for autonomy and freedom is not only the most creative force evolution ever produced but also simultaneously the most destructive, as can be observed in collective delusional ideas, mass murder, and environmental destruction. The danger of error, egoism, and destructive social behavior grows with increasing emancipation from natural bonds and social ties.

The challenge is that we can decide. We can decide between good and evil, cooperation or competition, egoism and altruism. And, we can decide in what type of society we want to live and which values we follow. Of course, not all options are available to everyone, but social realities are predominantly made by man.

Although the difference between animals and man appears to be gradual, the special combination of features of autonomy described here is so far advanced that, when summed, it is finally something new in nature and grows above and beyond nature. A point is reached at which differences in degree become significant enough to become differences in kind. There is, for example, a rather profound difference between the tool making of a chimpanzee using a stick to gather termites and that of humans building a jet aircraft.

This circumstance makes it possible for researchers to look in two directions: They can select those properties of man that are comparable to those of animals, and then they can say: "Look, man is not so different from animals!" Or, they can look in the other direction – from animals to man – and describe the ways man differs from animals and why he belongs to a realm of his own. Both directions can arise with arguments that can be proven to be true, but finally it is a matter of the preferred perspective.

I propose that this is a solution for two sharply divergent philosophical perspectives, dominating discussions of the nature of animal versus human minds (Gibson 2002). One, the Darwinian or continuity perspective, asserts that the emotional and intellectual differences between animals and humans are matters of degree rather than of kind. Human mental capacities would represent expansions of similar capacities found in more rudimentary forms in other animals. In contrast, the qualitative gap perspective assumes that qualitatively different mental processes underpin animal and human behaviors; hence, humans possess unique behavior not found in other species.

According to autonomy theory, both perspectives do apply: Many features of autonomy are prepared during prehuman evolution, but humans elaborate these biological prerequisites in such an amount that they become largely detached from the biological underpinnings and thus become new qualities that are not present among animals.

The concept presented here has the advantage that it neither reduces all human abilities to the biological underpinnings nor ignores man's evolutionary past. Thus, this point of view is suitable to build a bridge between nature and culture. Our biological past and the biological prerequisites of our existence accompany us. Biology is under the surface somewhere, but cultural impact and individual reasoning go beyond our biological roots in many respects. We even can act contrary to our physical needs, as in asceticism or with a hunger strike.

The point is not that man has dissociated himself from the biological roots as it is often formulated. Rather, the biological roots are the basis we constantly use and act on. The relative autonomy of our physical and physiological organization forms the prerequisite for all those features that are specifically human. Our nature is not determining, but rather enabling.

## 11.4 A Final Look at Animals

Let us again return to the animals: It is a great panorama we experience when we observe how these various components of autonomy are distributed among the animal kingdom. Because I have been living with the insights I am presenting in this book, I have a new perspective on many phenomena throughout the animal kingdom. To watch animals play is no longer just a nice observation but is rather a direct experience of how nature begins to grow above and beyond itself. To see a hunting cat with its dynamic, elegant, and precise movements and its sensory vigilance or to see the musk ox within its immense metabolic autonomy in a snowstorm on the Dovrefjell in Norway gives new insights into what nature is telling us. It is similar when birds of prey soar into the air, using the warm upward currents early in the morning. They "play" thanks to the emancipation they achieved by flying. The wonderful book by Walter Streffer (2009) opened my ears to moments of the enormous musical flexibility that the master singers among the birds can generate. Training with a horse, which makes its movement possibilities available to a skilled and sensitive rider, presenting its beauty and grace; experiencing the cooperation of a dog, which leads a blind person; or watching people working with dolphins: all of these encounters obtained a new dimension for me.

Experiencing the emotional life of animals by means of empathy, which is so impressively described by Bekoff (2007), acquires a new quality. Affection, friendship, joy, empathy, and sorrow and grief are abilities of highly autonomous animals, and we can share these emotions with them. I absolutely share Bekoff's appeal for a change in how we treat animals in industrial farming and scientific experimentation in face of this knowledge.

The behavior of animals fascinates people of all ages and of all cultures. People around the world are interested in what animals do, how they feel, and how they behave. There is so much that connects us with animals, which is precisely part of the fascination they evoke, also when we live together with our domestic animals. We finally recognize a part of ourselves in them. Bekoff (p. 491) writes: “In many ways ‘we are them’ and ‘they are us.’” The famous German ethologist Günther Tembrock, who died at the age of 92 in January 2011, expresses it even more strikingly: “All of animal nature is contained in human nature, but not all of human nature is contained in animal nature.”

In summary, humans exhibit a special combination of features of autonomy that allows for the generation of largely detached capacities in a whole series of attributes. First, we share many features of autonomy with all mammals: skin, which simultaneously closes relatively tight opposite to the environment and is highly flexible and light; endothermy combined with high aerobic capacity, enabling movements with endurance and to a large extent emancipated from variations in environmental temperatures; effectively stabilized fluid management, including refined processes for homeostasis and highly efficient renal functions; medium body size, which supports homeostatic functions but does not generate a larger burden for movements; and an extremely refined immune system.

In addition, there are features humans share with other primates:

- The high movability of arms, fingers, and legs;
- The sensibilization of hands and feet, thus becoming exceptionally sensitive for manipulating and exploring;
- The increase of functions the hands can carry out;
- Refinement of visual functions; and
- Profound internalization of embryonic development.

Many of the following features, generated during the evolution of primates, are elaborated to a special degree in humans:

- Man’s hands are completely freed from locomotory functions because of his upright posture.
- The hands gain nearly unlimited possibilities of flexibility and dexterity.
- Because of refined neuronal control, precise manipulations of the surroundings become possible.
- Upright posture and walking with stamina made humans able to traverse long distances and thus were the prerequisites for far-reaching migrations.
- Man became able to buffer seasonal changes either by techniques or by avoidance because of migration.
- Profound changes in mental abilities led to cultural activities from stone tools to artistic productions.
- Social learning is expanded by the ratchet effect.
- Understanding of time was developed.
- A full symbolic language was developed.
- The ability of self-control and willful, conscious behavior evolved.

# Chapter 12
# Conclusion and Implications

In the first chapter, I brought up the issue of the character of evolutionary innovations. The question is not only how the major innovations were generated but also what was qualitatively generated during the macroevolutionary transitions. Are evolved organisms in some consistent way different from more primitive ones? Are there some general, definable differences between prokaryotes and eukaryotes, between single-cell and multicellular organisms, between early chordates and amniotes? Are there some common features to be found within the respective innovations and transitions (McShea 1998; McShea and Simpson 2011)?

There might be several features involved, as discussed in Chap. 2. However, the feature of increasing autonomy is among them. In the previous chapters, it has been demonstrated that autonomy can be identified repeatedly during the major transitions. Increases in self-regulation, self-control, general stability, and flexibility of the individual organism belong to the essential changes that occurred frequently during these transitions. Thus, it can be postulated that autonomy plays a crucial role, at least during these transitions.

The discussion was based on two definitions in Chap. 3: First, general autonomy was defined, with the help of some extensive literature, as a feature of living systems. They are autonomous in the sense that they maintain themselves in form and function within time and achieve a self-determined flexibility. They establish metabolic networks, which are defended against perturbations from the environment, the ability to process energy-rich molecules, self-regulated processing and usage of energy, the processing and replication of information to escape from the tendency of the environment toward entropy, a dynamic stage of order (a situation far from equilibrium), a seclusion from the environment, and endogenous time sequences of processes.

The second definition covered the idea that during evolution the internal processes and functions can change in such a way that the organisms gain increased abilities to compensate for given perturbations and thus become more independent from environmental factors. Through these changes, they become more flexible and self-determined in many of their life processes, including behavior. This is summarized as increasing

B. Rosslenbroich, *On the Origin of Autonomy: A New Look at the Major Transitions in Evolution*, History, Philosophy and Theory of the Life Sciences 5,
DOI 10.1007/978-3-319-04141-4_12, © Springer International Publishing Switzerland 2014

autonomy. Increasing autonomy can also be expressed as an increase in possibilities of the organism.

A set of recourses that can be involved to change autonomous capacities is identified. These include changes in spatial separations from the environment, changes in homeostatic capacities and robustness, internalization of structures or functions, increases in body size, and changes in the flexibility within the environment, including behavioral flexibility. In the subsequent chapters, these features were examined in some detail with the help of physiological, morphological, molecular, and paleontological data. As these recourses are shown to be involved during the major transitions, it can be stated that increases in autonomy are a relevant principle of evolution and thus must be taken into account to understand the overall process of evolution.

According to the definition, increasing autonomy principally describes a characteristic of morphological and physiological features. It is a recurring pattern during evolutionary transitions, and it is not necessary to include any goal directedness in the process. When certain trends or patterns are identified, this does not mean that some future goals must be involved that push evolution toward autonomy. Autonomy is just the observable result of certain evolutionary changes.

The theory is not a specific model. Instead, it is a general phenomenological perspective on macroevolution, a framework for understanding its characteristics and the broader significance of research findings, and it offers orientation for further scientific research.

In principle, the events and functional changes described in the previous chapters are well known from comparative zoology and evolutionary studies. However, the crucial point is that the central feature is hardly ever noticed or made a subject of discussion. Therefore, this reinterpretation might offer new insights to generate a coherent picture of macroevolutionary events.

The identified set of recourses to gain autonomy can be involved in essentially different combinations. For example, some organisms rely more on separation from the environment, as can be studied with the shells of bivalves and brachiopods. Others rely more on flexibility, speed, and agility, as do squids (*Loligo*). Because the combination of these features can look different, there is a bewildering variety of forms. It seems as if on each level all combinations and possibilities are "experimented" with. Thus, there is evidently not something like a continuous linear trend. Rather, the record shows large radiation on each level involving different forms and combinations of features.

The Cambrian explosion is an example for this. After a long prelude, during which multicellularity was invented, its potential was elaborated in a diversity of body plans with a variety of different forms, organs, appendages, and lifestyles. Chapter 5 described how the set was involved during these early days of metazoans.

The multicellularity of animals itself also came from one line among others, which followed the path to close cooperation between cells. Rokas (2008) mentions 25 instantiations of multicellularity known today, from simple, undifferentiated bacterial filaments to macroscopic multicellular forms seen in animals, plants, and fungi. Multicellular forms observed in prokaryotes and single-cell eukaryotes are architecturally relatively simple, characterized by the presence, even in their most elaborate manifestations, of just a few distinct cell types. Of course, nobody knows – and probably will never know – which other attempts toward multicellularity were made before Cambrian times. However, one (or a few) of them did have the potential to set off the explosion.

Thus the concept takes into account that evolution reveals a wide range of divergences. However, basically some groups are able to achieve an organization that leads to changes in their general organism-environment relation. On each level, there were certain forms with the potential to generate further levels of autonomy. After reaching such a new level, it might be possible to radiate within the environment, to elaborate the possibilities of the respective body plan, and to continue progressive evolution. Only in the long run, looking back from today's perspective, it might look like some linear process lead to autonomy, while in reality on every level it might have been open regarding which organisms not only would survive but also would have the potential to generate a starting point for a new radiation on another level. In retrospect, we can follow the path toward insects, squids, mammals, and birds, but on each level – during the respective time – it might not have been clear which line would be able to successfully continue down the road. Most likely, it was not at all predetermined that *Pikaia*, *Yunnanozoon*, and their cousins would become the great-grandmothers of a line that would evolve quite extensive possibilities of autonomy. However, with the central axis, the chorda, they had a suitable prerequisite on which further evolution could build. Thus, the evolutionary process seems to be open toward the respective future.

Another example for such a starting point is the early mammals, which gained a new and apparently extremely potential organization, leading to the radiation we know from Tertiary times (Kümmell 2009, 2012; Kemp 2005). Single lines in other groups might have been able to generate an unusual autonomy, and we can observe it in the surprisingly intelligent octopus among the mollusks, a group that is otherwise not really known for especially intelligent and flexible members.

It seems to be characteristic of evolution that respective features of autonomy are generated several times independently from each other, often using (or perhaps "experimenting with") different building blocks but exposing comparable results. Thus convergence is a recurrent principle of evolution (Conway Morris 2003). Examples that have been discussed in the previous chapters are viviparity and placentation; endothermy in mammals and birds, together with the related metabolic complex; brain capacities in mammals and birds, together with behavioral flexibility; multicellularity; and environmental seclusion by way of different types of epidermal reinforcements.

The overview of the preceding chapters indicates some groups that include organisms with extensive possibilities of autonomy. To these belong some arthropods, some mollusks, and especially the line within chordates, which leads to mammals and birds. Within groups such as chondrichthyans, there are also forms with their own potential for autonomy, such as some sharks that might, for example, be more intelligent than has been previously assumed, or within the reptiles some dinosaurs that "tested" their own way.

A tension, described in the previous chapters, is that no organism can completely emancipate itself from the environment and thus always needs some environmental contact and exchange. Each individual organism has to balance the two requirements of relative autonomy and the necessary exchange functions. Thereby, each solution looks different. One example for this was the skin, which has to manage the balance between environmental seclusion and exchange simultaneously.

The same holds true for the general metabolic processes, which need strong self-regulation but at the same time are dependent on the supply of substances and energy from the outside world. Thus, there is an interesting tension between the biological autonomy and at the same time the dependence on the external medium. In fact, the living must operate within this tension.

Luisi (2003) describes that the interaction with the environment must be seen from the internal logic of the living system. The consequence of the interaction between an autonomous unit and a given molecule is not primarily dictated by the properties of the molecule but by the way in which this molecule is "seen" by the living organism. Thus, there is no particular nutrient value in sugar except when an organism is crossing the sugar gradient and its metabolism utilizes the molecules to permit the continuity of its identity. Actually, the compounds, which the living organism extracts from the environment to "create its own world," can be seen as something that the organism itself lacks, something that is missing and therefore must be obtained from the outside. Therefore, there is a specific complementarity between the organism and certain components of its environment. This has also been described as a sort of recognition of the environment the autonomous system performs (Luisi 2003; Thompson 2007).

Luisi (2003) explains that the complementary interactions with the environment accompany the organisms during their life and during their biological evolution. The environment has its own structural dynamics, and although independent of the organism, it does not prescribe or determine the changes in it. It induces a reaction in the organism, but the accepted changes are determined by the internal structure of the organism itself. It is the structure of the living system and its previous history of perturbations that determine what reactions the new perturbation will induce. These coupled interactions, accumulated over time, give a particular historical perspective to the living system: This becomes a historical product, the result of a long series of coupled interactions, the product of evolution. Accordingly, evolution is seen as a result of the maintenance of the internal structure of the organism. Because the dynamics of the environment might be erratic, the result, in terms of evolution, is a natural drift, determined primarily by the inner coherence and autonomy of the living organism.

Ruiz-Mirazo and Moreno (2012, p. 28) describe this in the following way: "Even the simplest living organism creates a set of preferential partitions of the world, converting interactions with their surrounding media into elementary norms or values. … And here is where the nature of living systems as autonomous agents, as inventors of worlds with meaning, becomes manifest. … So, from the origins of life to the origins of humanity we envision a complex series of transitions in which autonomous systems are, one way or another, involved."

A further principle, which would need more consideration than is possible in the present text, is the changes in the performance of sensory organs parallel to changes in autonomy. Generally, their performance is enhanced with increasing autonomy of the organism. Thus, increasing self-determination generates at the same time devices to gain more information and experiences from the surroundings. In the sense of Luisi, there are changes in what the organism perceives of its environment to make it part of its own world.

On each level, organisms also had to adapt to their environment to survive in the conventional sense of the term *adaptation*. Thus, in summary, there might be at least three factors that in different combinations most likely affect the huge variety of forms and functions:

1. Different combinations of the set of resources for autonomy.
2. Different solutions for the simultaneity of environmental seclusion, openness, and complementarity of interactions.
3. Modulation by means of adaptations.

I propose that these three factors are responsible for the diversity of living beings, which have passed by during evolution, regarding the many forms that have been existent on our planet, as well as for the variety we still see around us today.

An objection may be that this makes the notion of autonomy inexact. This is, however, what nature itself reveals, and it makes no sense to work with restricted perspectives just to satisfy certain scientific needs. It is more appropriate to listen to what nature is telling us, namely: Living beings are indeed often ambiguous, multidimensional, and inexact, and they seldom fit into one-dimensional models.

## 12.1 Questions for Further Research

Taking all of this into consideration, many questions arise for future research. Thus, it would be interesting to discover whether the features of autonomy can be adequately formalized to test them in empirical studies, perhaps analogous to the cladistic methods that also deal with qualitative characteristics. This would make empirical inquiries of selected evolutionary lines concerning their differential organism-environment relation possible.

Also, a thorough phylogenetic study of the distribution of features of autonomy should be possible, although it would first need some methodological development. Analysis on lower systematic levels might also become feasible.

Another line of inquiry may be the relation of autonomy to other "largest-scale trends" (McShea 1998), such as increasing complexity (McShea and Brandon 2010), increasing energy intensiveness (McCarthy and Enquist 2005; Milewski and Mills 2010; Vermeij 1999), increasing differentiation, increase in the number of cell types, and so on, as I discussed in Chap. 2 (see also Rosslenbroich 2006). It can be suspected that some of them are components of or result from autonomy or are at least in close correlation with it. Possibly, such analytic inquiries would also have feedback effects on the concept of autonomy itself.

The overview is established on a high systematic level because mainly comparisons between phyla and classes were consulted. It would be interesting to study these features on lower systematic levels in future work. With the framework proposed here, it should be possible to formulate suitable studies.

Which relation, however, does exist between adaptation and autonomy? This depends on how one understands adaptation. Mahner and Bunge (1997) claim that "the term 'adaptation' is highly ambiguous" and found eight meanings of the term in the literature. "The resulting confusion in the literature is increased by the fact that the term 'fitness' is occasionally used synonymously with one of the senses of 'adaptation'" (p. 160).

One widespread meaning of adaptation concerns organismic features, whose specific role has contributed to the selective success of their bearers. Thus, it is used as a term for the characters or traits observed in animals that are the result of selection. Alternatively, but in the same sense, it is defined as a process: the means by which natural selection adjusts the characters affecting fitness. Evolutionary adaptation then becomes almost synonymous with natural selection itself. This is, for example, a widespread use of the term in physiology. If one is convinced that natural selection is the only driving force of evolution, then every character finally becomes an adaptation.

If one accepts this meaning for a moment, there is no reason why an increase in autonomy should not be part of the fitness of organisms and thus adaptive. On the contrary, the stabilization of self-regulated functions and the autonomous buffering in regard to the influences of the environment lead to an increase in fitness.

In this sense, autonomy is contained in Darwin's theory. Organisms are increasingly stabilized in face of the vagaries of their environments and thus become fitter. This, for example, is associated with the fact that often organisms that are more autonomous have more prolonged life spans. This becomes evident if one realizes the difference on a large scale, comparing the life span of a bacterium or a protist to that of a mammal. However, there is again no simple trend in evolution but rather a rich diversity, as discussed for several features previously. A general pattern is, nonetheless, apparent.

However, as recent theories suggest (see Sect. 12.2), it is questionable that all these features were a product of natural selection. There is more involved in the process. Thus, the mixing of the notion of an environmental adaptation (changes to meet requirements of the environment) with the assumption that all features involved are a result of a selective success to cope with these requirements spoils the possibility to analyze those factors really involved.

Therefore, it is more important to understand adaptation as a notion that refers to the state of adjustment of an organism to the items in its environment (adaptation 5 according to Mahner and Bunge 1997, p. 161). This understanding of adaptation is often called "adaptedness." Adaptedness is a relational property of an organism or rather a property of the organism-environment system. Then, autonomy and adaptation become a central pair of this system. Both are dependent on each other: On the one hand, there is the organism, and on the other hand is the environment. The organism – even in its simplest form – always establishes its life function together

**Fig. 12.1** Features of autonomy and of adaptation in the case of the dolphin

with the generation of a boundary and thus produces its "being different" from the surrounding environment. To maintain this state, the organism not only needs regulatory and stabilizing functions on the one hand but also needs to react appropriately to cope successfully with the environmental influences. Self-assertion (autonomy) needs adaptations.

The polar bear possesses great autonomy with endothermy, flexible movement capacities, and flexible behavior, but to survive in the extreme climate, it needs adaptations such as the thick white fur and many others. The dolphin (Fig. 12.1) has the extensive regulation capacities and behavioral flexibility of mammals. These functions of autonomy are not prerequisites for life in water, as they are not present in fishes, and most of them can be traced back to the dolphin's phylogenetic history on land. At the same time, the dolphin has many adaptations to the aquatic environment, such as the blubber for isolation, secondary homodonty, fins and fluke, streamlined shape, nasal openings moved to the top of the head, and many more. This means that two elements are involved: (1) the individual biological integrity and (2) maintaining its autonomy while contending with the factors of the environment. Autonomy theory describes that element (1) changes during evolution, just as element (2) changes as well.

From the principle of niche construction (see Sect. 12.9), we learn that organisms not only adapt to a given environment but also might themselves severely influence the environment. Thus, the picture becomes more complicated: Organisms that maintain their autonomy are intertwined with given factors and with constructed factors of the environment. The organism can adapt to some factors as well as change others to maintain its autonomy. The organism as well as the environment can change during this process. There are mutual dynamic interactions between organismic autonomy and environments, thus again exhibiting features of a system.

The differentiation between autonomy and environmental adaptation allows a more detailed analysis of the organism-environment relation because there are functions for self-regulation and self-assertion, as well as for coping with environmental factors. Of course, this overlaps in many cases and is not mutually exclusive. In this sense, I propose that the notion has an underestimated heuristic potential.

## 12.2 The Present Situation of Evolutionary Theory

The following paragraphs summarize some of the most recent approaches to evolution. The intention is to show generally that a new image of evolution is in the making in the twenty-first century. I propose that the notion of increasing autonomy is worthy as a part of it.

To cite these authors does not mean that they take the principle of autonomy into consideration. However, some of the concepts or at least some formulations contain elements of it. This will show that, although the feature is not a part of these theories, it also is not a foreign body within the new developments and may add some new opportunities to them.

At the least, it might become clear that a new look at largest-scale trends, as I argued in Chap. 2, should be among the spectrum of new approaches. McShea and Simpson (2011) regard it as long overdue to study phenomena of evolution at the largest scale. With the new developments in the field, there is a new chance to do this.

Today, a lot is known about the course of evolution. The fossil record is being studied extensively and continues to reveal new findings and new insights. Often, new fossils correct some of the former interpretations, but there are no really large surprises coming from paleontology concerning the general picture of the evolutionary process. Apparently, the backbone of how the course of evolution is seen is near historical reality. Likewise, morphology and comparative physiology present a relatively robust image of the process. Although comparative genetics added new insights, its results were at least roughly consistent with the overall branching pattern that was derived from the classic disciplines. I admire the naturalists of the second half of the nineteenth century, who already knew so much from detailed observations in comparative morphology, embryology, and paleontology. Many of these observations have been confirmed by research during the twentieth century; others have been corrected or made more precise.

However, knowledge about the driving factors of evolution still appears to be incomplete. The claim that selection is the most important or even the only driving force of evolution has often been debated. Ever since Darwin's famous book, scientists have been arguing about whether and how the theory of selection works by way of small random changes. At Darwin's time, the notion of an evolutionary change within the organic world was in the air anyway, and Darwin's principle of divergence from a common descent enabled a burst of new research attempts and new insights. The driving factors he proposed, however, never did get rid of critical and challenging questions. Can competition between individuals with heritable differences in their ability to survive and reproduce lead to new features? Is natural selection the explanation of all evolutionary change? Where does all the hereditary variation, on which Darwin's theory depends, come from? What is the origin of major transitions and innovations? Can new species really be produced by natural selection?

At least with the beginning of the twenty-first century new doubts have been increasingly articulated regarding whether the theory of evolution is complete with its almost-exclusive emphasis on natural selection. These doubts did not come from opponents of the evolutionary theory or the Darwinian approach. Rather, they were formulated within a Darwinian framework. This is important, as there seems to be an "evolution of the evolutionary theory" under way (Wieser 1994; Pigliucci and Müller 2010).

Some critical remarks are typical: "There have been rumblings for some time to the effect that the neo-Darwinian synthesis of the early twentieth century is incomplete and due for a major revision. … The clamor to revise neo-Darwinism is becoming so loud that hopefully most practicing evolutionary biologists will begin to pay attention" (Pigliucci 2005, p. 565, 566). "In fact, there is a growing feeling that Darwinism is due for a major transformation" (Jablonka and Lamb 2005, p. 40). "I do not contest natural selection … but I do contest the notion that it alone sets the evolutionary sails" (Arthur 2004, p. 10). "The study of evolution is riddled with paradoxes. … This … is not to question the Darwinian formulation, but simply to insist we are engaged in unfinished business" (Conway Morris 2009a, p. 1314). "There is a post-Darwinian world waiting to be explored" (Conway Morris 2009b, p. R930).

What happened? Several new findings as well as new ideas are driving these changes. Concerning new concepts, it is increasingly realized that the synthetic theory of the twentieth century had little to say about the origin of macroevolutionary innovations. Concerning new findings, modern molecular biology shows that many of the old assumptions about the genetic system were incorrect. In this context, the disregard of the phenotype and the concentration mainly on the genotype, without knowing much about the relation between the two, is also being questioned.

The following paragraphs present a brief overview of some concepts that are presently under discussion and identify further considerations of autonomy among them.

## 12.3 New Concepts in Genetics

Modern genetics shows that there is something wrong with the assumption of the genome as a puzzle of accumulated random changes (Shapiro 2011). First, there is a great deal of conservation in the genome. Genetic components involved in basic cell functions are old building blocks, which changed little during evolution.

Second, it is becoming less clear what a gene really is. From one transcript, often several or even many different messenger RNAs can be assembled by way of the splicing process and thus are often the basis for many different proteins.

Another point is that there is seldom a one-to-one translation from genotype to phenotype. There are complicated self-sustaining feedback loop systems with multiple interacting genes and gene products, such as daily and seasonally cyclical/rhythmic changes in physiological states. To realize a phenotype, usually a cascade of interactions between DNA information and regulative functions is at work. Today, geneticists are beginning to think in terms of genetic networks composed of tens or hundreds of genes and gene products, which interact with each other and together affect the development of a particular trait. They recognize that the development of a trait does not depend, in the majority of cases, on a difference in a single gene. The process involves interactions among many genes, many proteins and other types of molecules, the whole organism, and the environment in which an individual develops. Thus, the "genes," whatever they will turn out to be, do not determine organisms, although they contain indispensable information within the network.

In their influential book, Jablonka and Lamb (2005) make a strong case about the new view of the genome, as recent discoveries challenge old ideas about what genes are: "No longer can the gene be thought of as an inherently stable, discrete stretch of DNA that encodes information for producing a protein, and is copied faithfully before being passed on. We now know that a whole battery of sophisticated mechanisms is needed to maintain the structure of DNA and the fidelity of its replication. Stability lies in the system as a whole, not in the gene. … And because the effect of a gene depends on its context, very often a change in a single gene does not have a consistent effect on the trait that it influences" (p. 7).

These new ideas about genes and genomes are having an increasing impact on evolutionary theory building. If a gene has meaning only in the context of the complex system of which it is a part, the standard way of thinking about evolution, in terms of changes in the frequency of one or more isolated genes, needs to be questioned. It is more appropriate to focus on changes in the frequency of alternative networks of interactions rather than on the frequencies of individual genes. Also, these ideas and data are a strong argument in favor of the systems model of Paul Weiss (see Chap. 3), not only as a model for single organisms but also as a framework for understanding evolutionary changes within these systems.

Jablonka and Lamb (2005) call for a renovation of evolutionary theory at the start of the twenty-first century. They argue for recognition of "evolution in four dimensions," rather than focusing on just one. In addition, to take into account the regulative character of genetic components, they argue for adding a perspective wherein three other inheritance systems also play essential roles in evolutionary change. These other systems are the epigenetic (organic systems outside, or in addition to, the DNA that can affect genetic expression, development, and biological function); behavioral (by means of social attention and learning); and symbolic inheritance systems (via language and other forms of symbolic communication). Epigenetic inheritance is found in all organisms, behavioral inheritance in most, and symbolic inheritance only in human beings.

Jablonka and Lamb state that some hereditary variations are nonrandom in origin, that some acquired information is inherited, and that evolutionary change can result from instructions as well as selection.

## 12.4 Epigenetic Inheritance

Most revolutionary are the findings of recent years about the mentioned further levels of inheritance beyond the level of DNA itself, which is called epigenetic inheritance (Jablonka and Raz 2009; Shapiro 2011). A person's liver cells, skin cells, and kidney cells look different and function differently, yet they all contain the same genetic information. With few exceptions, the differences between specialized cells are epigenetic, not genetic. They are the consequences of events that occurred during the developmental history of each type of cell and determined which genes are turned on and how their products act and interact. The remarkable thing about many specialized cells is that they not only can maintain their own particular phenotype for long periods of time but also can transmit it to daughter cells. When liver cells divide, the resulting cells are liver cells, not muscle cells or neurons. Thus, although their DNA sequences remain unchanged during development, cells acquire information that they can pass on to their progeny. This information is transmitted by way of what are known as epigenetic inheritance systems.

At present, the most well-known principles of epigenetic inheritance are chromatin marking systems. They involve the degree of condensation of chromatin (chromatin that is more condensed is less accessible to the factors needed for transcription), DNA methylation (methyl groups are attached to some bases, also regulating the access of transcription factors), and the modifications of the nucleosomal histones. Another process is that small RNA pieces are able to regulate post-transcriptional processes in the cell and thus contribute essentially to what is finally assembled.

Jablonka and Lamb (2005) argue that these principles of epigenetic inheritance are important in evolution as prerequisites to the evolution of complex organisms, where developmental decisions have to be transmitted to daughter cells and where the long-term maintenance of tissue functions depends on stable and transmissible cell phenotypes. They also argue that epigenetic variations can be transmitted not only in cell lineages but also between generations of organisms. Although genetically identical, organisms could have evolved because they passed on some of their epigenetic characteristics and thus provided an additional source of variation. Evolution can occur through the epigenetic dimension of heredity even if nothing is happening within the genetic dimension.

In addition, epigenetic variations might be generated at a higher rate than genetic ones, especially in changed environmental conditions, and several epigenetic variations might occur at the same time. Furthermore, they might not be blind to function because changes in epigenetic marks probably occur preferentially on genes, which are induced to be active by new conditions. This could increase the chances that a variation will be beneficial. The combination of these two properties – a high rate of generation and a good chance of being appropriate – means that adaptation by way of the selection of epigenetic variants might be rapid compared with changes via genetic change.

The discovery that some induced epigenetic states are inherited so phenotypic variants can persist even if the inducing conditions do not is adding a whole new dimension to the question of evolutionary innovations. It seems that at least short-term evolution does not depend so much on new mutations but on epigenetic changes. They might introduce new variants or just unveil genetic variants already present in a population. In addition, it is perhaps possible that heritable epigenetic variants can do a retaining job until genes catch up.

According to Jablonka and Lamb, this has general consequences on how we think about evolution. Because they give weight to the epigenetic, behavioral, and symbolic dimensions of heredity, evolutionary change does not have to wait for genetic changes. These can follow. Phenotypic modifications might come first. It also implies that evolution can be rapid because often an induced change will occur repeatedly and in many individuals simultaneously; therefore, it becomes likely that some levels can produce variations targeted toward special needs of the organism in its environment.

## 12.5 Developmental Plasticity

West-Eberhard's (2003) broad overview of developmental plasticity and evolution led her to suggest phenotypic plasticity as one of the key factors of evolution. Similar to other authors, she argues that reducing the processes of development as well as evolutionary change to the genomic level is not appropriate. She rather sees evolution as phenotypic change involving gene frequency change and not just as gene frequency change alone.

For her, the secret to understanding evolution is first to comprehend phenotypes, including their development and their responsiveness to the environment. The phenotype is characterized by plasticity, which allows for the evolution of variations. Only later might this be fixed within the genome. In this context, genes are followers rather than leaders in evolutionary change.

West-Eberhard considers environmental induction a major initiator of evolutionary change. Evolutionary novelties can result from the reorganization of preexistent phenotypes, and phenotypic plasticity can facilitate evolution by accommodation.

The term *accommodation* is central to her approach. She differentiates between phenotypic accommodation and genetic accommodation. Phenotypic accommodation is the adjustment, without genetic change, among variable aspects of the phenotype, following a novel or unusual – external or internal – input during development. Genetic accommodation is a genetic change that affects the regulation or form of a new trait.

Thus, an evolutionary change might follow several steps: First, a novel input – a mutation or an environmental change – affects some individuals or a population.

Because of inherent developmental plasticity, a phenotypic accommodation of the novel input occurs; consequently, a novel phenotype emerges. If the novel phenotype is advantageous, natural selection fixes it by stabilizing its appearance through an alteration of the genetic processes, so that genetic accommodation has occurred.

## 12.6 Facilitated Variation

As already described, increasing knowledge in comparative molecular biology produced a huge surprise: Where the synthetic theory expected the most variation, on the level of genes and their products, there is far-reaching conservation. This is the starting point for the discussion by Gerhart and Kirschner, who developed the theory of facilitated variation (Gerhart and Kirschner 1997; Kirschner and Gerhart 2005). Conservation means that even distantly related organisms use similar processes for cellular function, development, and metabolism. Each process, with many protein components working together, contributes to the phenotype. When a process is conserved, most of its protein components are conserved. Details of metabolism are the same in certain bacteria and humans, basic cell organization and functions are similar in yeast and humans, and some developmental strategies of fruit flies are strikingly similar to those of human beings.

There are also new features that had no forerunners in more ancestral organisms, so that organisms are a mixture of conserved and nonconserved processes. However, novelty in the organisms' physiology, anatomy, or behavior arises mostly by the use of conserved processes in new combinations, at different times and in different places and amounts, rather than by the invention of new procedures. The surprisingly small number of genes for humans and complex animal forms reflects the anatomical and physiological complexity that can be achieved by the reuse of genetic products. The conserved processes are fundamental cellular processes: They operate on many levels within the development and functioning of the organism. Gerhart and Kirschner call them "the core cellular processes" (Fig. 12.2).

Central to the argument of Gerhart and Kirschner is that these conserved processes facilitate rather than constrain evolutionary change. These processes have been conserved, they suggest, not only because changes would be lethal but also because they have repeatedly facilitated changes of certain kinds around them. Many of the conserved core processes have the capacity to be easily linked in new combinations. New linkages can occur with a minimum requirement on genetic change and hence can happen readily. They can arise with little or no change of the units themselves.

Thus, some genetic – and possibly much epigenetic – variation is needed to integrate conserved components into new heritable functions. Genetic changes might especially have occurred in regulatory regions of genomes. Gerhart and Kirschner call the increasing possibilities of integrating conserved functions into new combinations and the involved regulatory changes to give new outputs of the conserved processes "facilitated variation." This increased during evolution so that there was an "evolution of evolvability."

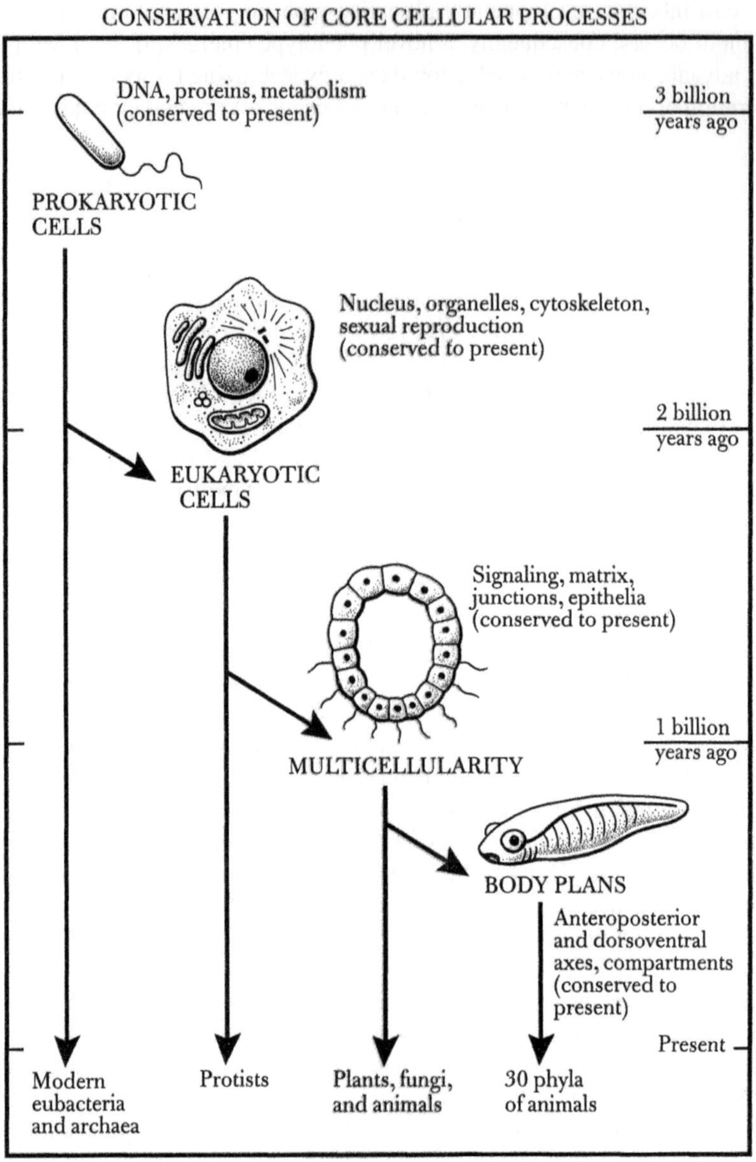


**Fig. 12.2** Conservation of core processes (From Kirschner and Gerhart 2005, p. 49)

One most obvious example of this principle on a systems level next higher than the genome and the organelles is the eukaryotic cell. It is essentially always the same cell type that delivers the building block for multicellular life. These cells might have many specializations, such as being neurons, liver cells, or epithelial cells, but they all have the same basic equipment. Novelty usually comes about by the deployment of existing cell behaviors in new combinations and to new extents, rather than in their drastic modification or the invention of completely new ones.

What changes on the cellular and molecular level, the argument of Gerhart and Kirschner continues, are regulatory components: small features of proteins, RNA or DNA, that allow a regulation of time, circumstance, and degree of activity in the processes. These are often involved in controlling the linkage and activity of the processes as well.

The eukaryotic cells within a multicellular organism especially gain functions for their regulation within the whole system. Multicellularity is characterized by the emergence of cell specializations within the same organism: complex physiology, complex spatial organization, embryonic development, and complex life cycles. All of these depend on an elaborate interaction of cells with each other via systems of signals and responses. Thus, the mostly conserved processes within the cell must have a regulatory linkage to extracellular and intracellular events. Regulatory linkage is defined as the complex association of conditions and responses. For example, cells in multicellular organisms acquire many signaling molecules at their membrane to make them receptive within their tissue and within the whole organism.

Within such multicellular systems, the organism's anatomy, physiology, and behavior are only remotely connected to the DNA sequence, Gerhart and Kirschner argue. In between, there are all these complex processes of growth, development, and metabolism. A change in the DNA sequence can therefore be only indirectly correlated with a change in anatomical and physiological characteristics of the organism. Gerhard and Kirschner maintain that currently our understanding of this connection is not sufficient enough for us to predict the phenotypic consequences of most genetic changes. We can identify genes that predispose a person to cancer, but we cannot draw a perfect correlation between the gene and the disease. Given the remote connection between the DNA and the phenotype, we have no way of knowing how often random DNA modification can produce useful results for the organism. Without an adequate understanding of how DNA changes are interpreted by the organism, we cannot recognize how important they are during evolution.

Further principles the authors identify as components of evolvability are "exploratory processes" (many complex processes are not determined by the genome but have self-organizing features according to their functionality, such as the fine wiring in the brain); "weak linkage" (features of information systems, where signals of low information content evoke complex, preprogrammed responses from the core process); and "compartmentation" (different compartments can develop different functions from the same basic organization).

In their view, the capacity for facilitating variation has itself evolved as the core processes of organisms have accumulated "more adaptive and robust behaviors." Evolution does not proceed on a random generation of dysfunctional phenotypes, which usually results in lethality and only by accident gives rise to an advantageous trait. Lethality is mostly an issue when genes are mutated that encode components of the conserved processes. These mutations are eliminated by selection in each generation. Exempting those, the population accumulates genetic variation because of the robustness of physiologically adaptable processes, and the individual generates phenotypic variation in response to genetic change or environmental change, which is predisposed to be less lethal. They accentuate the importance of the phenotype for evolutionary changes:

> "Evolutionary biologists … need to explain why organisms have changed. Evidence of conservation, to a first approximation, suggests that they are looking in the wrong place. The difference between birds and mammals is not going to be found in the structures of their muscles or nerves, their types of collagen or their microtubules. It will be found in their wings, their feathers, their sweat glands, their hair, and the organization of their brain cortex. Change has occurred principally in the organization of tissues and in the evolution of novel physiological and embryological mechanisms" (Gerhart and Kirschner 1997, p. 140).

Organisms are seen to be more actively involved in their evolution: "On the side of generating phenotypic variation, we believe the organism indeed participates in its own evolution, and does so with a bias related to its long history of variation and selection" (Kirschner and Gerhart 2005, p. 252).

On several occasions, Gerhard and Kirschner mention the principle of autonomy, for example, when they describe the essentials of multicellularity. Thus, their term *conditionality* has parallels to the term of autonomy:

> "Multicellularity … might seem just another step in the increasing complexity of eukaryotic cells, but the opportunities for evolution that accrued from multicellularity suggest that it ranks in evolutionary importance with the earlier emergence of eukaryotic cells themselves. Whereas single-celled eukaryotes had little control over their environment and evolved mostly in response to it, the cells of multicellular eukaryotes could largely produce their own intercellular conditions and respond to these, as they could to the external environment. The capacity of the cell to create its own conditions we will refer to as the cell's conditionality. Conditionality, we suggest here, has made possible the manifold cell differentiations and cell arrangements evolved by metazoa" (Gerhart and Kirschner 1997, p. 238). "The environment within a multicellular organism is protected from the external milieu and is consequently mostly under the organism's control. Thus cells of a multicellular organism can, through evolution, change both these extra cellular conditions as well as the contingent responses to these conditions" (Gerhart and Kirschner 1997, p. 45).

They regret that the capacity of metazoans to isolate the intercellular environment has not been discussed by theoreticians, although it has considerable bearing on evolvability. "Controlling the intercellular environment appears to be a major evolutionary innovation" (Gerhart and Kirschner 1997, p. 595). Thus, it is justified to include these two authors in Table 3.1.

It would be interesting to see whether increased evolvability is another feature of autonomy, as it adds a further dimension of flexibility to the components that have been described so far.

Their concept of "exploratory systems," which are especially available in sophisticated animal forms, also contains elements of flexibility and autonomy. These systems are robust and adaptable to the respective morphologies and physiological necessities. They are robust toward "noise" such as mistakes and injuries. Within certain limits, exploratory systems are even able to correct themselves. All this contributes to flexibility toward influences from inside and outside the organism and stabilizes its autonomy. It is also possible that they react flexibly toward functional evolutionary changes in other parts of the organism and in this way contribute to the evolvability of the organism itself. During integration of conserved processes and components, "the now-functioning processes and components would have undergone modification toward robustness, flexibility, compartmentation, exploratory behavior, and capacity for weak linkage. They could function well despite variable conditions both outside and inside. They could work in combination with other processes, and they were easily connected to other processes. Their capacity to buffer environmental and genetic variation increased" (Kirschner and Gerhart 2005, p. 258).

## 12.7 "Bernard Machines"

Turner (2007) contributes another piece to the jigsaw puzzle of evolution by pointing to the importance of physiology on the level of the organism, including its immediate environment, as it builds up homeostasis to generate stability and regulation.

One of his examples is the termite mound, which he studied extensively. The mound not only ventilates the air but also regulates the gaseous composition within the entire colony. The mound captures wind energy at a particular rate matched to the colony's metabolism, which makes it an organ of homeostasis. During the growth of the colony, the termites adjust the rate of the mound's wind-powered ventilation by continually adjusting its structure to keep pace with the colony's growing respiratory demand.

A simple principle underlies the complex process of building the chimney within the mound: The termites transport soil from areas of high carbon dioxide concentration to areas of low carbon dioxide concentration. In so doing, they maintain the structure of the mound as a whole in a way that allows it to perform its respiratory and regulatory functions.

Thus, a plan for the building does not exist within the termite so that it would know what to do at each and every point. The termite, rather, works according to the necessities of its direct environment, for which it is equipped with the necessary sensibility and the behavioral repertoire to operate appropriately at the respective point. For the mound as a whole, this sums up to a system that Turner calls a "form of embodied physiology" and "embodied homeostasis" (p. 27), as it is simultaneously structure and function.

Turner describes how he was contemplating the construction principle of the termite chimneys and – in the direct face of what he perceived – had to correct his thoughts about biological structures in a general manner:

> "I had been thinking about biological structures in entirely the wrong way. I had been subscribing to the conventional notion that a living structure is an object in which function takes place. That's all wrong, I came to see. A living structure is not an object, but is itself a process, just as much so as the function that takes place in it. Even the convenient dodge that structure and function are inextricably linked is wrong, I decided. That implies that structure and function are somehow distinct. … But living structures are not distinct from the function they support; they are themselves the function, no different in principle from the physiology that goes on there. In this sense, the mound is not a physical structure for the function of ventilation, it is itself the function of ventilation: it is embodied physiology" (p. 20).

Because of this embodied physiology, the mound also takes on a certain shape, a design, for which the intelligence is not in the individual termite. It is in the system as a whole: the termites together with their immediate environment. In this sense, self-organization can build up homeostasis and at the same time contribute to the phenotypic design of organisms.

Turner calls this principle "Bernard machines," named for Claude Bernard, mentioned in Chap. 3. He coined the term *milieu intérieur*: the equilibrium of the environment within the body, which Turner believes to be "physiology's core idea." Bernard machines are agents of homeostasis, devices that create environments and regulate them.

Further examples Turner refers to are tendons and muscle systems, arterial trees, and bones, which are typical samples from within single organisms that demonstrate similar general principles.

Evolution generated such self-organizing systems, so this principle appears to be an essential instrument for evolutionary change. Persistent environments are created by systems of Bernard machines, having a process-based form of heritable memory. Therefore, self-organizing systems might be of enormous significance in evolution and might in a certain sense also be intentional, as Turner assumes. Turner suggests that something analogous to intentional planning goes on in many other domains of the biological world when homeostatic machines, themselves endowed with a form of reactive plasticity, help to direct the proper functioning of organisms.

Turner (2007, p. 1) summarizes:

> "My thesis is quite simple: organisms are designed not so much because natural selection of particular genes has made them that way, but because agents of homeostasis build them that way. These agents' modus operandi is to construct environments upon which the precarious and dynamic stability that is homeostasis can be imposed, and design is the result."

Like some of the other theories I mentioned previously in this chapter, Turner focuses on the phenotype, the largest level in the model of Weiss. Turner's point is that Bernard machines have a kind of flexibility, which means that the action of a system cannot possibly be understood in terms of genetically specified rules. Some genetic prerequisites are necessary, of course, but then the system works on its own and builds physiological stability, a principle Gerhart and Kirschner (1997) call "exploratory systems."

With this concept, Turner comes close to what I describe as autonomy. This is already expressed by his term "Bernard machines" (although I personally dislike the term *machine*: Organisms are not machines, as some scientists believed since Descartes and some still believe today). With his theory of the milieu intérieur, Claude Bernard belonged to the first scientists pointing to the principle of organisms becoming independent from vagaries of the environment. Basically, what Turner describes are the means by which organisms gain autonomy.

## 12.8 Evo-Devo

The field of evolutionary developmental biology ("evo-devo") compares the processes of development of different organisms to study the evolutionary relationships between them and to discover how developmental processes evolved. It addresses the origin and evolution of embryonic development, how evolutionary modifications of developmental processes lead to the production of novel features, the role of developmental plasticity in evolution, and the developmental basis of homoplasy and homology (Carroll 2005a; Arthur 2004, 2010; Müller and Newman 2003; Pigliucci and Müller 2010; Gilbert et al. 1996; Müller 2007).

Since the 1990s, the contemporary field of evo-devo has gained impetus from the discovery that cascades of gene regulation are essential during the development of the embryo, and that small changes in the regulation networks might lead to extensive changes in phenotypes. Evo-devo demonstrates that during evolution developmental processes altered, creating novel structures from old gene networks. In other cases, conserved processes are involved in new structures and functions, so that unexpected "deep" homologies exist, as in *Hox* genes, for example.

The *Hox* gene cluster, or complex, as a classical example of evo-devo, belongs to the so-called tool kit genes. They are transcription factors containing the homeobox protein-binding DNA motif, which functions in patterning the body axis. By specifying the identity of particular body regions, *Hox* genes determine where limbs and other body segments will grow in a developing embryo or larva. Another example of such a toolbox gene is *Pax6/eyeless*, which induces eye formation in many animals, also when they are only distantly related to each other. Evo-devo also studies how development itself evolved.

Development of the embryo might generate variations by self-regulation, either spontaneously or induced by internal or external stimuli, including environmental ones. Thus, morphological form and complex structures such as body plans are not necessarily generated through random genetic changes and natural selection as an external factor but rather through describable internal changes within the system itself.

An important part of this approach is modularity: As has been long recognized, plants and animals are modular because they are organized in developmentally and anatomically distinct parts. Often, these parts are repeated, such as body segments or fingers, for example. This modularity is now found increasingly on different organizational levels, including the molecular level.

Findings suggest that the crucial distinction between different species, orders, or phyla might be caused less by differences in their genes than differences in spatial and temporal expression of conserved genes. The implication that large evolutionary changes in body morphology are associated with changes in gene regulation, rather than the evolution of new genes, suggested that changes in "switch" genes might play a major role in evolution and might be induced on different levels within the systemic hierarchy of the organism.

The proposal of West-Eberhard (2003) on developmental plasticity is also a focus of evo-devo and is derives from the recognition that phenotypes are not uniquely determined by their genotypes. Evolutionary changes in development might proceed by a "phenotype-first" route, with genetic change following, rather than initiating, the formation of phenotypic novelties.

The study of developmental processes is further extended by the more general idea of heterochrony. This term describes changes in the timing of a developmental process as relevant as a process of evolutionary change. Therefore, there can be dissociations of the development of different organs or body parts within an organism or also of the whole organism compared with the normal course of development and closely related species. Hypermorphosis, for example, involves a delay in the offset of a developmental process. Pedomorphosis (or juvenification) is a change in which the adults of a species retain traits previously seen in juveniles (McKinney and McNamara 1991; McNamara 1990; Gould 1977; Schad 1993). Heterochrony basically describes features of phenotypic plasticity, but there can also be heterochronic changes on the genetic level when the timing of gene expression is altered.

## 12.9 Niche Construction

Building on the work of Lewontin, Mayr, and Waddington, Odling-Smee formalized and proposed niche construction as another significant evolutionary principle (Odling-Smee et al. 2003; Odling-Smee 2010; Laland and Sterelny 2006; Jones 2005; Sterelny 2005). Niche construction, the building of niches by organisms and the mutual dynamic interaction between organisms and environments, was a long-accepted concept but has not been taken further into account as a major factor in evolution because adaptation was the predominant notion. Adaptation, in one version of its different meanings, considers the environment as a given factor with which the organism has to cope. The environment is only seen as the agent of selection, determining which variants survive and reproduce. However, Lewontin (2000) demonstrates that there is more in regard to the environment that complicates – or better enriches – the way in which we have to think about the role of the environment in evolution. Often, the organism itself is responsible for selecting its environment and for constructing some aspects of it. This is relevant for evolution as well. Lewontin has been stressing the importance of niche construction for years, but only recently have more biologists adopted his ideas and expanded them.

Niche construction creates feedback situations between organisms and their environment, and it is one source of the fit between organism and environment (Sterelny 2005). This, in turn, is supposed to influence evolutionary dynamics. It is expected that this includes a prominent role for phenotypes. Niche construction behavior might produce alterations that persist across generations and space, such that they, themselves, are a factor creating the pattern and strength of selection. In some cases, these changes are extreme: Living in mounds influences every aspect of termite morphology, physiology, and behavior (Turner 2007).

Organisms can influence the environment in ways that affect not only their own lives, but also the development and lives of their descendants, forming a sort of ecological inheritance. Organisms transmit to their offspring altered physical and selective environments and a niche choice as they affect their offspring's lives by choosing where they will live and breed (Jablonka and Lamb 2005; Sterelny 2005; Odling-Smee et al. 2003).

Even bacteria are ecological engineers because the products of their metabolism diffuse into the environment and transform it, changing the situation of their neighbors and descendants. These organisms contribute centrally to the conditions that make life possible. Other famous examples are the beaver, whose "inherited" dams provide the environment for new generations of beavers, or earthworms changing the properties of the soil in which they and their descendants will grow, develop, and be selected.

Differentiating all these relations with a much higher resolution will be a crucial advance beyond the externalist picture, in which a lineage is seen to accommodate to the environment. It will contribute to the problem of ecological theory, Sterelny (2005) points out, concerning the extent to which ecological communities are integrated systems rather than mere aggregates of individual agents that happen to live and die adjacent to one another. The concept does make it clear that the individual-aggregate conception of communities understates the range of potentially important and stabilizing interactions between organisms.

However, stabilizing ecosystems is as yet another form of autonomy, reaching beyond the single organism. Sterelny mentions the example of the forest, in which trees are important ecological engineers, physically stabilizing the soils on which they grow, moderating many physical impacts, and providing shelters, resources, and living space for hosts of animals and epiphytes. Especially, rain forests have been described as self-sustaining systems, thus showing clear features of self-regulation and self-adjustment.

Odling-Smee (2010) sees strong relevance of niche construction in human sciences, as humans also build ecological inheritance systems, including stabilized microenvironments, that incorporate both heritable material culture and heritable cultural knowledge (see also Sterelny 2006).

## 12.10 Increasing Diversity and Complexity

There is yet another approach that has to be included in a brief overview of novel theories on evolution. Because it has only been presented recently, it has not yet received much discussion. However, this new approach shows how manifold the aspects presently considered are.

McShea and Brandon (2010) argue that a spontaneous tendency toward increased diversity and complexity exists in evolution, one that is active whether natural selection is present or not. The theory focuses on diversity as well as individual complexity of the evolved organisms themselves. In the absence of selection and constraints, the theory predicts increasing diversity of a population (where the components are individuals). It also predicts increasing diversity of a species (made up of populations), of a clade (composed of species or lineages), and of all life (composed of clades). Further, it predicts that populations, species, and clades should become more diverse in every property – morphological, physiological, and behavioral – and at every level of organization, from macromolecules and cells to tissues and organs.

However, when selection or constraints operate, they might modulate the outcome of this tendency in profound ways, so that complexity might develop in relatively different ways and in some cases even decreases. This tendency toward complexity is presented as a background condition of evolving populations and organisms.

McShea and Brandon do not claim that their supposition represents a wholly new evolutionary principle, but rather that it is a unifying one.

## 12.11 Symbiosis and Horizontal Gene Transfer

The hypothesis that certain organelles of eukaryotic cells, in particular plant chloroplasts, evolved from bacteria had already been proposed by several researchers in the late nineteenth century. However, this relationship did not generate much interest for evolutionary theorists. In the 1960s, Lynn Margulis (Sagan) summarized the then-available data on the similarity between certain organelles and bacteria, in particular the striking discovery of organellar genomes, and came to the conclusion that not only chloroplasts but also the mitochondria evolved from endosymbiotic bacteria (Sagan 1967; Margulis 1993). The concept was heavily rejected in face of the predominance of ideas of gradual change in evolution. However, subsequent work, in particular phylogenetic analysis of both genes (i.e., those contained in the mitochondrial genome and those genes encoding proteins, which function in the mitochondria and apparently were transferred from the mitochondrial to the nuclear genome), turned the hypothesis on endosymbiosis into a well-supported theory.

The major evolutionary role assigned to effectively unique events like endosymbiosis is, of course, incompatible with gradualism. Nonetheless, it was finally incorporated into evolutionary thinking without further rumblings within the general theory. People returned to their former agenda without realizing that a fundamentally new, nongradualistic principle for the origin of evolutionary innovations had been introduced. It looks like a cuckoo's egg within the synthetic worldview.

Later, Margulis extended her theory of symbiogenesis and postulated similar relationships between organisms of rather different phyla or kingdoms as an essential driving force of evolution. Genetic variation, she proposes, occurs mainly as the result of transfer of nuclear information between bacterial cells or viruses and eukaryotic cells. Thus, symbiosis is considered as a central factor of the evolutionary process in general, stressing the importance of cooperative relationships between species (Margulis and Sagan 2002). Again, her theory is receiving heavy criticism, but the future will tell.

Similar to the hypothesis on endosymbiosis, there were other considerations under way during the second half of the twentieth century, which slowly undermined the view of gradual infinitesimal changes as the only material of evolution and mainly came from genetic research. Among these can be counted the neutral theory (Kimura), the concept of evolution by gene duplication (Ohno), the discovery of mobile elements ("jumping genes," i.e. genetic elements that were prone to frequently change their position in the genome, first described by McClintok), as well as unexpected insights into the genomic organization of viruses and bacteria (Koonin 2009).

Today, because of increasing knowledge from whole-genome sequencing, the extent of larger genomic changes becomes evident. Horizontal gene transfer between prokaryotes seems to be the rule rather than the exception (Doolittle 1999; Koonin 2009; Shapiro 2010, 2011; Boto 2010). The rate of horizontal gene transfer seems to differ for different genes depending on their functions. Eukaryotes are different from prokaryotes with respect to the role played by horizontal gene transfer in genome evolution. The genomes of eukaryotes exhibit more stability and thus robustness or autonomy. Especially, endosymbiosis made substantial contributions to the genomes of the host cells, as many genes of the symbiont were integrated into its nuclear DNA. Thus, there is no reasonable doubt that the gene complement of eukaryotes is a chimera comprising functionally distinct genes of archaeal and bacterial descent. Also, later in the evolution of plants and animals, there were several occurrences of horizontal gene transfer. This, however, is still under debate, at least concerning its extent and its significance for evolution.

Other substantial reorganizations such as gene and whole-genome duplications; large deletions, including loss of genes or groups of genes; and various types of genome rearrangements, which are becoming increasingly well documented today, contribute to a dynamic view of the genome.

The observation of extensively occurring horizontal gene transfer leads to a fundamental generalization: The genomes of all forms of life might be collections of genes with diverse evolutionary histories. Thus, it is possible that on the genetic level the tree of life is a sort of network rather than a branching tree (Koonin 2009). However, this possibly does not affect so much the usual tree-like representation of the history of life in regard to whole organisms because the role of the genome within organisms obviously has to be corrected in a profound manner.

Kitano and Oda (2006) connect their considerations about increasing robustness through evolution with these symbiotic events. In Chap. 3, I showed that the term *robustness* parallels my definition of autonomy. Kitano and Oda argue that biological robustness fosters evolvability, and that selection tends to favor individuals with robust traits; thus, robust systems progressively adapt to become more robust against the environment in which they are embedded. They suggest that, over evolutionary time, robustness against external perturbations was enhanced by adding diverse new functions to the input and output components of the organism, and that many of these new functions were gained through symbiosis.

They argue for a "self-extending symbiosis" as a process to further enhance robustness. Self-extending symbiosis refers to phenomenon by which evolvable robust systems continue to extend their system boundary by incorporating foreign biological forms (genes, microorganisms, etc.) to enhance their adaptive capability against environmental perturbations. Thus, robust evolvable systems have consistently extended themselves by incorporating "nonself" features into tightly coupled symbiotic states.

Looking at the history of evolutionary innovations, they describe some of the major innovations as the result of the acquisition of nonself into self at various levels, and that horizontal gene transfer facilitates evolution by exchanging genes between different species that have evolved within different contexts.

In early cases of symbiosis, such as during the generation of the eukaryotic cell, the relation between the host and the symbiont became close. In later cases, such as the acquisition of the gut flora in vertebrates, the mutual dependency is comparable, although the physiological relation is not that strong.

Kitano and Oda also argue in the sense of modern systems theory that different degrees of symbiosis add additional systemic layers within an organism. They see a general tendency in the continuous addition of external layers to the host system by means of symbiotic incorporation of foreign entities. They cite genomic studies that revealed that the bacterial flora even manipulates host gene expression to establish mutually advantageous partnerships, and that host gene expression changes according to the composition of microbes in the flora. Not only does the bacterial flora affect the host, but also the host affects the activity and composition of the flora through its immunological responses.

In this sense, Kitano and Oda see self-extending symbiosis as a fundamental process of significant evolutionary innovation that adds greater levels of robustness and functionalities to the species.

## 12.12 Patterns and Factors of Evolution

Looking at this outline of recent considerations on evolution, the overall situation becomes complicated. A multitude of processes comes into focus: Besides mutation there is developmental plasticity; heterochrony; facilitated variation; endosymbiosis; horizontal gene transfer; epigenetic inheritance and change, possibly including some inheritance of acquired characters; niche construction; and so on. It does not seem to be determinable which of these processes is the most important within evolution. Many authors assume that this is just a widened spectrum of variability, providing material for selection, which is still considered to be the most important driving force leading to adaptation. Others consider selection as a subordinated factor, solely functioning as a sieve, which can be passed only by viable forms, while creativity is more on the side of the variational processes.

However, this cannot be determined at present. There also seem to be different expectations regarding what should be explained by a hypothesis or by a theory: Is the description of a pattern already a valuable theory, or does it need to include some hypothesis about a driving force? To which extent do the cited theories include factors that function as causes in evolution? What is accepted as an explanation?

To differentiate possible types of answers, it might be helpful to distinguish between three levels of evolutionary analysis and explanation:

1. A first and basic level is the *description* of the evolutionary process and its events in different times. This includes those fields of paleontology that describe when and where certain organisms appear during evolution and what their functional, environmental, and phylogenetic background was. The comparative studies of phylogenetic relations, whether with classical morphological methods or aided by genetics and molecular biology to reconstruct certain parts of the genealogy, also have a predominantly descriptive character. This level collects information about *what* happened in the history of life.
2. A second level is the description of *patterns and processes*, which can be observed on the basis of the descriptions in level 1. Regularities, repeated events, and certain trends are studied and analyzed. On this level, *how* evolutionary processes come about is described. For example, molecular biology gives an account of how different organisms make use of the same building block to generate new phenotypes, or developmental biology describes how genetic elements are arranged to generate segmented organisms. These are for the time being descriptions of patterns, without asking about the driving factors involved because it might still not be known what really caused or induced the duplication of a gene and so on.
3. Answers for the question of *driving factors* might follow when more knowledge has been gathered on the first two levels.

A fairly similar differentiation has been indicated in a group report at a Dahlem workshop in Berlin, "Patterns and Processes in the History of Life" (Raup and Jablonski 1986): "We now reject simple notions of order such as special creation, foreordination, or that life is just the manifestation of the postulated law-like structure of the universe. But we are still left with the facts that there are perceived patterns and directional trends in the history of life. We must first recognize and define patterns in order to help us detect and identify underlying structure, cause, or both" (p. 47).

However, the third of the formulated levels mentioned itself contains a problem, which is a general one for biology: It is the question regarding the relevance of causality in biology. Biology inherited thinking in terms of linear causality from nineteenth century physics, which was the leading science of the time, so that life sciences also tried to describe organisms in cause-and-effect chains, the terms of classical physics. However, physics had to learn at the beginning of the twentieth century that linear causality is only a special case, mainly for macroscopic mechanics, and not a universal principle.

Nonetheless, causal thinking was partially successful in biology. First, it helped to overcome explanations of weak scientific value during the nineteenth century. Second, it was especially appropriate where mechanics is involved (movement of legs or wings), or where the behavior of fluids and gases are to be analyzed. Also, the dissection of molecular pathways into a series of single reactions is describable in causal sequences of chemical steps.

However, as soon as the biologist looks at functions and processes of larger entities in living systems and regards the interrelations between the components involved, the causal descriptions tend to be insufficient or even inappropriate. This corresponds with the attempts of systems biology to understand nonlinear processes of organic networks. The simplest principles of nonlinear processes are the feedback loop, as it is regularly found in physiology, and circular causality, as is the standard case in biochemical reaction cycles (Bechtel 2007). Extending this concept, several authors propose that organismic functions are more adequately described in the form of reciprocal interrelationships (Fuchs 2009a; Thompson 2007; Rosen 1991; Haken 1983).

Fuchs (2009a) differentiates a horizontal from a vertical circular interrelationship. Horizontal interrelationships are seen, for example, within the metabolism. Vertical interrelationships are the relations between the different system levels such as cells, organs, organisms, and so on. Fuchs describes both of them together as integral causality, thus extending well beyond simple linear causality.

Also, Mahner and Bunge (1997) discuss this problem from the perspective of systemic thinking, looking at the notion of levels of organization or levels of systems. They describe that a multilevel system, such as an organism, can be viewed either "bottom up" or "top down." Thus, the bottom-up standpoint takes the higher-level properties and laws of a system to be determined by, and hence reducible to, the properties and laws of its components: The parts determine the whole. This is the most widespread view in today's biology and also dominates large parts of medicine with profound consequences. The rival view is top down, according to which the higher-level properties and laws of a system determine its components: The whole determines its parts. Mahner and Bunge admit that the truth lies in a synthesis of both of these viewpoints, and that neither of these should be formulated in terms of causality. In living beings, there rather are functional relations or reciprocal causalities (for more detailed discussions of this topic, see Andersen et al. 2000; Bunge 1979; Noble 2006; Pigliucci 2013; Rosslenbroich 2011; Woese 2004; Wuketits 1981).

Thompson (2007) summarizes succinctly: "A whole not only arises from its parts, but the parts also arise from the whole. Part and whole co-emerge and mutually specify each other" (p. 38). These functional relations show mutual interdependencies of the components involved.

In Chap. 3, I introduced this principle according to the theory of Paul Weiss. Could it be that, analogous to these principles, also the evolutionary process has to be seen from a systems perspective? In the same sense, there seems to be a network of factors involved within the evolutionary process on the whole. Research is now beginning to understand increasingly about the components and factors involved and is perhaps just at the beginning of a new access to this system.

Then, it might be difficult to pinpoint single causes because this viewpoint predominantly deals with a network of interrelating, nonlinear factors in evolution. The third level of evolutionary explanation might find an adequate solution in the future by understanding this network of interrelated factors.

Shapiro (2011) holds the same view when he summarizes the significance of all these developments in evolutionary theory:

> The science of the 21st Century deals with the interactions between the multiple components of complex systems, ranging from aggregates of elementary particles ... to the behavior of the largest structures in the cosmos. This kind of science is fundamentally different from earlier periods, when the goal was to understand the unique property of each atomistic unit and then try to derive the behavior of large systems from a small set of interaction rules plus the character of their component parts. Today, a major focus in scientific inquiry is to understand how systems change over time, whether they are atoms, molecules, organisms, ecosystems, climates, galaxies, black holes, or universes. (p. 145)

Shapiro then states that the same shift in understanding the subject is on its way in evolutionary studies, and that it will have repercussions far beyond the life sciences.

There have been recurring proposals concerning a systems view of evolution (Riedl 2000; Wagner and Altenberg 1996; Shapiro 2011; Müller 2010; Noble 2013), and today the theoretical components of it are in the making. The main focus in present evolutionary research is increasingly on level 2 of the three levels described. Precisely this focus makes it possible to avoid speculations because the patterns that are found are interesting enough in themselves without decorating them with hypotheses on single primary causes or a selective scenario at any occasion (Gould and Lewontin 1979). Presumably, this work on patterns will serve as a basis for understanding the functional network of evolution.

Increasingly, insight is now being gained into patterns and processes that are involved in evolution, and autonomy is one of these patterns. As increases in autonomy are seen regularly during the major transitions, it must be an important pattern, and it can be expected that it is part of the network of factors involved. There must be something in evolution that drives essential changes toward increased autonomous capacities. As it is describable as a clear and recurring pattern during the major transitions, it is a necessary building stone of the modern theory of evolution.

When there is broader knowledge on level 2 investigations, the understanding of driving factors within the network (level 3) might become apparent on its own. However, there presumably will be a distinct difference to the physical principle of linear causation.

As we saw, several of the theories mentioned focus preeminently on phenotypes rather than on genes. Phenotypes are the entities that have to fulfill their function during life and the life cycle, and they have to survive and to reproduce. Of course, they rely not only on genetic prerequisites but also on many other prerequisites from the cells, tissues, and organs. Müller (2010) summarizes this point as the "epigenetic innovation theory," with *epigenetic* used here in its traditional meaning of "contextual development," and the constituent conditions for phenotypic evolution are seen as embodied in developmental systems.

According to Müller (2010, p. 323): "Epigenetic innovation theory represents a systems-oriented approach. It argues that genetic variation, natural selection, and environmental induction affect integrated developmental systems that generate specific phenotypic reactions when the canalized plasticity has reached its limits.

In other words, a developmental system falls into a new steady-state interaction among its components."

However, if phenotypes and developmental systems are the units of evolution, it is important to be able to describe their features and characteristics and to identify the patterns of their changes during evolution. The characteristics of phenotypes and the evolutionary patterns in question pertain to the autonomy of the individual organism, as it is described in the present book.

Phenotypes build on components, and their evolution increasingly integrated them into coherent systems, which in this manner gained in robustness and flexibility at the same time. Components and systems successful in evolution at lower levels become part of higher-level systems. Subsequent evolution uses these components as base units, be they nucleotides or body segments, building the next-higher level out of those units. For this to be coherent and heritable, the new level must be built of components that are themselves well defined and consistent. Each major evolutionary advance therefore conserves the details of the previous stage in an assemblage of components, from which more complex organization is constructed.

The critical functionality of each level lies in that level itself, not in the properties of the components that make it up. This is described by the theory of Weiss for an actual organism. Extending this view to evolution shows that this level itself is exposed to the environment with the challenge of establishing and maintaining stability and robustness in face of all the influences coming from the surroundings.

The summary of my hypothesis is the following: By means of the construction of system levels, a variety of functions for autonomy evolved and increased the capacities of self-determination, relative environmental independence, and flexibility of the individual organisms. Thus, autonomy is a central pattern in evolution, highly intertwined with other patterns and processes. This is a model that can be tested empirically and is suited to generate many new interesting and challenging questions about the organism-environment relations and their changes throughout the history of life. It integrates modern empirical knowledge on evolution, physiology, paleontology, and so on. Thus, it is not in opposition to modern fields of research but rather expands the scope of their interpretation to understand large-scale evolutionary processes.

# References

Abedin M, King N (2008) The premetazoan ancestry of cadherins. Science 319:946–948
Agnew ZK, Bhakoo KK, Puri BK (2007) The human mirror system: a motor resonance theory of mind-reading. Brain Res Rev 54(2):286–293
Ahlberg PE, Clak JA (2006) A firm step from water to land. Nature 440:747–749
Ahlberg PE, Clack JA, Blom H (2005) The axial skeleton of the Devonian tetrapod Ichthyostega. Nature 437:137
Alberts B, Johnson A, Lewis J, Raff M, Roberts K, Walter P (2002) Molecular biology of the cell, 4th edn. Taylor & Francis, London
Alcock J (1993) Animal behavior. An evolutionary approach. Sinauer, Sunderland
Alexander RM (1991) Wie Dinosaurier sich fortbewegten. Spektrum der Wissenschaft 14:82–89
Alibardi L (2003) Adaptation to the land: the skin of reptiles in comparison to that of amphibians and endotherm amniotes. J Exp Zool (Mol Dev Evol) 298B:12–41
Alibardi L (2009) Embryonic keratinization in vertebrates in relation to land colonization. Acta Zool 90(1):1–17
Alibardi L (2012) Perspectives on hair evolution based on some comparative studies on vertebrate cornification. J Exp Zool B Mol Dev Evol 318(5):325–343
Allman J (1999) Evolving brains. Freeman, New York
an der Heiden U, Roth G, Schwegler H (1985) Die Organisation der Organismen: Selbstherstellung und Selbsterhaltung. Funkt Biol Med 5:330–346
Andersen PB, Emmeche C, Finnemann NO, Christiansen PV (eds) (2000) Downward causation: minds, bodies, and matter. Aarhus University Press, Aarhus
Angelini F, Ghiara G (1991) Viviparity in squamates. In: Ghiara G (ed) Symposium on the evolution of terrestrial vertebrates. Mucchi, Modena, pp 305–334
Arnellos A, Moreno A, Ruiz-Mirazo R (2013) Organizational requirements for multicellular autonomy: insights from a comparative case study. Biol Philos. doi:10.1007/s10539-013-9387-x
Arnold WH (1989) Adaption und Emanzipation. Eine Betrachtung der Evolution am Beispiel der Entwicklung der Sprachorgane. In: Arnold WH (Hrsg.) Evolution. Interdisziplinäre Aspekte zur Evolutionsfrage. Urachhaus, Stuttgart, pp 143–173
Arthur W (2004) Biased embryos and evolution. Cambridge University Press, Cambridge
Arthur W (2010) Evolution – a developmental approach. Wiley, Chichester
Ayala FJ (1974) The concept of biological progress. In: Ayala JF, Dobzhansky T (eds) Studies in the philosophy of biology: reduction and related problems. University of California Press, Berkeley/Los Angeles, pp 339–355
Ball EE, Miller DJ (2010) Putting placozoans on the (phylogeographic) map. Mol Ecol 19:2181–2183

B. Rosslenbroich, *On the Origin of Autonomy: A New Look at the Major Transitions in Evolution*, History, Philosophy and Theory of the Life Sciences 5,
DOI 10.1007/978-3-319-04141-4, © Springer International Publishing Switzerland 2014

Ball EE, de Jong DM, Schierwater B, Shinzato C, Hayward DC, Miller DJ (2007) Implications of cnidarian gene expression patterns for the origins of bilaterality – is the glass half full or half empty? Integr Comp Biol 47(5):701–711. doi:10.1093/icb/icm028
Barandiaran X, Ruiz-Mirazo K (2008) Modelling autonomy: simulating the essence of life and cognition. BioSystems 91:295–304
Barnes RD (1986) Invertebrate zoology. Saunders, New York
Barrett HC (2012) A hierarchical model of the evolution of human brain specializations. Proc Natl Acad Sci U S A 109:10733–10740
Barrett L, Henzi SP, Lusseau D (2012) Taking sociality seriously: the structure of multi-dimensional social networks as a source of information for individuals. Philos Trans R Soc B Biol Sci 367:2108–2118. doi:10.1098/rstb.2012.0113
Bartholomew GA (1982) Body temperature and energy metabolism. In: Gordon MS (ed) Animal physiology: principles and adaptations. MacMillan Publishing, New York
Barton RA (2006) Primate brain evolution: integrating comparative, neurophysiological, and ethological data. Evol Anthropol 15(6):224–236. doi:10.1002/evan.20105
Bates LA, Byrne RW (2010) Imitation: what animal imitation tells us about animal cognition. Wiley Interdiscip Rev Cogn Sci 1(5):685–695. doi:10.1002/wcs.77
Bauwens D, Garland T (1995) Evolution of sprint in lacertid lizards: morphological, physiological, and behavioral covariation. Evolution 49:848–863
Bechtel W (2007) Biological mechanisms: organized to maintain autonomy. In: Boogerd F, Bruggeman FJ, Hofmeyr JHS, Westerhoff HV (eds) Systems biology: philosophical foundations. Elsevier, Amsterdam, pp 269–302
Bechtel W (2010a) The downs and ups of mechanistic research: circadian rhythm research as an exemplar. Erkenntnis 73:313–328. doi:10.1007/s10670-010-9234-2
Bechtel W (2010b) The cell: locus or object of inquiry? Stud Hist Philos Sci C Stud Hist Philos Biol Biomed Sci 41(3):172–182
Bekoff M (2004) Wild justice and fair play: cooperation, forgiveness, and morality in animals. Biol Philos 19:489–520
Bekoff M (2007) The emotional lives of animals. New World Library, Novato
Bekoff M, Allen C (1998) Intentional communication and social play: how and why animals negotiate and agree to play. In: Bekoff M, Byers JA (eds) Animal play. Evolutionary, comparative and ecological perspectives. Cambridge University Press, Cambridge, pp 97–114
Bekoff M, Byers JA (eds) (1998) Animal play. Evolutionary, comparative and ecological perspectives. Cambridge University Press, Cambridge
Bekoff M, Allen C, Burghardt G (eds) (2002) The cognitive animal: empirical and theoretical perspectives on animal cognition. MIT Press, Cambridge, MA
Bennett AF (1972) The effect of activity on oxygen consumption, oxygen debt, and heart rate in the lizards Varanus gouldii and Sauromalus hispidus. J Comp Physiol 79:259–280
Bennett AF (1989) Integrated studies of locomotor performance. In: Wake DB, Roth G (eds) Complex organismal functions: integration and evolution in vertebrates. Wiley, New York, pp 191–202
Bennett AF (1991) The evolution of aerobic capacity. J Exp Biol 160:1–23
Bennett AF, Ruben JA (1979) Endothermy and activity in vertebrates. Science 206:649–654
Bereiter-Hahn J (1996) Biologische Vorbedingungen für die Ermöglichung freier Willensentscheidungen. In: Fuchs G, Kessler H (Hrsg.) Gott, der Kosmos und die Freiheit. Echter Verlag, Würzburg, pp 31–56
Bereiter-Hahn J, Matoltsy AG, Richards KS (eds) (1984) Biology of the integument. Springer, Berlin
Bergquist PR (1978) Sponges. Huchinson University Library, London
Bernard C (1859) Lecons sur les Propriétés Physiologiques et les Altérations Pathologiques des Liquides de l'Organisme. Baillière, Paris
Bernard C (1878) Lecons sur les Phénomènes de la Vie Communs aux Animaux et aux Végétaux. Baillère, Paris

Bernard C (1974) Lectures on the phenomena common to animals and plants, vol 1. Thomas, Springfield
Bertalanffy Lv (1949) Vom Molekül zur Organismenwelt. Grundfragen der modernen Biologie. Athenaion, Potsdamm
Bertalanffy Lv (1973) General system theory. Foundations, development, applications. Penguin University Books, Harmondsworth
Bertschinger N, Olbrich E, Ay N, Jost J (2008) Autonomy: an information theoretic perspective. BioSystems 91:331–345
Beurlen K (1937) Die stammesgeschichtlichen Grundlagen der Abstammungslehre. G Fischer, Jena
Beurlen K (1949) Urweltleben und Abstammungslehre. Schwab, Stuttgart
Blackburn DG (1992) Convergent evolution of viviparity, matrotrophy, and specialisations for fetal nutrition in reptiles and other vertebrates. Am Zool 32:313–321
Blackburn DG (1995) Saltationist and punctuated equilibrium models for the evolution of viviparity and placentation. J Theor Biol 174:199–216
Blackburn DG, Flemming AF (2009) Morphology, development, and evolution of fetal membranes and placentation in squamate reptiles. J Exp Zool B Mol Dev Evol 312(6):579–589
Blackburn DG, Flemming AF (2010) Reproductive specializations in a viviparous African skink and its implications for evolution and conservation. Herpetol Conserv Biol 5(2):263–270
Blanckenhorn WU (2000) The evolution of body size: what keeps organisms small? Q Rev Biol 75(4):385–407
Boal JG, Dunham AW, Williams KT, Hanlon R (2000) Experimental evidence for spatial learning in octopuses (Octopus bimaculoides). J Comp Psychol 114:246–252
Boero F, Schierwater B, Piraino S (2007) Cnidarian milestones in metazoan evolution. Integr Comp Biol 47(5):693–700. doi:10.1093/icb/icm041
Bonner JT (1988) The evolution of complexity. Princeton University Press, Princeton
Bonner JT (1993) Life cycles. Reflections of an evolutionary biologist. Princeton University Press, Princeton
Bonner JT (1998) The origins of multicellularity. Integr Biol 1:27–36
Bonner JT (2003) On the origin of differentiation. J Biosci 28(4):523–528
Bonner JT (2004) Perspective: the size-complexity rule. Evolution 58(9):1883–1890
Boto L (2010) Horizontal gene transfer in evolution: facts and challenges. Proc R Soc B Biol Sci 277(1683):819–827
Bowler PJ (1983) Evolution: the history of an idea. University of California Press, Berkeley/Los Angeles
Bowler PJ (1989a) The invention of progress. The Victorians and the past. Basil Blackwell, Oxford/Cambridge
Bowler PJ (1989b) The non-Darwinian revolution. John Hopkins University Press, Baltimore
Brain CKB, Prave AR, Hoffmann KH, Fallick AE, Botha A, Herd DA, Sturrock C, Young I, Condon D, Allison SG (2012) The first animals: ca. 760-million-year-old sponge-like fossils from Namibia. S Afr J Sci 108:1–2
Bramble DM, Lieberman DE (2004) Endurance running and the evolution of Homo. Nature 432:345–352
Breuer T, Ndoundou-Hockemba M, Fishlock V (2005) First observation of tool use in wild gorillas. PLoS Biol 3(11):e380
Briggs DEG, Erwin DH, Collier FJ (1994) The fossils of the Burgess Shale. Smithonian Institution Press, Washington
Bromham L (2011) The small picture approach to the big picture: using DNA sequences to investigate the diversification of animal body plans. In: Calcott B, Sterelny K (eds) The major transitions in evolution revisited. MIT Press, Cambridge, MA, pp 271–298
Bronn HG (1853) Allgemeine Einleitung in die Naturgeschichte. Müller, Stuttgart
Bronn HG (1858) Morphologische Studien über die Gestaltungsgesetze der Naturkörper überhaupt und der organischen insbesondere. C.F. Winter, Leipzig/Heidelberg
Brown AD, Borowitzka LJ (1979) Halotolerance od Dunaliella. In: Levandowsky M, Hutner SH (eds) Biochemistry and physiology of protozoa, vol I. Academic, New York, pp 139–190

Brusca RC, Brusca GJ (2002) Invertebrates. Sinauer Associates, Sunderland
Buchsbaum D, Bridgers S, Weisberg DS, Gopnik A (2012) The power of possibility: causal learning, counterfactual reasoning, and pretend play. Philos Trans R Soc B Biol Sci 367:2202–2212. doi:10.1098/rstb.2012.0122
Bullock TH (1986) Suggestions for research on ethological and comparative cognition. In: Schusterman RJ, Thomas JA, Wood FG (eds) Dolphin cognition and behavior: a comparative approach. Lawrence Erlbaum Associates, Hillsdale, pp 207–219
Bullock TH (1993) How are more complex brains different? One view and an agenda for comparative neurobiology. Brain Behav Evol 41:88–96
Bullock TH (1995) Are the main grades of brains different principally in numbers of connections or also in quality? In: Breidbach O, Kutsch W (eds) The nervous systems of invertebrates: an evolutionary and comparative approach. Birkhäuser, Basel, pp 439–448
Bunge M (1979) Causality and modern science, 3rd edn. Dover, New York
Burghardt GM (1998) The evolutionary origins of play revisited: lessons from turtles. In: Bekoff M, Byers JA (eds) Animal play. Evolutionary, comparative and ecological perspectives. Cambridge University Press, Cambridge, pp 1–26
Burghardt G (2005) The genesis of animal play. Testing the limits. Bradford Books, Cambridge, MA
Buss LW (1987) The evolution of individuality. Princeton University Press, Princeton
Butler AB (2008) Brain evolution and comparative neuroanatomy. In: Encyclopedia of life sciences. doi:10.1002/9780470015902.a0000088.pub2. www.els.net
Butler AB, Hodos W (1996) Comparative vertebrate neuroanatomy. Wiley-Liss, New York
Butterfield NJ (2009) Modes of pre-Ediacaran multicellularity. Precambrian Res 173:201–211
Byrne RW (1991) Hand preferences in the skilled gathering tasks of mountain gorillas (Gorillia g. beringei). Cortex 27:521–546
Byrne RW (1993) Do larger brains mean greater intelligence? Behav Brain Sci 16:696–697
Byrne RW (1995) The thinking ape. Evolutionary origins of intelligence. Oxford University Press, Oxford
Calcott B, Sterelny K (2011) The major transitions in evolution revisited. MIT Press, Cambridge, MA
Call J, Tomasello M (2008) Does the chimpanzee have a theory of mind? 30 years later. Trends Cogn Sci 12:187–192
Calvin WH (1994) The emergence of intelligence. Sci Am 271:79–85
Campbell NA (1997) Biologie. Spektrum Verlag, Heidelberg/Berlin
Cannon WB (1932) The wisdom of the body. Norton, New York
Carrier DR (1987) The evolution of locomotor stamina in tetrapods: circumventing a mechanical constraint. Paleobiology 13:326–341
Carroll RL (1970) Quantitative aspects of the amphibian-reptilian transition. Form Funct 3:165–178
Carroll RL (1988) Vertebrate paleontology and evolution. Freeman, New York
Carroll RL (1991) The origin of reptiles. In: Schultze HP, Trueb L (eds) Origins of the higher groups of tetrapods – controversy and consensus. Cornell University Press, Ithaca, pp 331–353
Carroll RL (1997) Patterns and processes of vertebrate evolution. Cambridge University Press, Cambridge
Carroll RL (2005a) Between water and land. Nature 437:38–39
Carroll SB (2005b) Endless forms most beautiful. The new science of evo devo and the making of the animal kingdom. Norton, New York/London
Carroll SB, Grenier JK, Weatherbee SD (2005) From DNA to diversity. Molecular genetics and the evolution of animal design. Blackwell, Malden
Carter AM (2001) Evolution of the placenta and fetal membranes seen in the light of molecular phylogenetics. Placenta 22:800–807
Carter AM, Pijnenborg R (2011) Evolution of invasive placentation with special reference to non-human primates. Best Pract Res Clin Obstet Gynaecol 25(3):249–257
Cartwright P, Collins A (2007) Fossils and phylogenies: integrating multiple lines of evidence to investigate the origin of early major metazoan lineages. Integr Comp Biol 47(5):744–751. doi:10.1093/icb/icm071

Cavalier-Smith T (1987) The origin of eukaryote and archaebacterial cells. In: Lee JJ, Fredrick JF (eds) Endocytobiology III. Ann N Y Acad Sci 503:17–54

Cereijido M, Contreras RG, Shoshani L (2004) Cell adhesion, polarity, and epithelia in the dawn of metazoans. Physiol Rev 84:1229–1262

Chang C, Wu P, Baker RE, Maini PK, Alibardi L, Chuong CM (2009) Reptile scale paradigm: evo- devo, pattern formation and regeneration. Int J Dev Biol 53(5–6):813–826

Chen JY, Dzik J, Edgecombe GD (1995) A possible early Cambrian chordate. Nature 377:720–722

Chuong CM, Homberger DG (2003) Development and evolution of the amniote integument: current landscape and future horizon. J Exp Zool (Mol Dev Evol) 298B:1–11

Chuong CM, Wu P, Zhang FC, Xu X, Yu M, Widelitz RB, Jiang TX, Hou L (2003) Adaptation to the sky: defining the feather with integument fossils from mesozoic China and experimental evidence from molecular laboratories. J Exp Zool (Mol Dev Evol) 298B:42–56

Clack JA (2002) Gaining ground: the origin and early evolution of tetrapods. Indiana University Press, Bloomington

Clack JA (2007) Devonian climate change, breathing, and the origin of the tetrapod stem group. Integr Comp Biol 47(4):510–523

Clack JA, Ahlberg PE, Blom H, Finney SM (2012) A new genus of Devonian tetrapod from North-East Greenland, with new information on the lower jaw of Ichthyostega. Palaeontology 55(1):73–86

Clark RB (1964) Dynamics in metazoan evolution. The origin of the coelom and segments. Clarendon Press, Oxford

Clarke A, Pörtner HO (2010) Temperature, metabolic power and the evolution of endothermy. Biol Rev 85:703–727

Clavijo-Baque S, Bozinovic F (2012) Testing the fitness consequences of the thermoregulatory and parental care models for the origin of endothermy. PLoS One 7(5):e37069. doi:10.1371/journal.pone.0037069

Coates MI, Clack JA (1995) Romer's gap: tetrapod origins and terrestriality. Bulletin du Museum national d'Histoire naturelle Paris 4th series 17:373–388

Conway Morris S (1998) Early metazoan evolution: reconciling paleontology and molecular biology. Am Zool 38:1867–1877

Conway Morris S (2000) Evolution: bringing molecules into the fold. Cell 100:1–11

Conway Morris S (2003) Life's solution. Inevitable humans in a lonely universe. Cambridge University Press, Cambridge

Conway Morris S (2009a) The predictability of evolution: glimpses into a post-Darwinian world. Naturwissenschaften 96:1313–1337

Conway Morris S (2009b) Walcott, the Burgess Shale and rumours of a post-Darwinian world. Curr Biol 19(20):R927–R931

Conway Morris S, Caron JB (2012) Pikaia gracilens Walcott, a stem-group chordate from the Middle Cambrian of British Columbia. Biol Rev 87(2):480–512. doi:10.1111/j.1469-185X.2012.00220.x

Cornish-Bowden A (2006) Putting the systems back into systems biology. Perspect Biol Med 49:475–489

Cornish-Bowden A, Cardenas ML (2005) Systems biology may work when we learn to understand the parts in terms of the whole. Biochem Soc Trans 33:516–519

Cowels RB (1958) Possible origin of dermal temperature regulation. Evolution 12:347

Cowen R (1973) Respiration in metazoan evolution. Evolution 27:696

Crimes TP (1992) The record of trace fossils across the Proterozoic-Cambrian boundary. In: Lipps JH, Signor PW (eds) Origin and early evolution of the metazoa. Plenum Press, New York/London, pp 177–202

Daeschler E, Shubin NH, Jenkins FA (2006) A Devonian tetrapod-like fish and the evolution of the tetrapod body plan. Nature 440:757–763

Dan-Sohkawa M, Kaneko H, Noda K (1995) Paracellular, transepithelial permeation of macromolecules in the body wall epithelium of starfish embryos. J Exp Zool 271:264–272

Darwin C (1872) On the origin of species by means of natural selection or the preservation of favoured races in the struggle for life, Reprints of the 6th edn. In: The world's classics. Oxford University Press, London

de Waal FBM (2009) The age of empathy. Natures lessons for a kinder society. Harmony Books, New York

de Waal FBM (2012) The antiquity of empathy. Science 336(6083):874–876

Deacon TW (1990) Problems of ontogeny and phylogeny in brain-size evolution. Int J Primatol 11:237–282

Decety J, Svetlova M (2012) Putting together phylogenetic and ontogenetic perspectives on empathy. Dev Cogn Neurosci 2(1):1–24

DeDuve C (1991) Blueprint for a cell: the nature and origin of life. Neil Patterson, Burlington

Delius JD, Siemann M, Emmerton S, Xia L (2001) Cognitions of birds as products of evolved brains. In: Roth G, Wullimann MF (eds) Brain evolution and cognition. Wiley/Spektrum, New York/Heidelberg, pp 451–490

Delsuc F, Brinkmann H, Chourrout D, Phillippe H (2006) Tunicates and not cephalochordates are the closest living relatives of vertebrates. Nature 439:965–968

DeSalle R, Schierwater B (2007) Key transitions in animal evolution. Integr Comp Biol 47(5):667–669. doi:10.1093/icb/icm042

Dewel RA (2000) Colonial origin for eumetazoa: major morphological transitions and the origin of bilaterian complexity. J Morphol 243(1):35–74

Di Paolo EA (2004) Unbinding biological autonomy: Francisco Varela's contributions to artificial life. Artif Life 10:231–233

Di Paolo EA (2005) Autopoiesis, adaptivity, teleology, agency. Phenomenol Cogn Sci 4:429–452

Dickinson PS (2006) Neuromodulation of central pattern generators in invertebrates and vertebrates. Curr Opin Neurobiol 16(6):604–614

Dobzhansky T, Ayala FJ, Stebbins GL, Valentine JW (1977) Evolution. Freeman and Company, San Francisco

Donoghue PCJ, Antcliffe JB (2010) Origins of multicellularity. Nature 466:41–42

Doolittle WF (1999) Phylogenetic classification and the universal tree. Science 284:2124–2128

Doolittle WF (2000) Stammbaum des Lebens. Spektrum der Wissenschaft 23:52–57

Downes SM (2001) The ontogeny of information. Perspect Biol Med 44(3):464–469

Drack M, Apftaler W (2007) Is Paul Weiss' and Ludwig von Bertalanffy's system thinking still valid today? Syst Res Behav Sci 24(5):537–546

Drack M, Wolkenhauer O (2011) System approaches of Weiss and Bertalanffy and their relevance for systems biology today. Semin Cancer Biol 21:150–155

Dubbeldam JL (2001) Evolution of playlike behaviour and the uncoupling of neural locomotor mechanisms. Netherlands J Zool 51:335–345

Duboule D (2003) Time for chronomics? Science 301:277–338

Dudel J, Menzel R, Schmidt RF (eds) (2001) Neurowissenschaft. Vom Molekül zur Kognition. Springer, Berlin/Heidelberg/New York

Dunn CW, Hejnol A, Matus DQ, Pang K, Browne WE et al (2008) Broad phylogenomic sampling improves resolution of the animal tree of life. Nature 452:745–749

Eckert R (2000) Tierphysiologie. Thieme, Stuttgart/New York

Eibl-Eibesfeldt I (1999) Grundriß der vergleichenden Verhaltensforschung. Piper, München/Zürich

El Albani A, Bengtson S, Canfield DE, Bekker A, Macchiarelli R et al (2010) Large colonial organisms with coordinated growth in oxygenated environments 2.1 Gyr ago. Nature 466:100–104

Elinson RP, Beckham Y (2002) Development in frogs with large eggs and the origin of amniotes. Zoology 105:105–117

Elliot MG, Crespi BJ (2009) Phylogenetic evidence for early hemochorial placentation in eutheria. Placenta 30(11):949–967

Else PL, Hulbert AJ (1985) An allometric comparison of the mitochondria of mammalian and reptilian tissues: implications for the evolution of endothermy. J Comp Physiol 156:3–11

Emery NJ (2006) Cognitive ornithology: the evolution of avian intelligence. Philos Trans R Soc B Biol Sci 361:23–43

Enders AC, Carter AM (2012) The evolving placenta: convergent evolution of variations in the endotheliochorial relationship. Placenta 33(5):319–326

Erwin DH (2000) Macroevolution is more than repeated rounds of microevolution. Evol Dev 2:78–84

Erwin DH, Tweedt S (2012) Ecological drivers of the Ediacaran-Cambrian diversification of Metazoa. Evol Ecol 26(2):417–433

Erwin DH, Laflamme M, Tweedt SM, Sperling EA, Pisani D, Peterson KJ (2011) The Cambrian conundrum: early divergence and later ecological success in the early history of animals. Science 134:1091–1097. doi:10.1126/science.1206375

Fagen R (1981) Animal play behaviour. Oxford University Press, New York

Fagen R (1986) Play and behavioural flexibility. In: Smith PK (ed) Play in animals and humans. Blackwell, Oxford, pp 159–173

Falk D, Gibbson KR (eds) (2001) Evolutionary anatomy of the primate cerebral cortex. Cambridge University Press, Cambridge

Farmer CG (1999) Evolution of the vertebrate cardio-pulmonary system. Annu Rev Physiol 61:573–592

Farmer CG (2000) Parental care: the key to understanding endothermy and other convergent features in birds and mammals. Am Nat 155:326–334

Feder ME (1992) A perspective on environmental physiology of the amphibians. In: Feder ME, Burggren WW (eds) Environmental physiology of the amphibians. Chicago University Press, Chicago, pp 1–6

Ferner K, Mess A (2011) Evolution and development of fetal membranes and placentation in amniote vertebrates. Respir Physiol Neurobiol 178(1):39–50

Fetcho JR (1987) A review of the organization and evolution of motoneurons innervating the axial musculature of vertebrates. Brain Res Rev 12:143–280

Fetcho JR (1992) The spinal motor system in early vertebrates and some of its evolutionary changes. Brain Behav Evol 40:82–97

Filshie BK, Flower NE (1977) Junctional structures in hydra. J Cell Sci 23:151–172

Finlay BJ, Esteban GF (2009) Can biological complexity be rationalized? BioScience 59(4):333–340

Fiorito G, Scotto P (1992) Observational learning in Octopus vulgaris. Science 256:545–547

Fioroni P (1992) Allgemeine und vergleichende Embryologie der Tiere. Springer, Berlin/Heidelberg/New York

Franck D (1997) Verhaltensbiologie. Thieme, Stuttgart/New York

Freyer C, Renfree MB (2009) The mammalian yolk sac placenta. J Exp Zool B Mol Dev Evol 312(6):545–554

Fricke H, Reinicke O, Hofer H, Nachtigall W (1987) Locomotion of the coelacanth Latimeria chalumnae in its natural environment. Nature 329:331–333

Frolich LM (1997) The role of the skin in the origin of amniotes: permeability barrier, protective covering and mechanical support. In: Sumida SS, Martin KLM (eds) Amniote origins. Completing the transition to land. Academic, San Diego, pp 327–352

Fronius M, Clauss WG, Althaus M (2012) Why do we have to move fluid to be able to breathe? Front Physiol, 3 May, art. no. Article 146. doi:10.3389/fphys.2012.00146

Fuchs T (2009a) Das Gehirn – ein Beziehungsorgan. Eine phänomenologisch-ökologische Konzeption. Kohlhammer, Stuttgart

Fuchs T (2009b) Embodied cognitive neuroscience and its consequences for psychiatry. Poiesis Praxis 6(3–4):219–233. doi:10.1007/s10202-008-0068-9

Fuchs T (2012) The feeling of being alive. Organic foundations of self-awareness. In: Fingerhut J, Marienberg S (eds) Feelings of being alive. de Gruyter, Berlin, pp 149–166

Gaidos E, Dubuc T, Dunford M, Mcandrew P, Padilla-Gamiño J, Studer B, Weersing K, Stanley S (2007) The Precambrian emergence of animal life: a geobiological perspective. Geobiology 5(4):351–373

Gallese V, Fadiga L, Fogassi L, Rizzolatti G (1996) Action recognition in the premotor cortex. Brain 119:593–609

Gallup G (1997) On the rise and fall of self-conception in primates. Ann N Y Acad Sci 818:73–84

Gallup G (1998) Self-awareness and the evolution of social intelligence. Behav Process 42:239–247

Gans C (1989) Stages in the origin of vertebrates: analysis by means of scenarios. Biol Rev 64:221–268
Gärdenfors P (2003) How homo became sapiens: on the evolution of thinking. Oxford University Press, Oxford
Gee H (2006) Careful with that amphioxus. Nature 439:923–924
Gerhart J, Kirschner M (1997) Cells, embryos, and evolution. Toward a cellular and developmental understanding of phenotypic variation and evolutionary adaptability. Blackwell, Malden
Gibson KR (2002) Evolution of human intelligence: the roles of brain size and mental construction. Brain Behav Evol 59:10–20
Gilbert SF, Opitz JM, Raff RA (1996) Resynthesizing evolutionary and developmental biology. Dev Biol 173:357–372
Gillooly JF, Allen AP, Charnov EL (2006) Dinosaur fossils predict body temperatures. PLoS Biol 4:1467–1469
Giufra M, Zhang S, Jenett A, Menzel R, Srinivasan MV (2001) The concept of 'sameness' and 'difference' in an insect. Nature 410:930–933
Gleeson TT, Mitchell GS, Bennett AF (1980) Cardiovascular responses to graded activity in the lizards Varanus and Iguana. Am J Physiol 239:R174–R179
Goldenfeld N, Woese C (2007) Biology's next revolution. Nature 445:369
Goodall J (1986) The chimpanzees of Gombe: patterns of behavior. Harvard University Press, Cambridge, MA
Gordon IT, Whelan PJ (2006) Deciphering the organization and modulation of spinal locomotor central pattern generators. J Exp Biol 209(11):2007–2014
Gore J, Oudenaarden A (2009) The yin and yang of nature. Nature 457:271–272
Gould SJ (1977) Ontogeny and phylogeny. Harvard University Press/Belknap Press, Cambridge, MA
Gould SJ (1985) The paradox of the first tier: an agenda for paleobiology. Paleobiology 11(1):2–12
Gould SJ (1996) Full house. The spread of excellence from Plato to Darwin. Three Rivers Press, New York
Gould SJ (2002) The structure of evolutionary theory. The Belknap Press of Harvard University Press, Cambridge, MA/London
Gould J, Gould C (1994) Animal mind. Freeman, New York
Gould SJ, Lewontin RC (1979) The spandrels of San Marco and the Panglossian paradigm. Proc R Soc Lond B Biol Sci 205:581–598
Grant V (1985) The evolutionary process. A critical review of evolutionary theory. Columbia University Press, New York
Green CR, Bergquist PR (1979) Cell membrane specialisations in the porifera. Colloques internationaux du Cent Natl Res Sci 291:153–158
Greene JC (1986) The history of ideas revisited. Rev Synth 107:201–228
Greene JC (1991) Progress, science, and value: a biological dilemma. Biol Philos 6:99–106
Grell KG (1971) Trichoplax adhaerens F. E. Schulze und die Entstehung der Metazoen. Naturwissenschaftliche Rundschau 24:160–161
Grell KG (1981) Trichoplax adhaerens and the origin of Metazoa. Atti dei convegni Lincei 49:107–121
Grell KG, Ruthmann A (1991) Placozoa. In: Harrison FW, Westfall JA (eds) Microscopic anatomy of invertebrates, vol 2. Wiley-Liss, New York, pp 13–27
Griffin DR (1982) Animal mind – human mind. Dahlem Konferenzen Springer Verlag, Berlin
Griffin DR (1984) Animal thinking. Harvard University Press, Cambridge, MA
Griffin DR (2001) Animal minds: beyond cognition to consciousness. University of Chicago Press, Chicago
Griffin DR, Speck GB (2004) New evidence of animal consciousness. Anim Cogn 7(1):5–18
Griffith RW (1994) The life of the first vertebrates: a novel hypothesis suggests that our earliest vertebrate ancestors may have spent their lives partly in freshwater and partly at sea. Bioscience 44:408–417
Griffiths P (2009) The distinction between innate and acquired characteristics. In: Zalta EN (ed) The Stanford encyclopedia of philosophy. http://plato.stanford.edu/entries/innate-acquired/

Grillner S, Wallén P, Saitoh K, Kozlov A, Robertson B (2008) Neural bases of goal-directed locomotion in vertebrates – an overview. Brain Res Rev 57(1):2–12

Gruner HE (ed) (1993) Lehrbuch der Speziellen Zoologie. Band I: Wirbellose Tiere. Part 1: Einführung, Protozoa, Placozoa, Porifera. G. Fischer, Jena/Stuttgart/New York

Guillot C, Lecuit T (2013) Mechanics of epithelial tissue. Homeostasis and morphogenesis. Science 340:1185–1189

Günther K (1950) Phylogenetik, Teleologie und Freiheit. In: Grüneberg H, Ulrich W (eds) Moderne Biologie (Festschrift zum 60. Geburtstag von Hans Nachtsheim). F W Peters, Berlin, pp 94–130

Gutmann W (1981) Relationships between invertebrate phyla based on functional-mechanical analysis of the hydrostatic skeleton. Am Zool 21:63–81

Gutmann W (1985) The hydraulic principles of the chordate and vertebrate "bauplan". In: Duncker HR, Fleischer G (eds) Fortschritte der Zoologie, vol. 30: Vertebrate morphology. Fischer, Stuttgart/New York, pp 23–26

Haken H (1983) Synergetics: an introduction. Springer, Berlin

Hall BK (2003) Unlocking the black box between genotype and phenotype: cell condensations as morphogenetic (modular) units. Biol Philos 18:219–247

Hanlon RT, Messenger JB (1996) Cephalopod behaviour. Cambridge University Press, Cambridge

Hart B, Hart L (2004) Tool use by elephants. In: Bekoff M (ed) Encyclopedia of animal behavior, vol 3. Greenwood, Westport/London, pp 1071–1073

Hartman H (2002) Progenotes. In: Encyclopedia of life sciences. doi:10.1038/npg.els.0000463, www.els.net

Hartman J, Garvik B, Hartwell L (2001) Principles of the buffering of genetic variation. Science 291:1001–1004

Hassenstein B (1969) Aspekte der "Freiheit" im Verhalten von Tieren. Universitas 24:1325–1330

Hayes JP (2010) Metabolic rates, genetic constraints, and the evolution of endothermy. J Evol Biol 23:1868–1877

Hayes JP, Garland T (1995) The evolution of endothermy: testing the aerobic capacity model. Evolution 49(5):836–847

Heath JE (1968) Origins of thermoregulation. In: Drake ET (ed) Evolution and environment. Yale University Press, New Haven, pp 259–278

Heffner R, Masterton RB (1983) The role of the corticospinal tract in the evolution of human digital dexterity. Brain Behav Evol 23:165–183

Heilinger JC (Hrsg.) (2007) Naturgeschichte der Freiheit. de Gryter, Berlin/New York

Heinrich B (1999) Mind of the raven: investigations and adventures with wolf-birds. Harper Collins, New York

Heinrich B (2004) Corvids: the crow family. In: Bekoff M (ed) Encyclopedia of animal behavior, vol 1. Greenwood, Westport/London, pp 445–447

Heinrich B, Bugnyar T (2007) Just how smart are ravens? Sci Am (April):48

Heldmaier G, Neuweiler G (2003) Vergleichende Tierphysiologie. Vol. 1: Neuro- und Sinnesphysiologie. Springer, Berlin/Heidelberg/New York

Heldmaier G, Neuweiler G (2004) Vergleichende Tierphysiologie. Vol. 2: Vegetative Physiologie. Springer, Berlin/Heidelberg/New York

Hendrichs H (1988) Lebensprozesse und wissenschaftliches Denken. Zur Logik der Lebendigkeit und ihrer Erstarrung in den Wissenschaften. Karl Alber, Freiburg/München

Hennig E (1955) Der Begriff des Fortschritts im Werdegang des Lebens. Forsch Fortschr 29:289–290

Herman LM, Richards DG, Wolz JP (1984) Comprehension of sentences by bottle-nosed dolphins. Cognition 16:129–219

Herman LM, Abichandani SL, Elhajj AN, Herman EYK, Sanchez JK, Pack AA (1999) Dolphins (Tursiops truncatus) comprehend the referential character of the human pointing gesture. J Comp Psychol 113:1–18

Herman LM, Matus DS, Herman EYK, Ivancic M, Pack AA (2001) The bottlenosed dolphin's (Tursiops truncatus) understanding of gestures as symbolic representations of its body parts. Anim Learn Behav 29:250–264

Hertz PE, Huey PE, Garland T (1988) Time budgets, thermoregulation, and maximal locomotor performance: are reptiles Olympians or boy scouts? Am Zool 28:927–938
Hess J (1989) Familie 5. Berggorillas in den Virunga-Wäldern. Birkhäuser, Basel/Boston/Berlin
Heuts M (1953) Theorien und Tatsachen der Biologischen Evolution. Verhandlungen der Deutschen Zoologischen Gesellschaft (Zoologischer Anzeiger Suppl. Bd. 17) 409–429
Heyes C (2012a) New thinking: the evolution of human cognition. Philos Trans R Soc B Biol Sci 367:2091–2096. doi:10.1098/rstb.2012.0111
Heyes C (2012b) Grist and mills: on the cultural origins of cultural learning. Philos Trans R Soc B Biol Sci 367:2091–2096. doi:10.1098/rstb.2012.0120
Hildebrand M, Goslow GE (2001) Analysis of vertebrate structure. Wiley, New York
Hildebrandt G (1979) Rhythmical functional order and man's emancipation from the time factor. In: Schaefer KE, Hildebrandt G, Macbeth N (eds) Basis of an individual physiology. Futura Publishing Company, Mount Kisco, pp 15–43
Hildebrandt G, Moog R, Raschke F (Hrsg) (1987) Chronobiology and chronomedicine. Verlag Peter Lang, Frankfurt
Hill RW, Wyse GA, Anderson M (2008) Animal physiology. Sinauer Associates, Sunderland
Hillenius WJ, Ruben JA (2004) The evolution of endothermy in terrestrial vertebrates: who? when? why? Physiol Biochem Zool 77(6):1019–1042
Hillman SS (2001) Osmoregulation by vertebrates in terrestrial environments. In: Nature encyclopedia of life sciences. Nature Publishing Group, London. http://www.els.net
Hinegardner R, Engelberg J (1983) Biological complexity. J Theor Biol 104:7–20
Hogenesch JB, Herzog ED (2011) Intracellular and intercellular processes determine robustness of the circadian clock. FEBS Lett 585(10):1427–1434
Holland PWH (1998) Major transitions in animal evolution: a developmental genetic perspective. Am Zool 38:829–842
Hopson JA (1973) Endothermy, small size, and the origin of mammalian reproduction. Am Nat 107:446–452
Hotton N (1980) An alternative to dinosaur endothermy: the happy wanderers. In: Thomas RDK, Olson EC (eds) A cold look at the warm-blooded dinosaurs, AAAS selected symposium, 28. Westview Press, Boulder, pp 311–350
Hulbert AJ, Else PL (1989) Evolution of mammalian endothermic metabolism: mitochondrial activity and cell composition. Am J Physiol 256:R63–R69
Hutchison VH, Dupre RK (1992) Thermoregulation. In: Feder ME, Burggren W (eds) Environmental physiology of the amphibians. University of Chicago Press, Chicago, pp 206–249
Huxley J (1953) Evolution in action. Chatto & Windus, London
Huxley J (1974) Evolution, the modern synthesis. Allen & Unwin, London
Jablonka E, Lamb MJ (2005) Evolution in four dimensions. Genetic, epigenetic, behavioral, and symbolic variation in the history of life. MIT Press, Cambridge, MA
Jablonka E, Raz G (2009) Transgenerational epigenetic inheritance: prevalence, mechanisms and implications for the study of heredity and evolution. Q Rev Biol 84(2):131–176
Jablonski D (2007) Scale and hierarchy in macroevolution. Palaeontology 50:87–109
Jane SM, Ting SB, Cunningham JM (2005) Epidermal impermeable barriers in mouse and fly. Curr Opin Genet Dev 15(4):447–453
Jenkins FA, Goslow GE (1983) The functional anatomy of the shoulder of the savannah monitor lizard (Varanus exanthematicus). J Morphol 174:195–216
Jensen S (2003) The Proterozoic and earliest Cambrian trace fossil record; patterns, problems and perspectives. Integr Comp Biol 43:219–228
Jerison HJ (1973) Evolution of the brain and intelligence. Academic, New York
Jerison HJ (1991) Brain size and the evolution of mind: 59th James Arthur lecture on the evolution of the human brain. American Museum of Natural History, New York
Jerison HJ (2001) The evolution of neural and behavioral complexity. In: Roth G, Wullimann MF (eds) Brain evolution and cognition. Wiley/Spektrum, New York/Heidelberg, pp 523–553
Jerison HJ, Jerison I (1988) Intelligence and evolutionary biology. Springer, New York

Johanson D, Edgar B (2000) Lucy und ihre Kinder. Spektrum Akademischer Verlag, Heidelberg/Berlin

Jonas H (1966) The phenomenon of life: toward a philosophical biology. Harper and Row, New York

Jones D (2005) Niche construction. Nature 438:14–16

Jordan H (1908) Über Entwicklung vom physiologischen Standpunkte aus. Biologisches Zentralblatt 28:278–287

Jordan H (1913) Vergleichende Physiologie wirbelloser Tiere. Band 1. Fischer, Jena

Joyner M, Pedersen BK (2011) Ten questions about systems biology. J Physiol 589(5):1017–1030

Kacelnik A, Chappell J, Weir AAS, Kenward B (2004) Tool use and manufacture by birds. In: Bekoff M (ed) Encyclopedia of animal behavior, vol 3. Greenwood, Westport/London, pp 1067–1069

Kaminski J, Call J, Tomasello M (2008) Chimpanzees know what others know, but not what they believe. Cognition 109:224–234

Kämpfe L (1985) Evolution und Stammensgeschichte der Organismen. G. Fischer, Stuttgart

Kandel E, Schwartz J, Jesseil T (1995) Essentials of neural science and behavior. Appelton & Lange, Norwalk

Karasov WH, Diamond J (1985) Digestive adaptations for fuelling the cost of endothermy. Science 228:202–204

Karten HJ (1991) Homology and evolutionary origins of the “neocortex”. Brain Behav Evol 38:264–272

Kather R (2003) Was ist Leben? Philosophische Positionen und Perspektiven. Wissenschaftliche Buchgesellschaft, Darmstadt

Kauffman S (2003) Molecular autonomous agents. Philos Trans R Soc Lond A 361:1089–1099

Kemp TS (2005) The origin and evolution of mammals. Oxford University Press, Oxford

Kemp TS (2006) The origin of mammalian endothermy: a paradigm for the evolution of complex biological structure. Zool J Linn Soc 147:473–488

Kenward B, Weir AAS, Rutz C, Kacelnik A (2005) Tool manufacture by naive juvenile crows. Nature 433:121

Kiehn O (2006) Locomotor circuits in the mammalian spinal cord. Annu Rev Neurosci 29:279–306

Kimbrough Oller D, Griebel U (2008) Evolution of communicative flexibility. Complexity, creativity, and adaptability in human and animal communication, The Vienna series in theoretical biology. MIT Press, Cambridge

King N, Hittinger CT, Carroll SB (2003) Evolution of key cell signaling and adhesion protein families predates animal origins. Science 301:361–363

Kipp FA (1948) Höherentwicklung und Menschwerdung. Hippokrates, Stuttgart

Kipp FA (1949) Arterhaltung und Individualisierung in der Tierreihe. Verhandlungen der Deutschen Zoologen in Mainz 1949. Geest & Portig, Leipzig, pp 24–27

Kipp FA (1980) Die Evolution des Menschen im Hinblick auf seine lange Jugendzeit. Freies Geistesleben, Stuttgart

Kirk DL (1998) Volvox. Molecular-genetic origins of multicellularity and cellular differentiation. Cambridge University Press, Cambridge

Kirk DL (2005) A twelve-step program for evolving multicellularity and a division of labor. Bioessays 27:299–310

Kirschner MW, Gerhart JC (2005) The plausibility of life. Resolving Darwin's dilemma. Yale University Press, New Haven/London

Kitano H (2002a) Looking beyond the details: a rise in system-oriented approaches in genetics and molecular biology. Curr Genet 41:1–10

Kitano H (2002b) Systems biology: a brief overview. Science 295:1662–1664

Kitano H (2004) Biological robustness. Nat Rev Genet 5:826–837

Kitano H (2007) Towards a theory of biological robustness. Mol Syst Biol 3:137

Kitano H, Oda K (2006) Self-extending symbiosis: a mechanism for increasing robustness through evolution. Biol Theory 1(1):61–66

Klein RG (2009) The human career. Human biological and cultural origins, 3rd edn. University of Chicago Press, Chicago

Klein W, Owerkowicz T (2006) Function of intracoelomic septa in lung ventilation of amniotes: lessons from lizards. Physiol Biochem Zool 79:1019–1032

Knoll AH, Carroll SB (1999) Early animal evolution: emerging views from comparative biology and geology. Science 284:2129–2137

Knoll AH, Hewitt D (2011) Phylogenetic, functional, and geological perspectives on complex multicellularity. In: Calcott B, Sterelny K (eds) The major transitions in evolution revisited. MIT Press, Cambridge, MA, pp 251–270

Köchy K (1997) Ganzheit und Wissenschaft. Das historische Fallbeispiel der romantischen Naturforschung. Königshausen & Neumann, Würzburg

Koob TJ, Long JH (2000) The vertebrate body axis: evolution and mechanical function. Am Zool 40:1–18

Koonin EV (2009) Darwinian evolution in the light of genomics. Nucleic Acids Res 37(4): 1011–1034

Koteja P (2000) Energy assimilation, parental care and the evolution of endothermy. Proc R Soc Lond B 267:479–484

Koukkari WL, Sothern RB (2006) Introducing biological rhythms. Springer, New York

Kuhn-Schnyder E (1954) Der Ursprung der Säugetiere. Vierteljahreszeitschrift der Naturforschenden Gesellschaft in Zürich 99:165–197

Kuhn-Schnyder E (1967) Paläontologie als stammesgeschichtliche Urkundenforschung. In: Heberer G (ed) Die Evolution der Organismen. Ergebnisse und Probleme der Abstammungslehre, vol 1. G. Fischer, Stuttgart, pp 350–373

Kümmell S (2009) Die Digiti der Synapsida: Anatomie, Evolution und Konstruktionsmorphologie. Shaker, Aachen

Kümmell S (2012) Digital arcade in the autopodia of Synapsida: standard position of the digits and dorsoventral excursion angle of digital joints in the rays II–V. Palaeobiodivers Palaeoenviron 92(2):171–196. doi:10.1007/s12549-012-0076-6, Senckenberg Institut (Springer)

Kuypers HGJM (1981) Anatomy of the descending pathways. In: Brookhart JM, Mountcastle VB (eds) Handbook of physiology – the nervous system. Vol II: Motor control. American Physiology Society, Bethesda, pp 597–666

Lacalli T (2012) The Middle Cambrian fossil Pikaia and the evolution of chordate swimming. EvoDevo 3:12. doi:10.1186/2041-9139-3-12

Lacquaniti F, Ivanenko YP, Zago M (2012) Development of human locomotion. Curr Opin Neurobiol 22(5):822–828

Laland KN, Sterelny K (2006) Seven reasons (not) to neglect niche construction. Evolution 60:1751–1762

Landmann L (1986) The skin of reptiles: epidermis and dermis. In: Bereiter-Hahn J, Matoltsy AG, Sylvia Richards K (eds) Biology of the integument, Bd. 2. Springer, Berlin, pp 150–187

Lane N (2009) Life ascending. The ten great inventions of evolution. Profile Books, London

Lange E (1976) Ein Beitrag zur Frage “Was ist Höherentwicklung?”. Biologische Rundschau 14:206–214

Larhlimi A, Blachon S, Selbig J, Nikoloski Z (2011) Robustness of metabolic networks: a review of existing definitions. BioSystems 106(1):1–8

Lauder GV (2000) Function of the caudal fin during locomotion in fishes: kinematics, flow visualization, and evolutionary patterns. Am Zool 40:101–122

Laurin M (2004) The evolution of body size, Cope’s rule and the origin of amniotes. Syst Biol 53:594–622

Laurin M, Reisz RR (1997) A new perspective on tetrapod phylogeny. In: Sumida SS, Martin KLM (eds) Amniote origins. Completing the transition to land. Academic, San Diego, pp 9–59

Le Tensorer JM (2001) Ein Bild vor dem Bild? Die ältesten menschlichen Artefakte und die Frage des Bildes. In: Boehm G (Hrsg.) Homo Pictor. Colloquium Rauricum, vol. 7. Sauer, München/ Leipzig

Lechner J, Wieland F (1989) Structure and biosynthesis of prokaryotic glycoproteins. Annu Rev Biochem 58:173–194

Ledger PW (1975) Septate junctions in the calcareous sponge Sycon ciliatum. Tissue Cell 7:13–18
Lefebvre L, Whittle P, Lascaris E, Finkelstein A (1997) Feeding innovations and forebrain size in birds. Anim Behav 53:549–560
Lefebvre L, Gaxiola A, Dawson S, Timmermans S, Rosza L, Kabai P (1998) Feeding innovations and forebrain size in Australian birds. Behavior 135:1077–1097
Lefebvre L, Nicolakakis N, Boire D (2002) Tools and brains in birds. Behaviour 139:939–973
Lemon RN (2008) Descending pathways in motor control. Annu Rev Neurosci 31:195–218
Leslie AM (1987) Pretense and representation: the origins of 'theory of mind'. Psychol Rev 94:412–426
Lewis HM, Laland KN (2012) Transmission fidelity is the key to the build-up of cumulative culture. Philos Trans R Soc B Biol Sci 367:2171–2180. doi:10.1098/rstb.2012.0119
Lewitus E, Soligo C (2011) Life-history correlates of placental structure in eutherian evolution. Evol Biol 38(3):287–305
Lewontin R (1968) Evolution. In: Sills DL (ed) International encyclopedia of the social sciences, vol 5. The MacMillan Company/The Free Press, Chicago, pp 202–210
Lewontin R (2000) The triple helix. Gene organism and environment. Harvard University Press, Cambridge
Lillywhite HB, Maderson PFA (1988) The structure and permeability of integument. Am Zool 28:945–962
Lockley M (1991) Tracking dinosaurs – a new look at an ancient world. Cambridge University Press, Cambridge
Lockley M (1999) The eternal trail. A tracker looks at evolution. Perseus Books, Reading
Lockley M, Hunt AP (1995) Dinosaur tracks and other fossil footprints of the western United States. Columbia University Press, New York
Lorenz K (1935) Der Kumpan in der Umwelt des Vogels. Journal für Ornithologie 83(2–3): 137–215. doi:10.1007/BF01905355, 289–413
Love GD, Grosjean E, Stalvies C, Fike DA, Grotzinger JP, Bradley AS, Kelly AE, Bhatia M, Meredith W, Snape CE, Bowring SA, Condon DJ, Summons RE (2009) Fossil steroids record the appearance of Demospongiae during the Cryogenian period. Nature 457:718–721
Lovegrove BG (2012) The evolution of endothermy in Cenozoic mammals: a plesiomorphic-apomorphic continuum. Biol Rev 87:128–162. doi:10.1111/j.1469-185X.2011.00188.x
Lovejoy A (1982) The great chain of being: a study of the history of an idea. Harvard University Press, Cambridge, MA
Luckett WP (1993) Uses and limitations of mammalian fetal membranes and placenta for phylogenetic reconstruction. J Exp Zool 266:514–527
Luisi PL (2003) Autopoiesis: a review and a reappraisal. Die Naturwissenschaften 90:49–59
Mackie GO (1974) Locomotion, flotation and dispersal. In: Muscatine L, Lenhoff HM (eds) Coelenterate biology. Academic, London/New York, pp 313–357
Mackie GO (1984) Introduction to the diploblastic level. In: Bereiter-Hahn J, Matoltsy AG, Sylvia Richards K (eds) Biology of the integument. Springer, Berlin
Mackie GO (1990) The elementary nervous system revisited. Am Zool 30:907–920
MacPhail EM (1982) Brain and intelligence in vertebrates. Clarendon Press, Oxford
MacPhail EM (2001) Conservation in the neurology and psychology of cognition in vertebrates. In: Roth G, Wullimann MF (eds) Brain evolution and cognition. Wiley/Spektrum, New York/Heidelberg, pp 401–430
MacPhail EM, Bolhuis JJ (2001) The evolution of intelligence: adaptive specialisations versus general process. Biol Rev 76:341–364
Madara JL (1998) Regulation of the movement of solutes across tight junctions. Annu Rev Physiol 60:143–159
Madigan MT, Marinko JM, Parker J (2000) Brock. Biology of microorganisms. Prentice-Hall, Upper Saddle River
Magie CR, Martindale MQ (2008) Cell-cell adhesion in the Cnidaria: insights into the evolution of tissue morphogenesis. Biol Bull 214:218 232
Mahner M, Bunge M (1997) Foundations of biophilosophy. Springer, Berlin/Heidelberg

Maina JN (2000) Comparative respiratory morphology: themes and principles in the design and construction of the gas exchangers. Anat Rec 261(1):25–44

Maldonado M (2004) Choanoflagellates, choanocytes, and animal multicellularity. Invert Biol 123:1–22

Maravita A, Iriki A (2004) Tools for the body (schema). Trends Cogn Sci 8:79–86

Margulis L (1990) Kingdom Animalia: the zoological malaise from a microbial perspective. Am Zool 30:861–875

Margulis L (1991) Symbiosis as a source of evolutionary innovation: speciation and morphogenesis. MIT Press, Cambridge/London

Margulis L (1993) Symbiosis in cell evolution. Freeman, San Francisco

Margulis L, Sagan D (2002) Acquiring genomes. A theory of the origins of species. Basic Books, New York

Marino L (1998) A comparison of encephalization between odontocete cetaceans and anthropoid primates. Brain Behav Evol 51(4):230–238

Marino L (2002) Convergence of complex cognitive abilities in cetaceans and primates. Brain Behav Evol 59:21–32

Marler P (2004) Bird calls: their potential for behavioral neurobiology. Ann N Y Acad Sci 1016:31–44

Martin RD (1990) Primate origins and evolution: a phylogenetic reconstruction. Chapman and Hall, London

Martin W (1999) Mosaic bacterial chromosomes: a challenge en route to a tree of genomes. Bioessays 21:99–104

Martin W, Embley TM (2004) Early evolution comes full circle. Nature 431:134–137

Martin KLM, Nagy KA (1997) Water balance and the physiology of the amphibian to amniote transition. In: Sumida SS, Martin KLM (eds) Amniote origins. Completing the transition to land. Academic, San Diego, pp 399–423

Masel J, Siegal ML (2009) Robustness: mechanisms and consequences. Trends Genet 25(9):395–403

Masel J, Trotter MV (2010) Robustness and evolvability. Trends Genet 26(9):406–414

Maturana HR, Varela FJ (1987) The tree of knowledge: the biological roots of human understanding. Shambhala Press, Boston

Maynard Smith J, Szathmary EM (1995) The major transitions in evolution. Freeman, New York

Mayr E (1974) Behavior programs and evolutionary strategies. Am Sci 62:650–659

McCarthy MC, Enquist BJ (2005) Organismal size, metabolism and the evolution of complexity in metazoans. Evol Ecol Res 7(5):681–696

McCoy JW (1977) Complexity in organic evolution. J Theor Biol 68:457–458

McFarland D (2007) Animal behaviour: psychobiology, ethology and evolution. Pearson/Prentice Hall, Harlow

McGrew WC (2010) Chimpanzee technology. Science 328:579–580

McKinney ML, McNamara KJ (1991) Heterochrony: the evolution of ontogeny. Plenum Press, New York, 19

McMahon TA, Bonner JT (1983) On size and life. Scientific American Books, New York

McNab BK (1966) The metabolism of fossorial rodents: a study of convergence. Ecology 47:712–733

McNab BK (1978) The evolution of endothermy in the phylogeny of mammals. Am Nat 112:1–21

McNab BK, Auffenberg WA (1976) The effect of large body size on the temperature regulation of the Komodo dragon, Varanus komodoensis. Comp Biochem Physiol 55A:345–350

McNamara KJ (ed) (1990) Evolutionary trends. Belhaven Press, London

McShea DW (1991) Complexity and evolution: what everybody knows. Biol Philos 6:303–324

McShea DW (1996) Metazoan complexity and evolution: is there a trend? Evolution 50:477–492

McShea DW (1998) Possible largest-scale trends in organismal evolution: eight 'live hypotheses'. Annu Rev Ecol Syst 29:293–318

McShea DW (2002) A complexity drain on cells in the evolution of multicellularity. Evolution 56:441–452

McShea DW, Brandon RN (2010) Biology's first law: the tendency for diversity and complexity to increase in evolutionary systems. University of Chicago Press, Chicago
McShea DW, Simpson C (2011) The miscellaneous transitions in evolution. In: Calcott B, Sterelny K (eds) The major transitions in evolution revisited. MIT Press, Cambridge, MA, pp 19–33
Meckel JF (1821) System der vergleichenden Anatomie. Band I. Renger, Halle
Menkhorst E, Nation A, Cui S, Selwood L (2009) Evolution of the shell coat and yolk in amniotes: a Marsupial perspective. J Exp Zool B Mol Dev Evol 312:625–638
Menzel EW (1986) How can you tell if an animal is intelligent? In: Schusterman RJ, Thomas JA, Wood FG (eds) Dolphin cognition and behaviour: a comparative approach. Lawrence Erlbaum Associates, Hillsdale
Menzel R, Giufra M (2001) Cognitive architecture of a mini-brain: the honeybee. Trends Cogn Sci 5:62–71
Mesarovic M, Sreenath SN (2006) Beyond the flat earth perspective in systems biology. Biol Theory 1:33–34
Mesarovic MD, Sreenath N, Keene J (2004) Search for organizing principles: understanding in systems biology. Syst Biol 1:19–27
Mess A, Carter AM (2006) Evolutionary transformations of fetal membrane characters in Eutheria with special reference to Afrotheria. J Exp Zool B Mol Dev Evol 306:140–163
Mess A, Blackburn DG, Zeller U (2003) Evolutionary transformations of fetal membranes and reproductive strategies. J Exp Zool 299A:3–12
Michod R (1997) Evolution of the individual. Am Nat 150:S5–S21
Michod RE (2007) Evolution of individuality during the transition from unicellular to multicellular life. Proc Natl Acad Sci 104:8613–8618
Miklósi Á (2004) Imitation. In: Bekoff M (ed) Encyclopedia of animal behavior, vol 1. Greenwood, Westport/London, pp 299–305
Milewski AV, Mills AJ (2010) Does life consistently maximise energy intensity? Biol Rev 85(4):859–879
Mitic LL, Anderson JM (1998) Molecular architecture of tight junctions. Annu Rev Physiol 60:121–142
Mohawk JA, Green CB, Takahashi JS (2012) Central and peripheral circadian clocks in mammals. Annu Rev Neurosci 35:445–462
Moore JA (1990) A conceptual framework for biology, Part II. Am Zool 30:723–858
Moore BR (1993) Avian movement imitation and a new form of mimicry: tracing the evolution of complex learning behaviour. Behaviour 122:231–263
Moore BR (2004) The evolution of learning. Biol Rev 79:301–335
Moreno A, Etxeberria A, Umerez J (2008) The autonomy of biological individuals and artificial models. BioSystems 91:309–319
Moroz LL (2012) Phylogenomics meets neuroscience: how many times might complex brains have evolved? Acta Biol Hung 63(Suppl 2):3–19
Morris PJ (1993) The developmental role of the extracellular matrix suggests a monophyletic origin of the kingdom animalia. Evolution 47:152–165
Morris SC (2006) Darwin's dilemma: the realities of the Cambrian 'explosion'. Philos Trans R Soc B Biol Sci 361:1069–1083
Mrzljak L, Uylings HBM, van Eden CG, Judas M (1990) Neuronal development in human prefrontal cortex in prenatal and postnatal stages. In: Uylings HBM, van Eden CG, de Bruin JPC, Corner MA, Feenstra MGP (eds) The prefrontal cortex. Its structure, function and pathology. Elsevier, Amsterdam, pp 185–222
Müller WEG (2001) Review: how was metazoan threshold crossed? The hypothetical Urmetazoa. Comp Biochem Physiol A 129:433–460
Müller GB (2007) Evo-devo: extending the evolutionary synthesis. Nat Rev Genet 8:943–949
Müller GB (2010) Epigenetic innovation. In: Pigliucci M, Müller G (eds) Evolution – the extended synthesis. MIT Press, Cambridge, MA, pp 307–332
Müller GB, Newman SA (2003) Origination of organismal form – beyond the gene in development and evolutionary biology. MIT Press, Cambridge, MA

Murphy BF, Thompson MB (2011) A review of the evolution of viviparity in squamate reptiles: the past, present and future role of molecular biology and genomics. J Comp Physiol B Biochem Syst Environ Physiol 181(5):575–594
Natochin YV, Chernigovskaya TY (1997) Evolutionary physiology: history, principles. Comp Biochem Physiol A 118:63–79
Needham J (1931) Chemical embryology. Cambridge University Press, Cambridge
Nektaria N, Lefebvre L (2000) Forebrain size and innovation rate in european birds: feeding, nesting and confounding variables. Behaviour 137:1415–1429
Nespolo RF, Bacigalupe LD, Figueroa CC, Koteja P, Opazo JC (2011) Using new tools to solve an old problem: the evolution of endothermy in vertebrates. Trends Ecol Evol 26(8):414–423
Neuweiler G (2008) Und wir sind es doch – die Krone der Evolution. Wagenbach, Berlin
Nielsen C (1998) Morphological approaches to phylogeny. Am Zool 38:942–952
Nielsen C (2001) Animal evolution. Interrelationships of the living phyla. Oxford University Press, Oxford
Nitecki MH (ed) (1988) Evolutionary progress? The University of Chicago Press, Chicago
Nitecki HM (1990) Evolutionary innovations. The University of Chicago Press, Chicago/London
Noble D (2006) The music of life. Biology beyond genes. Oxford University Press, Oxford
Noble D (2008) Claude Bernard, the first systems biologist, and the future of physiology. Exp Physiol 93:16–26
Noble D (2011) Neo-Darwinism, the modern synthesis and selfish genes: are they of use in physiology? J Physiol 589(5):1007–1015
Noble D (2013) Physiology is rocking the foundations of evolutionary biology. Exp Physiol 98(8):1235–1243. doi:10.1113/expphysiol.2012.071134
Northcutt RG (2012) Evolution of centralized nervous systems: two schools of evolutionary thought. Proc Natl Acad Sci USA 109(Suppl 1):10626–10633
Nowell A (2014) Childhood, play and the evolution of cultural capacity in Neandertals and modern humans. In: Haidle M, Conard N, Bolus M (eds) The nature of culture, Vertebrate paleobiology and paleoanthropology series. Springer (in press) zitiert in Kap. 10 und 11 (mehrfach)
Nowell A, Davidson I (eds) (2010) Stone tools and the evolution of human cognition. University Press of Colorado, Boulder
Ochman H, Lawrence JG, Groisman EA (2000) Lateral gene transfer and the nature of bacterial innovation. Nature 405:299–304
O'Connor MP, Dodson P (1999) Biophysical constraints on the thermal ecology of dinosaurs. Paleobiology 25(3):341–368
Odling-Smee J (2010) Niche inheritance. In: Pigliucci M, Müller GB (eds) Evolution. The extended synthesis. MIT Press, Cambridge, MA, pp 175–207
Odling-Smee FJ, Laland KN, Feldman MW (2003) Niche construction: the neglected process in evolution. Princeton University Press, Princeton
O'Malley MA, Dupré J (2005) Fundamental issues in systems biology. Bioessays 27:1270–1276
O'Malley MA, Dupré J (2007) Size doesn't matter: towards a more inclusive philosophy of biology. Biol Philos 22:155–191. doi:10.1007/s10539-006-9031-0
Ortega JC, Bekoff M (1987) Avian play: comparative evolutionary and developmental trends. Auk 104:338–341
Ortiz RM (2001) Osmoregulation in marine mammals. J Exp Biol 204(11):1831–1844
Overhage P (1957) Der "Biologische Aufstieg" und seine Kriterien. Acta Biotheor 12:80–115
Overhage P (1963) Die Evolution des Lebendigen – das Phänomen. Herder, Freiburg/Basel/Wien
Overton J (1997) Paul Alfred Weiss. In: Office of the Home Secretary, National Academy of Sciences (eds) Biographical memoirs, vol. 72. pp 373–386. www.nap.edu/readingroom/books/biomems/pweiss.html
Oyama S (2000a) The ontogeny of information: developmental systems and evolution. Duke University Press, Durham
Oyama S (2000b) The evolution's eye. A systems view of the biology-culture divide. Duke University Press, Durham
Oyama S, Griffiths PE, Gray RD (2001) Cycles of contingency: developmental systems and evolution. MIT Press, Cambridge, MA

Özbek S, Balasubramanian PG, Chiquet-Ehrismann R, Tucker RP, Adams JC (2010) The evolution of extracellular matrix. Mol Biol Cell 21(24):4300–4305

Packard MJ, Seymour RS (1997) Evolution of the amniote egg. In: Sumida SS, Martin KLM (eds) Amniote origins. Completing the transition to land. Academic, San Diego

Packard GC, Tracy RC, Roth JJ (1977) The physiological ecology of reptilian eggs and embryos, and the evolution of viviparity within the class Reptilia. Biol Rev 52:71–105

Paladino EV, O'Conner MP, Spotila JR (1990) Metabolism of leatherback turtles, gigantothermy, and thermoregulation of dinosaurs. Nature 344:858–860

Papentin F (1980) On order and complexity, I. General considerations. J Theor Biol 87:421–456

Parker ST, Mitchell RW, Boccia ML (eds) (1994) Self-awareness in animals and humans: developmental perspectives. Cambridge University Press, Cambridge, MA

Passingham RE (1975) Changes in the size and organization of the brain in man and his ancestors. Brain Behav Evol 11:73–90

Pearce JM (1997) Animal learning and cognition. Psychology Press, Exeter

Pepperberg IM (1990) Cognition in an African gray parrot (Psittacus erithacus): further evidence for comprehension of categories and labels. J Comp Psychol 104:41–52

Pepperberg IM (1994a) Numerical competence in an African gray parrot (Psittacus erithacus). J Comp Psychol 108:36–44

Pepperberg IM (1994b) Vocal learning in African grey parrots: effects of social interaction. Auk 111:300–313

Perry SF (1985) Functional anatomy and evolution of reptilian lungs. In: Duncker HR, Fleischer G (eds) Fortschritte der Zoologie, Band 30: Vertebrate morphology. Fischer, Stuttgart/New York, pp 379–382

Perry SF, Carrier DR (2006) The coupled evolution of breathing and locomotion as a game of leapfrog. Physiol Biochem Zool 79(6):997–999

Perry SF, Wilson RJA, Straus C, Harris MB, Remmers JE (2001) Which came first, the lung or the breath? Comp Biochem Physiol A Mol Integr Physiol 129:37–47

Petras JM (1968) Corticospinal fibers in new world and old world simians. Brain Res 8:206–208

Pflumm W (1996) Biologie der Säugetiere. Parey, Berlin

Pierce SE, Clack JA, Hutchinson JR (2012) Three-dimensional limb joint mobility in the early tetrapod Ichthyostega. Nature 486:523–526

Pigliucci M (2005) Expanding evolution. Nature 435:565–566

Pigliucci M (2013) Between holism and reductionism: a philosophical primer on emergence. Biol J Linn Soc. doi:10.1111/bij.12060

Pigliucci M, Müller G (2010) Evolution – the extended synthesis. MIT Press, Cambridge, MA

Plessner H (1975) Die Stufen des Organischen und der Mensch. De Gruyter, Berlin

Plotnik JM, de Waal F, Reiss D (2006) Self-recognition in an Asian elephant. Proc Natl Acad Sci U S A 103(45):17053–17057. doi:10.1073/pnas.0608062103

Pough FH (1980) The advantages of ectothermy for tetrapods. Am Nat 115:92–112

Pough FH (1983) Amphibians and reptiles as low-energy systems. In: Aspey WP, Lusick SI (eds) Behavioral energetics: the cost of survival in vertebrates. Ohio State University Press, Columbus, pp 141–188

Preston SD, de Waal FBM (2002) Empathy: its ultimate and proximate bases. Behav Brain Sci 25:1–72

Prior H, Schwarz A, Güntürkün O (2008) Mirror-induced behavior in the Magpie (Pica pica): evidence of self-recognition. PLoS Biol 6:e202. doi:10.1371/journal.pbio.0060202

Prud'homme B, Gompel N, Carroll SB (2007) Emerging principles of regulatory evolution. Proc Natl Acad Sci 104(Suppl 1):8605–8612

Putnam NH, Butts T, Ferrier DEK et al (2008) The amphioxus genome and the evolution of the chordate karyotype. Nature 453:1064–1072

Quiring H (1931) Die Grundlagen des Artaufstiegs. Forsch Fortschr 7:161–162

Rapp F (1992) Fortschritt. Entwicklung und Sinngehalt einer philosophischen Idee. Wissenschaftliche Buchgesellschaft, Darmstadt

Raup DM, Jablonski D (eds) (1986) Patterns and processes in the history of life. Report of the Dahlem Workshop on Patterns and Processes in the History of Life, Berlin 1985. Springer, Berlin/Heidelberg

Reader SM, Laland KN (2002) Social intelligence, innovation, and enhanced brain size in primates. Proc Natl Acad Sci U S A 99:4436–4441

Rehkämper G, Frahm HD, Mann MD, Vogt O (2001) Evolutionary constraints of large telencephala. In: Roth G, Wullimann MF (eds) Brain evolution and cognition. Wiley/Spektrum, New York/ Heidelberg, pp 265–293

Rehmann-Sutter C (2002) Genetics, embodiment and identity. In: Grunwald A, Gutmann M, Neumann-Held EM (eds) On human nature. Anthropological, biological, and philosophical foundations. Springer, Berlin/Heidelberg/New York, pp 23–50

Reichholf JH (1992a) Der schöpferische Impuls. Eine neue Sicht der Evolution. Deutsche Verlagsanstalt, Stuttgart

Reichholf JH (1992b) Erfolgsprinzip Fortbewegung. Die Evolution des Laufens, Fliegens, Schwimmens und Grabens. Der Deutsche Taschenbuch Verlag, München

Reiss D, Marino L (2001) Mirror self-recognition in the bottlenose dolphin: a case of cognitive convergence. Proc Natl Acad Sci 98:5937–5942

Remane A (1967) Die Geschichte der Tiere. In: Heberer G (ed) Die Evolution der Organismen. Ergebnisse und Probleme der Abstammungslehre, Band I. G. Fischer, Stuttgart

Renfree MB (1991) Marsupial mammals: enigma variations on a reproductive theme. Verhandlungen der deutschen Zoologischen Gesellschaft 84:153–167

Rensch B (1959) Evolution above the species level. Columbia University Press, New York

Rensch B (1973) Gedächtnis, Begriffsbildung und Planhandlungen bei Tieren. Parey, Berlin

Richards KS (1984) Introduction, intercellular junctions and cilia. In: Bereiter-Hahn J, Matoltsy AG, Richards KS (eds) Biology of the integument, vol 1. Springer, Berlin, pp 1–16

Richards RJ (1992) The meaning of evolution: the morphological construction and ideological reconstruction of Darwin's theory. University of Chicago Press, Chicago

Richardson RC (2007) Evolutionary psychology as maladapted psychology. MIT Press, Cambridge, MA

Richerson PJ, Boyd R (2005) Not by genes alone. How culture transformed human evolution. University of Chicago Press, Chicago/London

Riedl R (2000) Strukturen der Komplexität. Eine Morphologie des Erkennens und Erklärens. Springer, Berlin/Heidelberg

Rieger RM (1984) Evolution of the Cuticle in the lower eumeatzoa. In: Bereiter-Hahn J, Matoltsy AG, Sylvia Richards K (eds) Biology of the integument. Springer, Berlin, pp 389–399

Rieger RM (1994) Evolution of the "lower" Metazoa. In: Bengtson S (ed) Early life on earth. Columbia University Press, New York, pp 475–488

Rieger RM, Lombardi J (1987) Ultrastructure of coelomic lining in echinoderm podia: significance for concepts in the evolution of muscle and peritoneal cell. Zoomorphology 107:191–208

Rieger RM, Rieger M (1976) Fine structure of the archiannelid cuticle and remarks on the evolution of the cuticle within the spiralia. Acta Zool (Stockholm) 57:53–68

Rieger RM, Weyrer S (1998) The evolution of the lower Metazoa: evidence from the phenotype. In: Müller WE (ed) Molecular evolution: towards the origin of Metazoa. Springer, Berlin

Rieger RM, Haszprunar G, Schuchert P (1991) On the origin of the bilateria: traditional views and recent alternative concepts. In: Simonetta A, Conway Morris S (eds) The early evolution of Metazoa and the significance of problematic taxa. Cambridge University Press, Cambridge, pp 107–112

Ristau CA (ed) (1991) Cognitive ethology: the minds of other animals. Lawrence Erlbaum Associates, Hillsdale

Rivera MC, Lake JA (2004) The ring of life provides evidence for a genome fusion origin of eukaryotes. Nature 431:152–155

Roach NT, Venkadesan M, Rainbow MJ, Lieberman DE (2013) Elastic energy storage in the shoulder and the evolution of high-speed throwing in Homo. Nature 498:483–487. doi:10.1038/nature12267

Robson D (2013) Old schooled: you never stop learning like a child. New Scient 2918:20–24

Robson SL, Wood B (2008) Hominin life history: reconstruction and evolution. J Anat 212:394–425. doi:10.1111/j.1469-7580.2008.00867.x

Roche H, Blumenschine RJ, Shea JJ (2009) Origins and adaptations of early Homo: what archeology tells us. In: Grine FE, Fleagle JG, Leakey RE (eds) The first humans – origin and early evolution of the genus Homo. Springer, Berlin, pp 135–147

Rokas A (2008) The origins of multicellularity and the early history of the genetic toolkit for animal development. Annu Rev Genet 42:235–251

Romer AS (1957) Origin of the amniotic egg. Sci Mon 85:57–63

Romer AS (1967) Major steps in vertebrate evolution. Science 158:1629–1638

Romer AS, Parsons TS (1977) The vertebrate body. Saunders, Philadelphia

Rosen R (1991) Life itself. A comprehensive inquiry into the nature, origin, and fabrication of life. Columbia University Press, New York

Rosenberg A, McShea DW (2008) Philosophy of biology. A contemporary introduction. Routledge, New York

Rosslenbroich B (2006) The notion of progress in evolutionary biology – the unresolved problem and an empirical suggestion. Biol Philos 21:41–70

Rosslenbroich B (2007) Autonomiezunahme als Modus der Makroevolution. Galunder, Nümbrecht

Rosslenbroich B (2009) The theory of increasing autonomy in evolution: a proposal for understanding macroevolutionary innovations. Biol Philos 24:623–644

Rosslenbroich B (2011) Outline of a concept for organismic systems biology. Semin Cancer Biol 21(3):156–164. doi:10.1016/j.semcancer.2011.06001

Roth G (1981) Biological systems theory and the problem of reductionism. In: Roth G, Schwegler H (Hrsg.) Self-organizing systems. An interdisciplinary approach. Campus Verlag, Frankfurt/New York, pp 106–120

Roth G (2001) The evolution of consciousness. In: Roth G, Wullimann MF (eds) Brain evolution and cognition. Wiley/Spektrum, New York/Heidelberg, pp 555–582

Roth G, Dicke U (2005) Evolution of the brain and intelligence. Trends Cogn Sci 9:250–257

Roth G, Wullimann MF (2001) Brain evolution and cognition. Wiley/Spektrum, New York/Heidelberg

Rothchild I (2003) The yolkless egg and the evolution of eutherian viviparity. Biol Reprod 68:337–357

Roy K (2008) Dynamics of body size evolution. Science 321:1451–1452

Ruben JA (1976) Aerobic and anaerobic metabolism during activity in snakes. J Comp Physiol 109:147–157

Ruben JA (1995) The evolution of endothermy in mammals and birds: from physiology to fossils. Annu Rev Physiol 57:69–95

Ruben JA (1996) Evolution of endothermy in mammals, birds and their ancestors. In: Johnston IA, Bennett AF (eds) Animals and temperature: phenotypic and evolutionary adaptation. Cambridge University Press, Cambridge, pp 347–376

Ruben JA, Jones TD (2000) Selective factors associated with the origin of fur and feathers. Am Zool 40:585–596

Ruben JA, Bennett AF, Hisaw FL (1987) Selective factors in the origin of the mammalian diaphragm. Paleobiology 13:54–59

Ruben JA, Jones TD, Geist NR (2003) Respiratory and reproductive paleophysiology of dinosaurs and early birds. Physiol Biochem Zool 76:141–164

Ruiz-Mirazo K, Moreno A (2004) Basic autonomy as a fundamental step in the synthesis of life. Artif Life 10:235–259

Ruiz-Mirazo K, Moreno A (2012) Autonomy in evolution: from minimal to complex life. Synthese 185:21–52. doi:10.1007/s11229-011-9874-z

Runnegar B (1982) Oxygen requirements, biology and phylogenetic significance of the late Precambrian worm Dickinsonia, and the evolution of the burrowing habit. Alcheringa 6:223–239

Ruppert EE, Barnes RD (1994) Invertebrate zoology. Saunders College Publishers, Fort Worth

Ruppert EE, Carle KJ (1983) Morphology of metazoan circulatory systems. Zoomorphology 103:193–208

Ruse M (1996) Monad to man. The concept of progress in evolutionary biology. Harvard University Press, Cambridge, MA

Russon AE, Galdikas BMF (1993) Imitation in ex-captive orangutans. J Comp Psychol 107:147–161
Russon AE, Galdikas BMF (1995) The nature of imitation in rehabilitant orangutans. In: Russon AE, Bard KA, Parker ST (eds) Reaching into thought; The minds of the great apes. Cambridge University Press, Cambridge
Ruthmann A (2006) Placozoa. In: Westheide W, Rieger R (eds) Spezielle Zoologie. Part 1: Einzeller und Wirbellose Tiere, 2nd edn. Spektrum Akademischer Verlag, Heidelberg
Saetzler K, Sonnenschein C, Soto AM (2011) Systems biology beyond networks: generating order from disorder through self-organization. Semin Cancer Biol 21:165–174
Sagan L (1967) On the origin of mitosing cells. J Theor Biol 14:255–274
Santos LR (2004) Tool use. In: Bekoff M (ed) Encyclopedia of animal behavior, vol 3. Greenwood, Westport/London, pp 1065–1067
Satoh N, Tagawa K, Takahashi H (2012) How was the notochord born? Evol Dev 14(1):56–75
Saunders PT, Ho MW (1976) On the increase in complexity in evolution. J Theor Biol 63:375–384
Saunders PT, Ho MW (1981) On the increase in complexity in Evolution II. The relativity of complexity and the principle of minimum increase. J Theor Biol 90:515–530
Savage-Rumbaugh ES (1986) Ape language: from conditioned response to symbol. Columbia University Press, New York
Savage-Rumbaugh ES, Lewin R (1994) Kanzi: the ape at the brink of the human mind. Wiley, New York
Savage-Rumbaugh ES, McDonald K (1988) Deception and social manipulation in symbol-using apes. In: Byrne RW, Whiten A (eds) Machiavellian intelligence: social expertise and the evolution of intellect in monkeys, apes and humans. Clarendon Press, Oxford, pp 224–237
Savage-Rumbaugh ES, Shanker SG, Taylor TJ (1998) Apes, language and the human mind. Oxford University Press, Oxford
Sawyer R, Knapp LW (2003) Avian skin development and the evolutionary origin of feathers. J Exp Zool (Mol Dev Evol) 298B:57–72
Sawyer RH, Knapp LW, O'Guin M (1986) The skin of birds: epidermis, dermis and appendages. In: Bereiter-Hahn J, Matoltsy AG, Sylvia Richards K (eds) Biology of the integument. Springer, Berlin, pp 195–238
Schad W (1977) Man and mammals: toward a biology of form. Waldorf Press, Garden City
Schad W (1992) Der Heterochronie-Modus in der Evolution der Wirbeltierklassen und Hominiden. Dissertation, Universität Witten/Herdecke
Schad W (1993) Heterochronical patterns of evolution in the transitional stages of vertebrate classes. Acta Biotheor 41:383–389
Schad W (1997) Die Zeitintegration als Evolutionsmodus. Habilitation, Universität Witten/Herdecke
Scheiba B (1990) Schwimmen, Laufen, Fliegen. Die Bewegung der Tiere, Urania/Leipzig
Schierwater B, DeSalle R (2007) Can we ever identify the Urmetazoan? Integr Comp Biol 47(5):670–676. doi:10.1093/icb/icm040
Schierwater B, Eitel M, Jakob W, Osigus H-J, Hadrys H et al (2009) Concatenated analysis sheds light on early metazoan evolution and fuels a modern "Urmetazoon" hypothesis. PLoS Biol 7(1):e1000020. doi:10.1371/journal.pbio.1000020
Schlichter D (1984) Cnidaria: permeability, epidermal transport and related phenomena. In: Bereiter-Hahn J, Matoltsy AG, Sylvia Richards K (eds) Biology of the integument, vol 1. Springer, Berlin, p 79
Schmalhausen II (1949) Factors of evolution. The University of Chicago Press, Chicago
Schmuck R, Linsenmair KE (1997) Regulation of body water balance in reed frogs (Superspecies Hyperolius viridiflavus and Hyperolius maramoratues: Amphibia, Anura, Hyperoiliidae) living in unpredictably varying savannah environments. Comp Biochem Physiol 118A:1335–1352
Schrenk F (1998) Die Frühzeit des Menschen. Der Weg zum Homo sapiens. C.H. Beck, München
Schweitzer MH, Marshall CL (2001) A molecular model for the evolution of endothermy in the theropod-bird lineage. J Exp Zool 291:317–338

Seebacher F, Grigg GC, Beard LA (1999) Crocodiles as dinosaurs: behavioural thermoregulation in very large ectotherms leads to high and stable body temperatures. J Exp Biol 202:77–86
Seed A, Byrne RW (2010) Animal tool-use. Curr Biol 20(23):R1032–R1039
Seilacher A (1999) Biomat-related lifestyles in the Precambrian. Palaios 14:86–93
Seilacher A, Grazhdankin D, Legouta A (2003) Ediacaran biota: the dawn of animal life in the shadow of giant protists. Paleontol Res 7:43–54
Shapiro R (1987) Origins. A skeptic's guide to the creation of life on earth. Bantam, New York
Shapiro JA (1998) Thinking about bacterial populations as multicellular organisms. Annu Rev Microbiol 52:81–104
Shapiro JA (2010) Mobile DNA and evolution in the 21st century. Mobile DNA 1(4). doi:10.1186/1759-8753-1-4
Shapiro JA (2011) Evolution: a view from the 21st century. FT Press Science, Upper Saddle River
Shapiro JA, Dworkin M (1997) Bacteria as multicellular organisms. Oxford University Press, Oxford/New York
Shea N (2012) New thinking, innateness and inherited representation. Philos Trans R Soc B Biol Sci 367:2234–2244. doi:10.1098/rstb.2012.0125
Sheng G, Foley AC (2012) Diversification and conservation of the extraembryonic tissues in mediating nutrient uptake during amniote development. Ann N Y Acad Sci 1271(1):97–103
Sherwood C, Subiaul F, Zawidzki TW (2008) A natural history of the human mind: tracing evolutionary changes in brain and cognition. J Anat 212:426–454. doi:10.1111/j.1469-7580.2008.00868.x
Shettleworth SJ (1998) Cognition, evolution and behavior. Oxford University Press, New York
Shoemaker VH, Nagy K (1977) Osmoregulation in amphibians and reptiles. Annu Rev Physiol 39:449–471
Shoemaker VH, Hillman SS, Hillyard SD, Jackson DC, McClanahan LL, Withers PC, Wygoda ML (1992) Exchange of water, ions, and respiratory gases in terrestrial amphibians. In: Feder ME, Burggreen WW (eds) Environmental physiology of the amphibians. University of Chicago Press, Chicago, pp 125–150
Shu D, Zhang X, Chen L (1996) Reinterpretation of Yunnanozoon as the earliest known hemichordate. Nature 380:428–430
Shu DG, Luo HL, Conway Morris S, Zhang XL, Hu SX, Chen L, Han J, Zhu M, Chen LZ (1999) Lower Cambrian vertebrates from south China. Nature 402:42–46
Shubin N (2008) Your inner fish. A journey into the 3.5-billion-year history of the human body. Pantheon Books, New York
Shubin NH, Marshall CR (2000) Fossils, genes, and the origin of novelty. Paleobiology 26(Suppl 4):324–340
Shultz S, Nelson E, Dunbar RIM (2012) Hominin cognitive evolution: identifying patterns and processes in the fossil and archaeological record. Philos Trans R Soc B Biol Sci 367:2130–2140. doi:10.1098/rstb.2012.0115
Sibley R, McFarland D (1976) On the fitness of behavior sequences. Am Nat 110:601–617
Siewing R (1969) Lehrbuch der vergleichenden Entwicklungsgeschichte der Tiere. Parey Hamburg, Berlin
Siewing R (1980) Das Archicoelomatenkonzept. Zoologische Jahrbücher, Abteilung für Anatomie und Ontogenie der Tiere 103:439–482
Simpson GG (1971) The meaning of evolution. Yale University Press, New Haven/London
Simpson GG (1973) The concept of progress in organic evolution. Soc Res 41:28–51
Singer W (2006) The evolution of culture from a neurobiological perspective. In: Levinson SC, Jaisson P (eds) Evolution and culture. MIT Press, Cambridge, pp 181–201
Sinke AP, Deen PMT (2011) The physiological implication of novel proteins in systemic osmoregulation. FASEB J 25(10):3279–3289
Sjölander S (1993) Some cognitive breakthroughs in the evolution of cognition and consciousness, and their impact on the biology of language. Evol Cogn 3:1–10
Slobodkin LB, Rapoport A (1974) An optimality strategy of evolution. Q Rev Biol 49:181–200
Smith H (1953) From fish to philosopher. Little Brown, Boston

Smolker R (2004) Tool use by dolphins. In: Bekoff M (ed) Encyclopedia of animal behavior, vol 3. Greenwood, Westport/London, pp 1069–1070
Sol D, Timmermanns S, Lefebvre L (2002) Behavioural flexibility and invasion success in birds. Anim Behav 63:495–502
Sonea S (1991) Bacterial evolution without speciation. In: Margulis L, Fester R (eds) Symbiosis as a source of evolutionary innovation: speciation and morphogenesis. MIT Press, Cambridge
Sonea S, Panisset M (1983) A new bacteriology. Jones and Bartlett, Boston
Sonnenschein C, Soto A (1999) The society of cells: cancer and control of cell proliferation. Taylor & Francis, New York
Soto A, Sonnenschein C (2012) Is systems biology a promising approach to resolve controversies in cancer research? Cancer Cell Int 12:12. doi:10.1186/1475-2867-12-12
Soto AM, Sonnenschein C, Maini PK, Noble D (2011) Systems biology and cancer. Prog Biophys Mol Biol 106(2):337–339
Spencer H (1864) Illustrations of universal progress. D. Appleton, New York
Spotila JR, O'Connor MP, Dodson P, Paladino FV (1991) Hot and cold running dinosaurs: size, metabolism and migration. Mod Geol 16:203–227
Spotila JR, O'Connor MP, Bakken GS (1992) Biophysics of heat and mass transfer. In: Feder ME, Burggren W (eds) Environmental physiology of the amphibians. University of Chicago Press, Chicago, pp 59–80
Srivastava M, Simakov O, Chapman J, Fahey B, Gauthier M et al (2010) The Amphimedon queenslandica genome and the evolution of animal complexity. Nature 466:720–727
Stamp Dawkins M (1993) Through our eyes only? The search for animal consciousness. Freeman/Spektrum, Oxford/New York/Heidelberg
Starck D (1978) Das evolutive Plateau Säugetier. Sonderband des Naturwissenschaftlichen Vereins in Hamburg 3:7–33
Starck D (1982) Vergleichende Anatomie der Wirbeltiere auf evolutionsbiologischer Grundlage, (3 Bände 1978–1982). Springer, Berlin/Heidelberg/New York
Steiner R (1964) The philosophy of freedom. Rudolf Steiner Press, London
Stelling J, Sauer U, Szallasi Z, Doyle FJ III, Doyle J (2004) Robustness of cellular functions. Cell 118:675–685
Sterelny K (2005) Made by each other: organisms and their environment. Biol Philos 20:21–36
Sterelny K (2006) The evolution and evolvability of culture. Mind Lang 21(2):137–165
Sterelny K (2011) Evolvability reconsidered. In: Calcott B, Sterelny K (eds) The major transitions in evolution revisited. MIT Press, Cambridge, MA
Sterelny K (2012) Language, gesture, skill: the co-evolutionary foundations of language. Philos Trans R Soc B Biol Sci 367:2141–2151. doi:10.1098/rstb.2012.0116
Sterelny K, Griffiths PE (1999) Sex and death. An introduction to philosophy of biology. University of Chicago Press, Chicago/London
Stewart JR (1997) Morphology and evolution of the egg of oviparous amniotes. In: Sumida SS, Martin KLM (eds) Amniote origins. Completing the transition to land. Academic, San Diego, pp 291–326
Storch V, Welsch U (1989) Evolution- Tatsachen und Probleme der Abstammungslehre. Deutscher Taschenbuch Verlag, München
Storch V, Welsch U, Wink M (2001) Evolutionsbiologie. Springer, Berlin/Heidelberg
Streffer W (2003) Magie der Vogelstimmen. Die Sprache der Natur verstehen lernen. Freies Geistesleben, Stuttgart
Streffer W (2009) Klangsphären. Motive der Autonomie im Gesang der Vögel. Freies Geistesleben, Stuttgart
Striedter G (2005) Principles of brain evolution. Sinauer, Sunderland
Suchantke A (2002) Metamorphose. Kunstgriff der Evolution. Freies Geistesleben, Stuttgart
Suki B (2012) The major transitions of life from a network perspective. Front Physiol, 3 APR, art. no. Article 94. doi:10.3389/fphys.2012.00094

Swanson DL, Thomas NE, Liknes ET, Cooper SJ (2012) Intraspecific correlations of basal and maximal metabolic rates in birds and the aerobic capacity model for the evolution of endothermy. PLoS One 7(3):e34271. doi:10.1371/journal.pone.0034271

Szathmary EM, Maynard Smith J (1995) The major evolutionary transitions. Nature 374:227–232

Taigen TL (1983) Activity metabolism of anuran amphibians: implications for the origin of endothermy. Am Nat 121:94–109

Taigen TL, Emerson SB, Pough FH (1982) Ecological correlates of anuran exercise physiology. Oecologia 52:49–56

Taylor M, Saayman H (1973) Imitation behaviour by Indian ocean bottlenose dolphins (*Tursiops aduncus*) in captivity. Behaviour 44:286–298

Taylor EW, Campbell HA, Levings JJ, Young MJ, Butler PJ, Egginton S (2006) Coupling of the respiratory rhythm in fish with activity in hypobranchial nerves and with heartbeat. Physiol Biochem Zool 79:1000–1009

Taylor AH, Medina FS, Holzhaider JC, Hearne LJ, Hunt GR, Gray RD (2010) An investigation into the cognition behind spontaneous string pulling in New Caledonian crows. PLoS One 5:e9345

ten Donkelaar HJ (2001) Evolution of vertebrate motor systems. In: Roth G, Wullimann MF (eds) Brain evolution and cognition. Wiley/Spektrum, New York/Heidelberg, pp 77–112

Tepass U, Tanentzapf G, Ward R, Fehon R (2001) Epithelial cell polarity and cell junctions in Drosophila. Annu Rev Genet 35:747–748

Thomas RDK, Reif WE (1993) The skeleton space: a finite set of organic designs. Evolution 47:341–360

Thompson E (2007) Mind in life. Biology, phenomenology, and the sciences of mind. Harvard University Press, Cambridge, MA/London

Thompson E, Varela F (2001) Radical embodiment: neural dynamics and consciousness. Trends Cogn Sci 5:418–425

Thomson KE (1992) Macroevolution: the morphological problem. Am Zool 32:106–112

Tomasello M (2006) Uniquely human cognition is a product of human culture. In: Levinson SC, Jaisson P (eds) Evolution and culture. MIT Press, Cambridge, pp 203–217

Tomasello M (2009) Why we cooperate. MIT Press, Cambridge MA/London

Tomasello M, Call J (1997) Primate cognition. Oxford University Press, Oxford

Tresch MC, Saltiel P, D'Avella A, Bizzi E (2002) Coordination and localization in spinal motor systems. Brain Res Rev 40(1–3):66–79

Trinkaus E (1992) Evolution of human manipulation. In: Jones S, Martin R, Pilbeam D (eds) The Cambridge encyclopedia of human evolution. Cambridge University Press, Cambridge, pp 346–349

Turner JS (2007) The Tinkerer's accomplice. How design emerges from life itself. Harvard University Press, Cambridge, MA/London

Tyler S (2003) Epithelium: the primary building block for metazoan complexity. Integr Comp Biol 43:55–63

Uexküll J (1973) Theoretische Biologie. Suhrkamp, Frankfurt am Main

Valentine JW (2002) Prelude to the Cambrian explosion. Annu Rev Earth Planet Sci 30:285–306

Valentine JW (2004) On the origin of phyla. University of Chicago Press, Chicago/London

Valentine JW, Collins AG, Meyer CP (1993) Morphological complexity increase in metazoans. Paleobiology 20:131–142

Varela FJ (1979) Principles of Biological Autonomy. North Holland, New York/Oxford

Varela FJ (1981) Autonomy and autopoiesis. In: Roth G, Schwegler H (eds) Self-organizing systems. An interdisciplinary approach. Campus Verlag, Frankfurt/New York, pp 14–23

Vasilakos K, Kimura N, Wilson RJA, Remmers JE (2006) Lung and buccal ventilation in the frog: uncoupling coupled oscillators. Physiol Biochem Zool 79:1010–1018

Vermeij GJ (1987) Evolution and escalation – an ecological history of life. Princeton University Press, Princeton

Vermeij GJ (1996) Animal origins. Science 274:525–526

Vermeij GJ (1999) Inequality and the directionality of history. Am Nat 153:243–253

Visalberghi E, Fragaszy DM (1990) Do monkeys ape? In: Parker ST, Gibson KR (eds) Language and intelligence in monkey and apes. Cambridge University Press, Cambridge, pp 247–273

Waddington CH (1961) The nature of life. Allen & Unwin, London

Wagner A (2012) The role of robustness in phenotypic adaptation and innovation. Proc R Soc B 279:1249–1258. doi:10.1098/rspb.2011.2293

Wagner GP, Altenberg L (1996) Perspective: complex adaptations and the evolution of evolvability. Evolution 50:967–976

Wagner G, Chiu CH, Laubichler M (2000) Developmental evolution as a mechanistic science: the inference from developmental mechanisms to evolutionary processes. Am Zool 40:819–831

Wainright PO, Hinkle G, Sogin ML, Stickel SK (1993) Monophyletic origins of the Metazoa, an evolutionary link with Fungi. Science 260:340–342

Wake DB (1986) Directions in the history of life. In: Raup DM, Jablonsky D (eds) Patterns and processes in the history of life. Springer, Berlin, pp 47–67

Wang J, Yu X, Guo A, Hao Y, Wang D (2012) Research advances in Cetacean osmoregulation. Acta Theriol Sin 32(2):156–167

Weingarten M (1993) Organismen – Objekte oder Subjekte der Evolution? Philosophische Studien zum Paradigmawechsel in der Evolutionsbiologie. Wissenschaftliche Buchgesellschaft, Darmstadt

Weir AAS, Chappell J, Kacelnik A (2002) Shaping of hooks in New Caledonian crows. Science 297:981

Weiss PA (1963) The cell as unit. J Theor Biol 5:389–397

Weiss PA (1968) From cell to molecule. In: Weiss PA (ed) Dynamics of development: experiments and inferences. Selected papers on, developmental biology. Academic, New York, pp 24–95

Weiss PA (1969) The living system: determinism stratified. In: Koestler A, Smythies JR (eds) Beyond reductionism. New perspectives in the life sciences. Hutchinson, London, pp 3–55

Weiss PA (1971) The basic concept of hierarchic systems. In: Weiss PA (ed) Hierarchically organized systems in theory and practice. Hafner Publishing Company, New York, pp 1–43

Weiss PA (1973) The science of life: the living system – a system for living. Futura, New York

Weiss PA (1977) The system of nature and the nature of systems: empirical holism and practical reductionism harmonized. In: Schaefer KE, Hensel H, Brady R (Hrsg.) Toward a man-centered medical science. A new image of man in medicine, vol. 1. Futura Publishing Company, Mount Kisco, pp 17–63

Weissenfels N (1983) Bau und Funktion des Süsswasserschwammes *Ephydatia fluviatilis* L. (Porifera) X. Der Nachweis des offenen Mesenchyms durch Verfütterung von Bäckerhefe (*Saccharomyces cerevisiae*). Zoomorphologie 103:15–23

Weizsäcker Vv (1986) Der Gestaltkreis. Theorie der Einheit von Wahrnehmen und Bewegen. Thieme, Stuttgart

West-Eberhard MJ (2003) Developmental plasticity and evolution. Oxford University Press, Oxford

Westheide W, Rieger R (eds) (2006) Spezielle Zoologie. Part 1: Einzeller und Wirbellose Tiere, 2nd edn. Spektrum Akademischer Verlag, Heidelberg

Westheide W, Rieger R (eds) (2010) Spezielle Zoologie. Part 2: Wirbel- oder Schädeltiere, 2nd edn. Spektrum Akademischer Verlag, Heidelberg

Whiten A, Erdal D (2012) The human socio-cognitive niche and its evolutionary origins. Philos Trans R Soc B Biol Sci 367:2119–2129. doi:10.1098/rstb.2012.0114

Whittamore JM (2012) Osmoregulation and epithelial water transport: lessons from the intestine of marine teleost fish. J Comp Physiol B Biochem Syst Environ Physiol 182(1):1–39

Wicken JS (1979) The generation of complexity in evolution: a thermodynamic and information-theoretical discussion. J Theor Biol 77:349–365

Wicken JS (1987) Evolution, thermodynamics, and information. Oxford University Press, Oxford

Wieser W (Hrsg.) (1994) Die Evolution der Evolutionstheorie. Von Darwin zur DNA. Spektrum Akademischer Verlag, Heidelberg

Wieser W (1998) Die Erfindung der Individualität oder Die zwei Gesichter der Evolution. Spektrum, Heidelberg/Berlin

Williams GC (1966) Adaptation and natural selection. A critique of some current evolutionary thought. Princeton University Press, Princeton

Willmer P (1990) Invertebrate relationships. Patterns in animal evolution. Cambridge University Press, Cambridge

Willmer P, Stone G, Johnston I (2000) Environmental physiology of animals. Blackwell, London

Withers PC (1992) Comparative animal physiology. Brooks Cole, Pacific Grove

Witteveen J (2011) The softening of the modern synthesis. Julian Huxley: Evolution: the modern synthesis; The definitive edition. In: Pigliucci M, Müller GB (eds) Evolution – the extended synthesis (book review). Acta Biotheor 59:333–345

Woese CR (2000) Interpreting the universal phylogenetic tree. Proc Natl Acad Sci U S A 97: 8392–8396

Woese CR (2002) On the evolution of cells. Proc Natl Acad Sci U S A 99:8742–8747

Woese C (2004) A new biology for a new century. Microbiol Mol Biol Rev 68:137–186

Wolf YI, Rogozin IB, Grishin NV, Koonin EV (2002) Genome trees and the tree of life. Trends Genet 18:472–479

Wolpert L (1990) The evolution of development. Biol J Linn Soc 39:109–124

Wörheide G, Dohrmann M, Erpenbeck D, Larroux C, Maldonado M, Voigt O, Borchiellini C, Lavrov DV (2012) Deep phylogeny and evolution of sponges (Phylum Porifera). Adv Mar Biol 61:1–78

Wu P, Hou L, Plikus M, Hughes M, Scehnet J, Suksaweang S, Widelitz RB, Jiang TX, Chuong CM (2004) Evo-devo of amniote integuments and appendages. Int J Dev Biol 48(2–3):249–270

Wuethrich B (1999) Stunning fossil shows breath of a dinosaur. Science 283:468

Wuketits FM (1981) Biologie und Kausalität: Biologische Ansätze zur Kausalität, Determination u. Freiheit. Parey Verlag, Berlin/Hamburg

Xiao S, Kaufman AJ (eds) (2006) Neoproterozoic geobiology and paleobiology. Springer, Dordrecht

Xiao S, Yuan X, Knoll AH (2000) Eumetazoan fossils in terminal proterozoic phosphorites? Proc Natl Acad Sci U S A 97:13684–13689

Yigael Y (2000) The paradox of evolution. World Future 56:155–178

Yin C, Bengtson S, Yue Z (2004) Silicified and phosphatized Tianzhushania, spheroidal microfossils of possible animal origin from the neoproterozoic of south China. Acta Palaeontol Polon 49:1–12

Young JZ (1991) Computation in the learning system of cephalopods. Biol Bull 180:200–208

# Author Index




B. Rosslenbroich, *On the Origin of Autonomy: A New Look at the Major Transitions in Evolution*, History, Philosophy and Theory of the Life Sciences 5,
DOI 10.1007/978-3-319-04141-4, © Springer International Publishing Switzerland 2014

# Subject Index



B. Rosslenbroich, *On the Origin of Autonomy: A New Look at the Major Transitions in Evolution*, History, Philosophy and Theory of the Life Sciences 5,
DOI 10.1007/978-3-319-04141-4, 

## T

MIX
Papier aus verantwortungsvollen Quellen
Paper from responsible sources
FSC® C105338

If you have any concerns about our products,
you can contact us on
**ProductSafety@springernature.com**

In case Publisher is established outside the EU,
the EU authorized representative is:
**Springer Nature Customer Service Center GmbH**
**Europaplatz 3, 69115 Heidelberg, Germany**

Printed by Libri Plureos GmbH
in Hamburg, Germany